Data Analysis for Direct Numerical Simulations of Turbulent Combustion

Heinz Pitsch · Antonio Attili
Editors

Data Analysis for Direct Numerical Simulations of Turbulent Combustion

From Equation-Based Analysis to Machine Learning

Springer

Editors
Heinz Pitsch
Institute for Combustion Technology
RWTH Aachen University
Aachen, Germany

Antonio Attili
Institute for Combustion Technology
RWTH Aachen University
Aachen, Germany

ISBN 978-3-030-44717-5 ISBN 978-3-030-44718-2 (eBook)
https://doi.org/10.1007/978-3-030-44718-2

This Springer imprint is published by the registered company Springer Nature Switzerland AG
The registered company address is: Gewerbestrasse 11, 6330 Cham, Switzerland

Preface

This book is a graduate-level review of the theory and application of methods for the analysis of large datasets generated in experiments and simulations of reactive turbulent flows. The book originates from the "Combustion-DNS Strategy & Data Analysis Workshop" organized by the editors in Sorrento, Italy, in May 2018. The book consists of a collection of chapters describing traditional and innovative methods to extract global and local features from massive datasets and to inform the development of reduced models for the simulation of turbulent combustion. A number of different methodologies are presented, including analysis based on flame topology, dissipation elements, explosive modes (CEMA), computational singular perturbation (CSP), high-order tensors, and dynamic mode decomposition (DMD). In addition, techniques based on machine learning, which are steadily gaining popularity in the fluid and combustion communities, are presented in a series of chapters on evolutionary algorithms, data assimilation, principal component analysis (PCA), and artificial (ANN) and convolutional (CNN) neural networks.

This book is primarily intended for graduate-level engineering students and researchers interested in the analysis of large-scale data of reactive flows, but it can be also useful in other fields, including general fluid mechanics, applied mathematics, and physics.

Aachen, Germany
December 2019

Antonio Attili
Heinz Pitsch

Contents

About the Editors

Prof. Heinz Pitsch received his PhD from the RWTH Aachen University in 1998, where he is now a Full Professor and Director of the Institute for Combustion Technology. He has received numerous honours and awards, including an ERC Advanced Grant, and Fellow Awards of the American Physical Society and the International Combustion Institute. He has served on the board of directors of the International Combustion Institute since 2014 and has been the chair of the German section of the Institute since 2017. Professor Pitsch has over 200 ISI-listed, peer-reviewed journal publications to his credit.

Dr. Antonio Attili received his PhD from Sapienza University of Rome in 2009 and he is now Lecturer in Computational Reactive Flows in the School of Engineering at the University of Edinburgh, United Kingdom. Before that, he was a research scientist at the Institute for Combustion Technology, RWTH Aachen University, and at KAUST, Saudi Arabia. He cochaired and organized several workshops, including the Combustion-DNS Strategy and Data Analysis Workshop. Dr Attili has received several fellowship, including a European Space Agency and AVIO Groups Graduate Research Fellowship in 2007. He has authored and coauthored over 50 research papers published in journals and conference proceedings.

Chapter 1
Analysis of Flame Topology and Burning Rates

Shrey Trivedi, Girish V. Nivarti, and R. Stewart Cant

Abstract Datasets generated using Direct Numerical Simulation (DNS) are used to investigate the influence of local flame surface topology on global flame propagation. A mathematical framework based on Morse theory is presented and is shown to lead to a classification of all possible types of flame surface topology. A similar mathematical approach is shown to provide insight into the behaviour of the surface density function (SDF) and the displacement speed in the vicinity of flame pinch-off and pocket burnout events. DNS data for a pair of colliding premixed turbulent hydrogen–air flames is used to identify and locate topological points of interest and to determine their frequencies of occurrence on the flame surface. Further analysis of the dataset is carried out to evaluate terms of the SDF balance equation and the displacement speed in the presence of flame–flame interactions. Considerable insight is gained into the underlying mechanisms of flame propagation.

1.1 Introduction

Direct Numerical Simulation (DNS) of premixed turbulent flames has yielded increasingly large amounts of statistical data and has led to significant progress in understanding the interactions that occur between the fluid mechanics of the turbulence and the mainly chemical and diffusive structure of the flame. Recent DNS at ever-higher turbulence intensities [1, 2] has served to reinforce the notion that the inner structure of the flame is highly resistant to penetration by the turbulence, even if the leading-edge preheat zone and trailing-edge equilibration zone do become

S. Trivedi · G. V. Nivarti · R. S. Cant (✉)
Department of Engineering, University of Cambridge, Trumpington Street, Cambridge, United Kingdom
e-mail: rsc10@cam.ac.uk

S. Trivedi
e-mail: st634@cam.ac.uk

G. V. Nivarti
e-mail: gvn22@cam.ac.uk

H. Pitsch and A. Attili (eds.), *Data Analysis for Direct Numerical Simulations of Turbulent Combustion*, https://doi.org/10.1007/978-3-030-44718-2_1

disrupted. The steep gradients of temperature and species mass fraction within the central part of the flame are retained, and give rise to a thin and sheet-like reaction zone. This observation provides further support for the well-known flamelet concept for premixed flames [3], and provides a basis for the delineation of regimes of premixed turbulent combustion [4] based on the ratio of turbulence intensity to laminar flame speed and the ratio of turbulence integral length scale to laminar flame thickness. It appears to be difficult to reach a regime of combustion in which the interior structure of the flame does not survive in some recognisable form, even under somewhat extreme experimental conditions of strong turbulence [5].

Given the robustness of the interior flame structure, it is helpful to make use of the sheet-like nature of the flame to describe and analyse its behaviour. In particular, the local curvature of the flame sheet, along with the local hydrodynamic strain field, is known to affect the local propagation speed and hence the rate of conversion of reactants into products. Flame curvature occurs in association with many different geometrical configurations of the three-dimensional flame surface. For a single flame surface, it is possible to define a curvature shape factor [6, 7] to classify the local flame curvature into distinct spherical, cylindrical and spherical-saddle types, and to quantify the frequency of occurrence of each type on the flame surface. Further analysis of DNS data provides insight into the influence of each type of curvature on the propagation of the flame. For example, cusps can be formed by Huygens propagation of the flame [8] in regions of initially moderate curvature concave to the reactants, and this can lead to very high flame speeds close to the tip of the cusp [9].

Cases which involve flame–flame interaction provide a much greater range of possible geometrical configurations. Some of these can lead to phenomena such as reactant pocket formation and burnout, which can produce very rapid changes in the total surface area of the flame and hence can have a significant effect on the global rate of propagation. Hence it is useful to be able to identify and classify such configurations and also to investigate their local and global effects on flame propagation.

Previous work on this topic has included topological analysis of scalar field structure in non-reacting turbulent flow [10, 11] and in flames [12]. The topology of flame surfaces has been investigated in MILD combustion [13] using an approach based on Minkowski functionals [14]. Pocket formation resulting from flame pinch-off has been analysed in detail using Morse theory in two dimensions [15, 16] and this analysis has been extended recently to three dimensions [17]. An approach based on the analysis of critical points has proved useful in identifying all the principal types of flame configuration that can occur during flame–flame interaction and in quantifying the frequency of occurrence of each type [18] using a DNS dataset for hydrogen flames. The same approach has been extended to hydrocarbon flames, with emphasis on changes with increasing turbulence intensity [19]. The present chapter provides a description of this approach together with some recent DNS results.

1.2 Theory of Flame–Flame Interactions

The scalar field within a premixed flame can be described using a single reaction progress variable c which rises monotonically from zero in the unburned reactants to unity in the fully burned products. A suitable definition of c using (for example) the mass fraction Y_α of a major species is

$$c = \frac{Y_\alpha - Y_{\alpha R}}{Y_{\alpha P} - Y_{\alpha R}} \tag{1.1}$$

where R denotes reactants and P denotes products. Since the gradient of c is large within the reaction zone of the flame, the isosurfaces of c are close together and parallel to each other. Hence a single isosurface of c taken at a value $c = c^*$ within the reaction zone can be used as a reliable marker for the flame surface.

1.2.1 Flame Topology

The topology of the flame surface thus defined can be studied using an approach based on the Morse theory of critical points [20]. When two flame surfaces interact, the gradient of the progress variable at the point of flame–flame interaction becomes zero. Such points at which $\nabla c = 0$ are defined as critical points.

The Taylor expansion of $c(\underline{x})$ around a critical point located at $\underline{x} = \underline{x}_c$ is given by

$$c(\underline{x} + \underline{x}_c) = c(\underline{x}_c) + \frac{1}{2}\underline{x}^T . \underline{\underline{H}} . \underline{x} + \ldots \tag{1.2}$$

where there is no linear term in $\underline{x}$ since $\nabla c = 0$. The Hessian matrix $\underline{\underline{H}}\left(c(\underline{x}_c)\right)$ describes the curvature of the progress variable field in the vicinity of the critical point. The eigenvalues of the Hessian λ_1, λ_2 and λ_3 are all real and give the curvatures along the three orthogonal principal axes defined by the corresponding eigenvectors $(\underline{e}_{\lambda_1}, \underline{e}_{\lambda_2}, \underline{e}_{\lambda_3})$. Assuming that the orientation of these eigenvectors relative to the laboratory coordinate system is not of interest, the eigenvalues completely define the local topology to second order accuracy in $\underline{x}$.

It is helpful to make use of a spherical coordinate system. Taking $\lambda_1 > \lambda_2 > \lambda_3$ without loss of generality, it is possible to define angles ϕ and θ together with a scaling parameter κ, according to

$$\theta = \frac{6}{\pi} \arctan \frac{(\lambda_1 - 2\lambda_2 + \lambda_3)/6^{1/2}}{(\lambda_1 - \lambda_3)/2^{1/2}} \tag{1.3}$$

$$\phi = \frac{2}{\pi} \arctan \frac{(\lambda_1 + \lambda_2 + \lambda_3)\cos(\theta\pi/6)/3^{1/2}}{(\lambda_1 - \lambda_3)/2^{1/2}} \tag{1.4}$$

$$\kappa = \left(\lambda_1^2 + \lambda_2^2 + \lambda_3^2\right)^{1/2} \tag{1.5}$$

The angle ϕ gives the latitude relative to the pole vector taken as $(\underline{e}_{\lambda_1}, \underline{e}_{\lambda_2}, \underline{e}_{\lambda_3})$ while the angle θ gives the latitude relative to the meridional vector $(\underline{e}_{\lambda_1}, 0, -\underline{e}_{\lambda_3})$. Both angles are scaled within a range of plus or minus unity and may be interpreted as shape factors. The scaling factor κ provides a measure of the local mean curvature of the surface.

The usefulness of the shape factors is illustrated in Fig. 1.1 which shows examples of all of the possible topologies of flame–flame interaction, plotted as a function of ϕ and θ.

In each example shown, the blue colour indicates a low value of reaction progress variable while the red colour indicates a high value, thus allowing for the direction of flame propagation to be inferred, i.e. towards low values of c. The plot is divided from left to right into four main regions separated by curved lines. In the leftmost region, all three eigenvalues are negative and the spherical topology corresponds to product pockets (denoted as PP). Here the flame is seen from the reactant side (low c) and is propagating outwards. In the next region (second from left) one eigenvalue has become positive and the cylindrical topology now describes tunnel formation (TF). Again the propagation is outwards as seen from the reactant side. The third region has two positive eigenvalues and the cylindrical topology describes tunnel closure (TC). The flame is seen from the product side (high c) with inward propagation. Finally, in the rightmost region all three eigenvalues are positive and the spherical topology corresponds to reactant pockets (RP), as seen from the product side and propagating inwards.

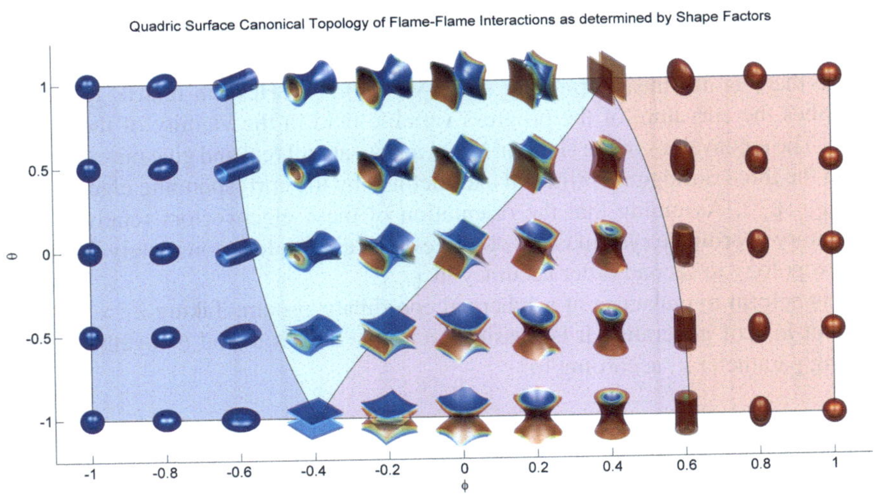

Fig. 1.1 The set of all possible flame–flame interaction topologies as determined by the shape factors. From [18]

1.2.2 Flame Surface Area at Critical Points

The reaction progress variable $c(\underline{x})$ in the vicinity of a critical point also provides the basis for a mathematical analysis of the local propagation behaviour of the flame [15]. Close to a critical point located at $\underline{x} = \underline{x}_0$, it is possible using Morse's lemma [20] to write a series expansion for c corresponding to each of the four principal configurations identified above. Using scaled local coordinates $\underline{x} = (x, y, z)$ corresponding to small deviations from the critical point, and taking the critical point $\underline{x}_0 = (0, 0, 0)$ as the origin, the expansions are

$$\text{(PP)} \quad c(x, y, z) = c_0 - \frac{1}{2}x^2 - \frac{1}{2}y^2 - \frac{1}{2}z^2 \tag{1.6}$$

$$\text{(RP)} \quad c(x, y, z) = c_0 + \frac{1}{2}x^2 + \frac{1}{2}y^2 + \frac{1}{2}z^2 \tag{1.7}$$

$$\text{(TF)} \quad c(x, y, z) = c_0 + \frac{1}{2}x^2 - \frac{1}{2}y^2 - \frac{1}{2}z^2 \tag{1.8}$$

$$\text{(TC)} \quad c(x, y, z) = c_0 - \frac{1}{2}x^2 + \frac{1}{2}y^2 + \frac{1}{2}z^2 \tag{1.9}$$

where $c_0 = c(\underline{x}_0)$ is the value of the progress variable at the critical point. The radial distance r from the critical point is given by $r = (x^2 + y^2 + z^2)^{1/2}$.

It can be seen from these expansions that a product pocket (PP) is described as a local maximum of progress variable with a high value of c in the centre surrounded in all directions by lower values. Hence propagation is radially outwards in all directions. By contrast, a reactant pocket (RP) consists of a local minimum, with a low value of c in the centre and higher values all around. Propagation is radially inwards.

Tunnel formation (TF) corresponds to a local saddle point with a minimum in c on the central axis of the tunnel (taken as the x-axis) combined with a maximum in the other two directions. Here, propagation is outwards in the radial direction of the cylinder. Again by contrast, tunnel closure (TC) consists of a local saddle point with a maximum in c on the central axis combined with a minimum in the other two directions. Propagation is radially inwards towards the central axis of the cylinder.

For each configuration it is straightforward to evaluate the normal vector defined as $\underline{n} = -\nabla c/|\nabla c|$ (pointing into the reactants) together with the curvature given by $\kappa = \nabla.\underline{n}$ (equal to twice the geometrical mean curvature) and these are listed in Table 1.1.

Some insight into the behaviour of the flame close to a critical point can be obtained using the surface density function σ defined as the magnitude of the gradient of c, i.e. $\sigma = |\nabla c|$ [21]. This quantity is related to the flame surface density (FSD) Σ [22] which has been found useful in modelling for Large Eddy Simulation (LES) [23, 24]. Note that for all four configurations of interest in the present context, the SDF may be evaluated as $\sigma = r$.

A balance equation for σ can be derived using the balance equation for progress variable, and takes the form [15]:

Table 1.1 Normal vector $\underline{n}$ and curvature κ for the four principal configurations. The radius r is given by $r = (x^2 + y^2 + z^2)^{1/2}$

	Normal vector	Curvature
PP	$(x/r, y/r, z/r)$	$\kappa = 2/r$
RP	$(-x/r, -y/r, -z/r)$	$\kappa = -2/r$
TF	$(-x/r, y/r, z/r)$	$\kappa = 2x^2/r^3$
TC	$(x/r, -y/r, -z/r)$	$\kappa = -2x^2/r^3$

$$\frac{\partial \sigma}{\partial t} + u_k \frac{\partial \sigma}{\partial x_k} = -n_i n_k \sigma \frac{\partial u_i}{\partial x_k} + n_k \frac{\partial}{\partial x_k}(D\kappa\sigma) + n_k \frac{\partial}{\partial x_k}\left[\frac{w}{\rho} + \frac{n_j}{\rho}\frac{\partial}{\partial x_j}(\rho D \sigma)\right] \tag{1.10}$$

The first term on the right-hand side describes surface area production due to hydrodynamic straining while the second term describes dissipation of surface area due to diffusive effects. The last term is the kinematic restoration term which contains two contributions arising respectively from reaction and normal diffusion.

The SDF dissipation term can be decomposed into two contributing terms as

$$n_k \frac{\partial}{\partial x_k}(D\kappa\sigma) = n_k(\kappa\sigma)\frac{\partial D}{\partial x_k} + n_k D \frac{\partial(\kappa\sigma)}{\partial x_k} \tag{1.11}$$

while the SDF normal restoration term may be decomposed into five contributing terms according to

$$\begin{aligned} n_k \frac{\partial}{\partial x_k}\left[\frac{n_j}{\rho}\frac{\partial}{\partial x_j}(\rho D\sigma)\right] &= D n_k \frac{\partial n_j}{\partial x_k}\frac{\partial \sigma}{\partial x_j} + D n_k n_j \frac{\partial^2 \sigma}{\partial x_k \partial x_j} + n_k \frac{\partial n_j}{\partial x_k}\frac{\sigma}{\rho}\frac{\partial}{\partial x_j}(\rho D \sigma) \\ &+ n_k n_j \frac{\partial D}{\partial x_k}\frac{\partial \sigma}{\partial x_j} + n_k n_j \frac{\partial}{\partial x_k}\left[\frac{\sigma}{\rho}\frac{\partial}{\partial x_j}(\rho D)\right] \end{aligned} \tag{1.12}$$

Of the four principal configurations, the tunnel closure (TC) configuration is chosen for detailed analysis due to its association with flame pinch-off which on physical grounds is likely to have a strong influence on the destruction of flame surface area. Using the expansion (Eq. 1.9) for progress variable and the associated results for the normal vector, curvature and SDF in the TC configuration, the two terms contributing to the SDF dissipation term may be written as

$$n_k(\kappa\sigma)\frac{\partial D}{\partial x_k} = -\frac{2x^2}{r^3}\left(x\frac{\partial D}{\partial x} + y\frac{\partial D}{\partial y} + z\frac{\partial D}{\partial z}\right) \tag{1.13}$$

$$n_k D \frac{\partial(\kappa\sigma)}{\partial x_k} = -8D\left(\frac{x^2y^2 + x^2z^2}{r^5}\right) \tag{1.14}$$

The critical point is approached by taking the limit $r \to 0$. Assuming that the diffusion coefficient D and its first derivatives remain well-behaved and that $r \sim x, y, z$

as $r \to 0$, the first term (l.h.s. of Eq. 1.13) remains bounded. The second term (l.h.s. of Eq. 1.14) however becomes singular, behaving as $1/r$ for $r \to 0$, and therefore the total dissipation will become unbounded as the tunnel closes.

The kinematic restoration term may be treated in a similar manner. First, it is straightforward to show that the reactive restoration term remains bounded at the critical point, and that the last three terms contributing to the SDF normal restoration term (see Eq. 1.12) also remain bounded. Substituting for the normal vector components and SDF for the TC configuration, the first two of the contributing terms may be evaluated together as

$$Dn_k \frac{\partial n_j}{\partial x_k} \frac{\partial \sigma}{\partial x_j} + Dn_k n_j \frac{\partial^2 \sigma}{\partial x_k \partial x_j} = 4D\left(\frac{x^2y^2 + x^2z^2}{r^5}\right) + 4D\left(\frac{x^2y^2 + x^2z^2}{r^5}\right) \tag{1.15}$$

It is clear that these terms become singular as the critical point is approached, behaving again as $1/r$ for $r \to 0$. Hence the kinematic restoration becomes unbounded as the tunnel closes.

The occurrence of singularities in the SDF equation points to the likelihood of very rapid variations in the flame surface area close to a tunnel closure event. Nevertheless it can be seen that the two singular terms cancel each other exactly in the limit. The balance equation for σ remains bounded (as it must), and indeed $\sigma \to 0$ as $r \to 0$. These findings are consistent with the previous two-dimensional analysis [15]. Similar singular behaviour may be observed in the corresponding terms of the SDF equation for the tunnel formation case.

This mathematical result has clear implications for the physical behaviour of the flame. A flame pinch-off event is necessarily preceded by a tunnel closure event causing mutual annihilation of the converging local flame surfaces on very short time scales. The emerging singularity in the kinematic restoration term will be balanced by the emerging singularity in the dissipation term, thus retaining a physically realisable flame surface as the pocket detaches and emerges as a distinct entity after the pinch-off.

For the reactant pocket and product pocket configurations, singularities in the σ equation are not observed as the critical point is approached. Hence the flame surface area remains physically well-behaved for these events.

1.2.3 Displacement Speed at Critical Points

Further insight into flame propagation may be gained by considering the displacement speed S_d which is the speed of advance of an isosurface of progress variable in the direction of the normal vector $\underline{n}$ relative to the surrounding material. The displacement speed is defined by the expression

$$\rho S_d \sigma = w + \frac{\partial}{\partial x_k}\left(\rho D \frac{\partial c}{\partial x_k}\right) \tag{1.16}$$

Here the right-hand side may be decomposed into contributing terms describing the effects of reaction, normal diffusion and tangential diffusion (curvature) effects according to

$$S_r = \frac{w}{\rho\sigma} \tag{1.17}$$

$$S_n = -\frac{n_k}{\rho}\frac{\partial}{\partial x_k}(\rho D) - \frac{n_k D}{\sigma}\frac{\partial \sigma}{\partial x_k} \tag{1.18}$$

$$S_t = -D\kappa \tag{1.19}$$

For the tunnel closure configuration, substitution of the expressions for $\underline{n}$, κ and σ close to the critical point indicates that only the first term $S_n^{(1)}$ in the expression for S_n remains bounded as $r \to 0$. The second term $S_n^{(2)}$ in the same expression may be combined with the curvature term as

$$S_n^{(2)} + S_t = -D\left[\frac{x^2}{r^3} - \frac{y^2}{r^3} - \frac{z^2}{r^3}\right] + 2D\frac{x^2}{r^3} \sim \frac{D}{r} \tag{1.20}$$

Hence the combined term becomes singular as $r \to 0$ assuming that D is non-zero in the limit. The reactive term also becomes singular assuming that the reaction rate w remains non-zero in the limit. Thus the displacement speed becomes unbounded during tunnel closure.

For the reactant pocket configuration the displacement speed again exhibits singular behaviour. Substituting for $\underline{n}$, κ and σ close to the critical point, the expression for $S_n^{(2)} + S_t$ becomes

$$S_n^{(2)} + S_t = D\left[\frac{x^2}{r^3} + \frac{y^2}{r^3} + \frac{z^2}{r^3}\right] + \frac{2D}{r} \sim \frac{D}{r} \tag{1.21}$$

which (given that D remains well-behaved) is unbounded as $r \to 0$. Note that the nature of the singularity ensures that the quantity $S_d\sigma$ remains bounded for both tunnel closure and reactant pocket configurations. Hence the SDF equation and the balance equation for the progress variable remain well-behaved.

These results imply that rapid flame propagation will occur during tunnel closure or the final burnout of a reactant pocket. Such events lead to rapid local destruction of flame surface area, and are likely to have a significant influence on the overall surface area and hence the overall flame speed.

Similar results may be obtained for the displacement speed in the product pocket and tunnel formation configurations also. However, these events originate due to straining within the flow field rather than from flame propagation. Once formed, these configurations expand due to propagation, i.e. the radius r from the critical point increases. Hence, the impact on overall flame speed is not expected to be as rapid or as significant.

1.3 Analysis Using DNS Data

The theory of flame behaviour near critical points as developed in the previous section can be used to underpin an investigation of flame topology using DNS data. Quantities of interest include the frequency of occurrence of each of the principal types of flame topology, and the magnitudes of the terms in the SDF equation close to a tunnel closure or reactant pocket configuration.

1.3.1 DNS Dataset for the Current Study

The study of Griffiths et al. [18] made use of the DNS dataset of Hawkes et al. [25] which contains two initially planar lean premixed hydrogen–air flames propagating towards each other in a shear-driven turbulent planar jet flow. The mixture was preheated to 700 K and the equivalence ratio was 0.7. The 9-species 21-step reaction mechanism of Li et al. [26] was used. The dataset was produced using the DNS code S3D [27] which employs an eighth-order explicit finite difference scheme for spatial derivatives and fourth-order explicit Runge–Kutta time-stepping. Two different cases were run for Damköhler numbers denoted as Da− (Da = 0.13) and Da+ (Da = 0.54).

The computational domain was a cuboid of physical size $16H \times 20H \times 12H$, where H is the slot-jet width, discretised using a mesh of $2400 \times 1600 \times 1800$ points. For the Da+ and Da− cases, the value of H was 5.4 and 2.7 mm, the mesh spacing was 36 and 18 µm, and the integral length scale was 1.8 and 0.9 mm, respectively. Note that the Da− case has half the domain length in all directions compared to the Da+ case. Periodic boundary conditions were applied in the streamwise and spanwise directions, with outflow conditions in the direction of mean flame propagation.

For both cases the flames are within the thin reaction zones regime in which the smaller scales of turbulence penetrate into the preheat zone but the reactants and products remain separated by a continuous flame surface. No broken reaction zones were observed. A representative subset of the temperature field for the Da+ case is shown in Fig. 1.2.

1.3.2 Flame–Flame Interaction Statistics

Subsets of the full hydrogen–air dataset were analysed to obtain statistics of the flame–flame interactions. The subsets were taken at a single instant of time when the turbulent burning velocity is close to its maximum value. Enough critical points were found within each subset to yield a good statistical sample at reasonable computational cost.

A reaction progress variable c was defined based on the mass fraction of H_2, and critical points of the progress variable field were captured using Newton's method

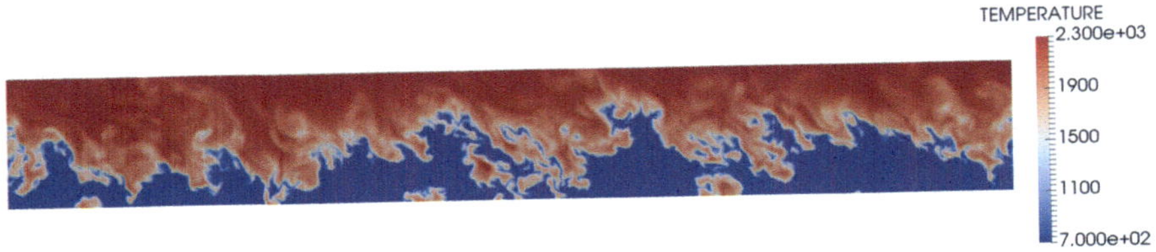

Fig. 1.2 Temperature field in a subset of the Da+ dataset from Hawkes et al. [25]

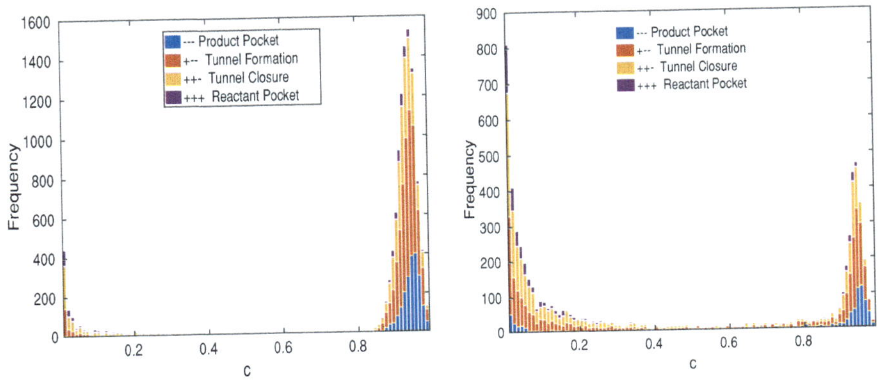

Fig. 1.3 Histograms showing the number of flame–flame interactions for Da+ (left) and Da− (right) cases

based on a tri-quintic interpolation stencil [18]. For each critical point, the data collected included the location in physical and progress variable space, together with the eigenvalues of the Hessian matrix. The shape factors were calculated using Eqs. (1.3)–(1.5).

The resulting flame–flame interaction statistics evaluated for the hydrogen–air dataset are presented in Fig. 1.3 for both Da+ (left) and Da− (right) cases. Different colours are used to indicate the different types of topology (see legend). The general shape of the histogram differs for the two cases but shows a common trend. Both cases show a significant number of critical points, and hence flame–flame interactions, at the leading edge of the flame ($0.01 < c < 0.2$) but their frequency of occurrence decreases in the middle of the flame ($0.2 < c < 0.8$). There is an increase in the number of flame–flame interactions at the trailing edge ($c > 0.8$).

It can been seen in the histograms that the cylindrical topological events (TC and TF) occur more frequently than the spherical (RP and PP) events. This is consistent with the findings of Pope et al. [6] and Cant et al. [7] for the prevalence of spherical and cylindrical curvatures.

The histograms shown in Fig. 1.3 are consistent with those presented by Griffiths et al. [18] who used a progress variable based on H_2O as well as a larger subset of the DNS data.

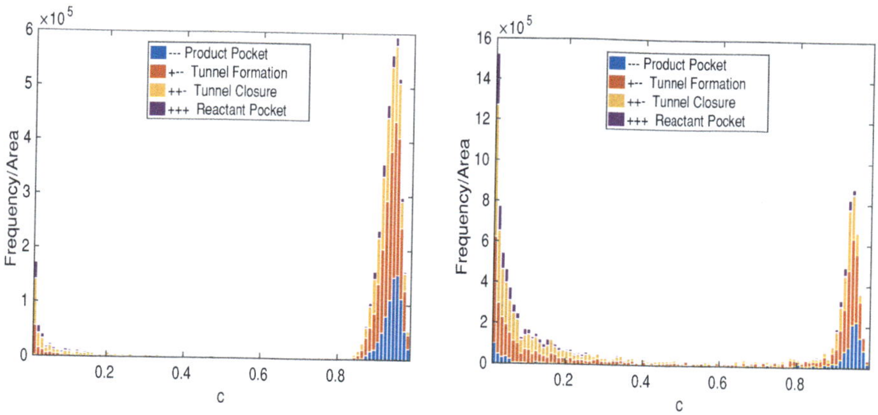

Fig. 1.4 Histograms showing the number of flame–flame interactions per unit surface area for Da+ (left) and Da− (right) cases

It is important to emphasise that the computational domain for the Da− case is half the size in all three dimensions compared to the Da+ case. Therefore the cross-sectional area differs by a factor of four. To make a direct comparison between these different cases, the number of interactions per unit area is computed for the isosurface at $c = 0.5$ and the results are presented in Fig. 1.4.

It is evident from Fig. 1.4 that the number of flame–flame interactions per unit surface area at the trailing edge is qualitatively similar in magnitude for both Da cases, whereas they are larger in magnitude at the leading edge.

1.3.3 Surface Density Function and Displacement Speed

The same hydrogen–air DNS dataset [25] as described above was used to analyse flame propagation behaviour close to critical points according to the theory. For this analysis, a reaction progress variable is defined using the mass fraction of O_2. The critical points are captured in the same manner, and the progress variable gradient close to the critical point is used to evaluate the flame normal vector $\underline{n}$, curvature κ and surface density function σ. Hence the dissipation and kinematic restoration terms in the SDF balance equation can be computed, along with the contributing terms in the definition of the displacement speed.

1.3.3.1 Surface Density Function

Here the focus is on the behaviour of the SDF during tunnel closure events. Several small subsets were separated from the complete Da+ dataset on the basis of locally

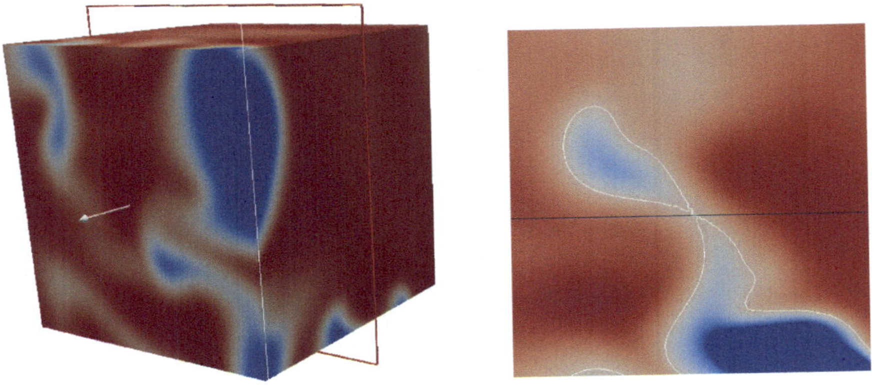

Fig. 1.5 Example of a tunnel closure event. Colour scale runs from blue in reactants to red in products. From [17]

high frequency of occurrence of the tunnel closure configuration. The subsets ranged in size from 30^3 points to 80^3 points. Each subset was scanned to locate and classify all the critical points according to the analysis of Griffiths et al. [18]. Critical points with the tunnel closure configuration were selected for further analysis.

An example of tunnel closure is shown in Fig. 1.5. The full three-dimensional topology is complicated (Fig. 1.5, left-hand picture) and is hard to visualise in general, and so a two-dimensional slice was extracted and is shown in Fig. 1.5 in the right-hand picture. In both pictures the colours depict the variation of the progress variable field running from blue in reactants to red in products. In the right-hand picture the white outline represents the isosurface of progress variable c closest to its value c_0 at the critical point, while the black line is the closest DNS mesh line. The location of the critical point is evident, and it is clear that a pocket is about to be pinched off. In general the geometry of the progress variable field surrounding each instance of tunnel closure was found to be quite different in each case despite the identical topology (by definition) close to the critical point.

Four further examples of tunnel closure are illustrated in Fig. 1.6 (left column) which shows two-dimensional slices taken close to the critical point. Again the white outlines represent the contour of the progress variable that corresponds as closely as possible to the critical point location captured on each plane. The corresponding plots of the SDF dissipation and kinematic restoration terms are shown in the right column of Fig. 1.6. These terms were evaluated at solution points along the DNS mesh line (again shown as a straight black line on each slice) that lies closest to the critical point. This avoids the need to interpolate, thus retaining accuracy despite the strong local variation in the quantities of interest.

It is clear from Fig. 1.6a that in all cases shown both the dissipation term (blue line) and the kinematic restoration term (red line) reach their highest magnitudes close to the critical point. Moreover, the two terms have similar magnitudes but opposite signs, exhibiting a tendency to cancel each other out on a local basis as expected from the theory.

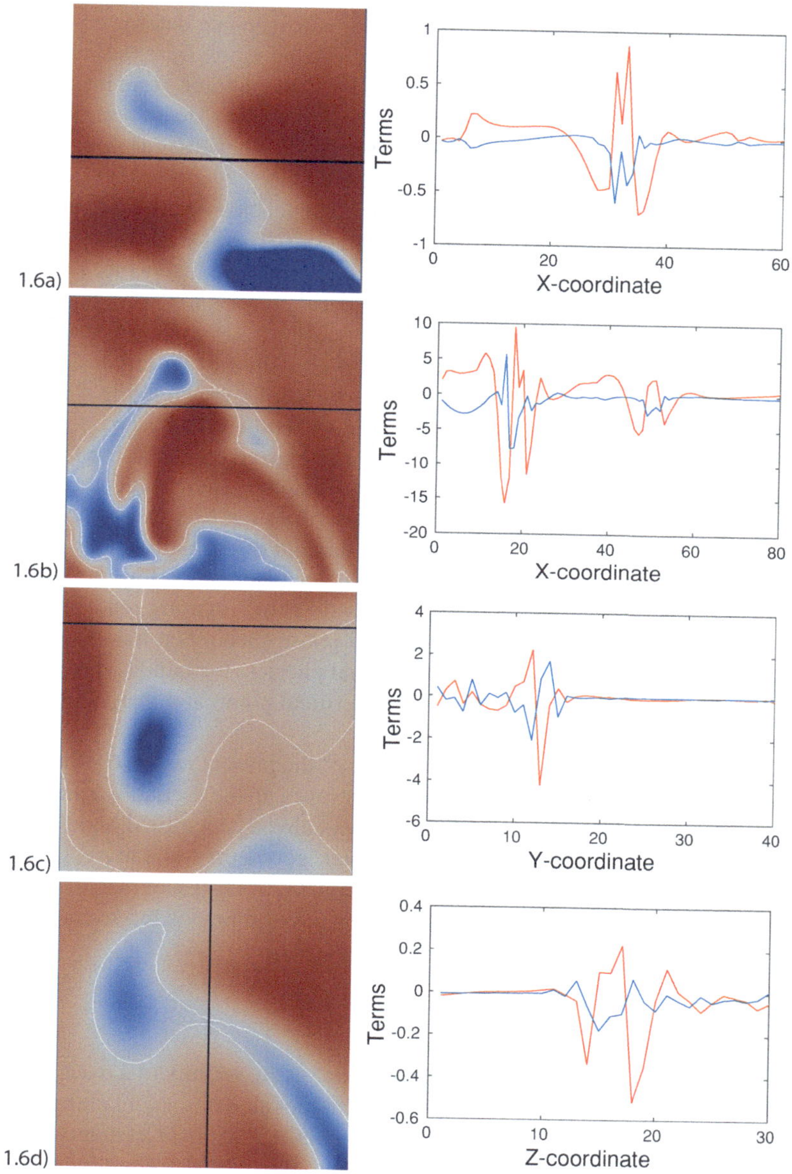

Fig. 1.6 Examples of tunnel closure events (left column) together with (right column) the corresponding SDF equation dissipation term (blue line) and kinematic restoration term (red line). From [17]

Figure 1.6b shows a different subset of the hydrogen–air dataset containing a different tunnel closure configuration. Again, the same terms were evaluated along the mesh line passing closest to the location of the critical point. Here it can be seen that there are two locations of interest close to the same line. Again, both SDF terms have large magnitudes close to each critical point and again there is a strong tendency for the terms to cancel each other at both locations.

Figure 1.6c, d show further subsets containing tunnel closure configurations, with planar views taken in different coordinate orientations. The SDF terms are evaluated along a y-coordinate line in Fig. 1.6c and along a z-coordinate line in Fig. 1.6d. Again the results show high magnitudes close to the critical point along with a tendency for the different terms to cancel each other.

The present three-dimensional results agree well with the two-dimensional analysis of Kollmann and Chen [15]. Nevertheless, the results shown in Fig. 1.6 indicate that the dissipation and kinematic restoration terms do not cancel each other exactly. This is not surprising, since exact cancellation is expected only at the location of the critical point. The departure from exact cancellation is a consequence of the fact that the mesh line used for evaluation of the terms does not pass exactly through the location of the critical point. Nevertheless the method of evaluation does guarantee that the critical point lies within a minimum distance corresponding to one-half of the DNS mesh spacing. Given that the results do show a strong tendency for the dissipation and kinematic restoration terms to oppose each other, it appears that the mesh line is close enough to capture the essential physics. It should be noted that the theory is valid only in a region "close" to the critical point, and that the actual size of this region is unknown.

It would be helpful to confirm the generality of the present results for the terms of the SDF equation by undertaking a statistical survey of all the available critical points. Given the large number of tunnel closure events already identified and the widely different sizes, spatial orientations and complex geometries already encountered, this would be a major task. An automated procedure making use of machine learning techniques is almost certainly required, and this development is left to future work.

1.3.3.2 Reactant Pocket Burnout

Investigation of the displacement speed behaviour in the vicinity of a reactant pocket was carried out using a further subset of the hydrogen–air dataset containing a critical point surrounded by a single isolated reactant pocket. The displacement speed was evaluated on progress variable isosurfaces close to the critical point. Linear interpolation in each coordinate direction was used to determine the location of each isosurface, together with the corresponding value of the displacement speed. Figure 1.7 (top) shows a slice through the reactant pocket showing contours of displacement speed (colour) and progress variable (white lines).

Figure 1.7 (bottom) shows the surface-averaged values of the displacement speed S_d on each progress variable isosurface, plotted against the corresponding value of the progress variable c. It can be seen that S_d is large for small values of the

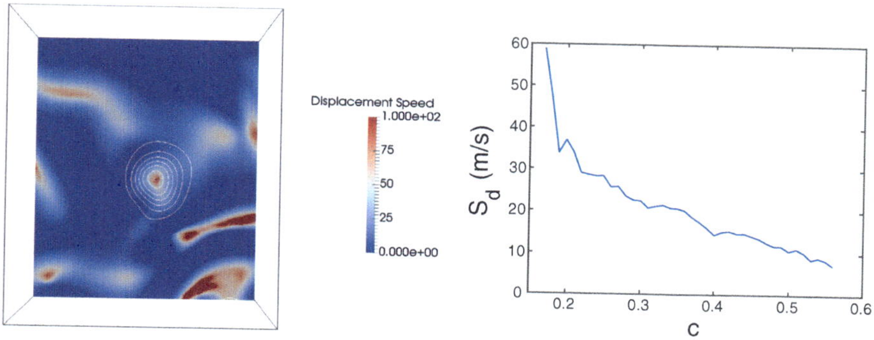

Fig. 1.7 (Left) Displacement speed on progress variable isosurfaces close to a critical point during reactant pocket burnout. (Right) Variation of surface-averaged displacement speed S_d with progress variable c close to reactant pocket burnout

progress variable in the reactant pocket close to the critical point, and falls rapidly with increasing progress variable away from the critical point. This behaviour is consistent with the theory, and confirms that the final burnout of a reactant pocket occurs very quickly.

1.4 Conclusions

The investigation of flame surface topology in premixed turbulent flames using DNS data offers a novel approach to understanding the influence of local flame geometry on global flame propagation behaviour. The interaction of two flame surfaces, or of a single flame surface with itself, is governed by a set of kinematic mechanisms which can induce rapid changes in the flame surface area and hence the turbulent flame speed.

A theoretical approach based on the Morse theory of critical points [20] has been presented and used to investigate the possible kinematic mechanisms. Firstly, the formulation developed by Griffiths et al. [18] has been described together with its application to the identification and extraction of critical points in premixed turbulent flame datasets produced from DNS. This formulation allows for all possible topologies to be divided into four general categories based on the eigenvalues of the Hessian matrix evaluated from the reaction progress variable field in the vicinity of a critical point. These categories are named as reactant pockets, product pockets, tunnel formation and tunnel closure. The pockets are essentially spherical while the tunnels are essentially cylindrical, and a map showing all categories has been constructed. Secondly, the progress variable field close to the critical point has been synthesised for each category using a three-dimensional extension of a previous two-dimensional theoretical analysis of Kollmann and Chen [15]. This approach permits the evaluation of key terms in the balance equation for the surface density function (SDF), and hence provides insight into flame propagation. The dissipation term and the kinematic restoration term are both found to become unbounded at the critical point in the tunnel closure configuration. Exact cancellation of these terms is found to occur at the critical point, thus avoiding singular behaviour in the flame surface area, but implying that flame pinch-off events resulting from tunnel closure are likely to be very rapid. Using a similar analysis technique, the displacement speed is found to become unbounded at the critical point for the reactant pocket configuration, with the implication that the final stage of pocket burnout is likely to be very fast. These results are consistent with the previous two-dimensional theory [15].

The predictions of the Morse theory were confirmed using a DNS dataset for a pair of premixed turbulent hydrogen–air flames [25]. The four different categories of flame topology were identified in the DNS dataset and their frequencies of occurrence were evaluated at two different Damköhler numbers. The majority of critical points were found to exist close to either the leading edge or the trailing edge of the flame, with fewer critical points occurring in the middle of the flame. There was a higher frequency of occurrence of tunnel formation and tunnel closure than for reactant pockets and product pockets. This is consistent with previous results which have indicated a prevalence of cylindrical curvature over spherical curvature on premixed turbulent flame surfaces [6].

Analysis of critical points from the same DNS dataset was used to confirm the behaviour of the dissipation and kinematic restoration terms in the SDF equation for the tunnel closure configuration. Strong variations in the magnitude of both terms were observed close to the critical point, and the two terms were found to have opposite signs in line with the theory. Similar analysis was carried out for the displacement speed in a reactant pocket, and large values of displacement speed were found close to the critical point.

The behaviour associated with each category of critical point, combined with knowledge of the frequency of occurrence of each category, is helpful in understanding the manner in which flame–flame interactions can affect overall flame propagation. Future work will involve improvements to the analysis techniques to improve accuracy, generality and throughput. In terms of flame physics, it will be necessary to investigate the influence of fuel type, fuel–air mixture strength and the effects of thermodiffusive and other inherent instability mechanisms. Finally, the insight gained will be invaluable in the construction of models of flame propagation.

References

1. H. Wang, E.R. Hawkes, J.H. Chen, B. Zhou, Direct numerical simulations of a high Karlovitz number laboratory premixed jet flame - an analysis of flame stretch and flame thickening. J. Fluid Mech. **815**, 511–536 (2017)
2. A.Y. Poludnenko, E.S. Oran, The interaction of high-speed turbulence with flames: global properties and internal flame structure. Combust. Flame **157**, 995–1011 (2010)
3. K.N.C. Bray, J.B. Moss, A unified statistical model of the premixed turbulent flame. Acta Astron. **4**, 291–320 (1977)
4. N. Peters, *Turbulent Combustion* (Cambridge University Press, Cambridge, 2000)
5. T.M. Wabel, A.W. Skiba, J.E. Temme, J.F. Driscoll, Measurements to determine the regimes of premixed flames in extreme turbulence. Proc. Combust. Inst. **36**, 1809–1816 (2017)

6. S.S. Girimaji, S.B. Pope, Propagating surfaces in isotropic turbulence. J. Fluid Mech. **234**, 247–277 (1992)
7. N. Chakraborty, R.S. Cant, Statistical behavior and modeling of the flame normal vector in premixed turbulent flames. Num. Heat Trans, Part I: Appl. **50**, 623–643 (2006)
8. W.T. Ashurst, Geometry of premixed flames in three–dimensional turbulence, in *Proceedings of the 1990 Summer Program*. Center for Turbulence Research, Stanford University & NASA Ames (1990), pp. 245–253
9. A.Y. Poludnenko, E.S. Oran, The interaction of high-speed turbulence with flames: turbulent flame speed. Combust. Flame **158**, 301–326 (2011)
10. M. Yoda, L. Hesselink, M.G. Mungal, Instantaneous three-dimensional concentration measurements in the self-similar region of a round high-Schmidt-number jet. J. Fluid Mech. **279**, 313–350 (1994)
11. L. Wang, N. Peters, The length-scale distribution function of the distance between extremal points in passive scalar turbulence. J. Fluid Mech. **554**, 457–475 (2006)
12. Y. Shim, S. Tanaka, M. Tanahashi, T. Miyauchi, Local structure and fractal characteristics of H_2-air turbulent premixed flame. Proc. Combust. Inst. **33**, 1455–1462 (2011)
13. Y. Minamoto, N. Swaminathan, R.S. Cant, T. Leung, Morphological and statistical features of reaction zones in MILD and premixed combustion. Combust. Flame **161**, 2801–2814 (2014)
14. H. Minkowski, Volumen und Overfläche. Math. Ann. **57**, 447–495 (1903)
15. W. Kollmann, J.H. Chen, Pocket formation and the flame surface density equation. Proc. Combust. Inst. **27**, 927–934 (1998)
16. J.H. Chen, T. Echekki, W. Kollmann, The mechanism of two-dimensional pocket formation in lean premixed methane-air flames with implications to turbulent combustion. Combust. Flame. **116**, 15–48 (1999)
17. S. Trivedi, R.A.C. Griffiths, H. Kolla, J.H. Chen, R.S. Cant, Topology of pocket formation in turbulent premixed flames. Proc. Combust. Inst. **37**, 2619–2626 (2019)
18. R.A.C. Griffiths, J.H. Chen, H. Kolla, R.S. Cant, W. Kollmann, Three-dimensional topology of turbulent premixed flame interaction. Proc. Combust. Inst. **35**, 1341–1348 (2015)
19. S. Trivedi, G.V. Nivarti, R.S. Cant, Flame self-interactions with increasing turbulence intensity. Proc. Combust. Inst. **37**, 2443–2449 (2019)
20. J. Milnor, *Morse Theory* (Princeton University Press, Princeton, 1963)
21. L. Vervisch, E. Bidaux, K.N.C. Bray, W. Kollmann, Surface density function in premixed turbulent combustion modeling, similarities between probability density function and flame surface approaches. Phys. Fluids **7**, 2496–2503 (1995)
22. R.S. Cant, S.B. Pope, K.N.C. Bray, Modelling of flamelet surface-to-volume ratio in turbulent premixed combustion. Proc. Combust. Inst. **23**, 809–815 (1990)
23. E.R. Hawkes, R.S. Cant, A flame surface density approach to Large-Eddy Simulation of premixed turbulent combustion. Proc. Combust. Inst. **28**, 51–58 (2000)
24. M. Boger, D. Veynante, H. Boughanem, A. Trouvé, Direct numerical simulation analysis of flame surface density concept for large eddy simulation of turbulent premixed combustion. Proc. Combust. Inst. **27**, 917–925 (1998)
25. E.R. Hawkes, O. Chatakonda, H. Kolla, A.R. Kerstein, J.H. Chen, A petascale direct numerical simulation study of the modelling of flame wrinkling for large eddy simulations in intense turbulence. Combust. Flame **159**, 2690–2703 (2012)
26. J. Li, Z. Zhao, A. Kazarov, F.L. Dryer, An updated comprehensive kinetic model of hydrogen combustion. Int. J. Chem. Kinet. **36**, 566–575 (2004)
27. J.H. Chen, A. Choudhary, B. de Supinski, M. DeVries, E.R. Hawkes, S. Klasky, W.K. Liao, K.L. Ma, J. Mellor-Crummey, N. Podhorszki, R. Sankaran, S. Shende, C.S. Yoo, Terascale direct numerical simulations of turbulent combustion using S3D. Comput. Sci. Discov. **2**, 015001 (2009)

Chapter 2
Dissipation Element Analysis of Inert and Reacting Turbulent Flows

Dominik Denker, Antonio Attili, and Heinz Pitsch

Abstract Dissipation elements provide a procedure for compartmentalizing scalar fields into physically meaningful sub-units which provides a direct measure for turbulent scales. Furthermore, dissipation elements enable a variety of additional ways of investigating non-local effects in reacting and non-reacting turbulent flows. After the underlying physical ideas of dissipation elements are explained and a parameterization of dissipation elements is defined, the method of detecting dissipation elements with gradient trajectories is explained and physical and numerical prerequisites are presented. Common characteristics of dissipation elements are interpreted and compared for a large range of selected reacting and non-reacting flow configurations. To provide the reader with a degree of familiarity, dissipation element statistics are then related to more commonly used methods of obtaining statistics. The additional benefit of using the dissipation element analysis in free shear flows is highlighted by using it as an alternative way of identifying turbulent core regions. Next, a dissipation element-based procedure for the local investigation of the turbulence–combustion interaction in the context of non-premixed flames is presented. The chapter is concluded with the application of a dissipation element statistics-based modeling procedure for computational fluid dynamics of a passenger car diesel engine, employing the previously gained insight into the structure of turbulent scalar fields.

2.1 An Introduction to Dissipation Elements

The sheer volume of data obtained from direct numerical simulations (DNS) poses a challenge in itself for any analysis of turbulence and turbulent combustion. If the nondeterministic behavior of turbulence is added on top, the complexity is increased to a point where novel methods of analysis are inevitably required to gain deeper insight into this physical phenomenon that defies complete comprehension regardless of the best efforts of many generations of researchers. A desirable quality of a method

D. Denker (✉) · A. Attili · H. Pitsch
Institut für Technische Verbrennung, Templergraben 64, 52056 Aachen, Germany
e-mail: ddenker@itv.rwth-aachen.de

H. Pitsch and A. Attili (eds.), *Data Analysis for Direct Numerical Simulations of Turbulent Combustion*, https://doi.org/10.1007/978-3-030-44718-2_2

for the analysis of turbulent flows is the ability to break down the complexity of the entire turbulent domain into smaller structures which are ideally simple enough for an intuitive consideration. At the same time, this method of analysis needs to be robust and efficient enough to cope with the vast amount of data generated by modern large-scale DNS. A potential method for the analysis for turbulent flows that fulfills these requirements is the dissipation element (DE) analysis.

In this chapter, the definition of DEs is introduced and a brief motivation for their use in turbulent flows will be given. Additionally, physical and numerical requirements of the analyzed flows will be discussed. Identification of turbulent scales by means of DEs in DNS is shown in Sect. 2.2. DE parameter statistics will be related to more commonly used conventional statistics in Sect. 2.3, followed by specific applications for inert flows and reacting flows in Sects. 2.4 and 2.5. Finally, an outlook on possible DE modeling procedures is given in Sect. 2.6.

2.1.1 *Definition of Dissipation Elements*

The inherent challenge of analyzing turbulent flows stems from the complexity introduced by the chaotic and multiscale nature of turbulence. A way to overcome this challenge and gain a deeper insight is to compartmentalize the entire turbulent field and its scales into simpler sub-units. If the criteria for compartmentailization are chosen sensibly, the analysis of these sub-units is far less complex and the comprehension of more detailed structures and correlations is possible.

Other scale-decomposing methods exist including the Fourier transform, the wavelet transform and proper orthogonal decomposition, and these methods have their individual advantages for linear problems. However these methods are less valuable for the understanding of the nonlinear interactions in turbulence or have limited applicability in flow configurations which display high levels of anisotropy.

Another desirable characteristic of a compartmentalization method for turbulent flows is the space-filling and non-overlapping nature of the sub-units. To ensure that all physical effects in the turbulent domain are considered, all material points must be uniquely assigned to a sub-unit. While systematic approaches for the space-filling decomposition exist, such as the velocity gradient tensor Q and pressure p-based approach [1], these methods suffer from the arbitrariness introduced from the setting of threshold levels. Finally, the method to decompose turbulent domains should ideally yield results that are easy to interpret in a physical way, as well as the potential for a model developement.

A physically motivated method for such a compartmentalization of scalar fields in turbulent flows that meets all mentioned requirements is the dissipation element (DE) analysis [2]. DEs are ensembles of material points in which the underlying scalar behaves monotonically, i.e., the space in between the turbulence-introduced extremal points. To illustrate the identification of DEs in a 1D setting, a schematic profile of a scalar ϕ in physical space is shown in Fig. 2.1. An intuitive way for a quick estimation of scales in turbulent profiles or signals is to observe the local extremal points and

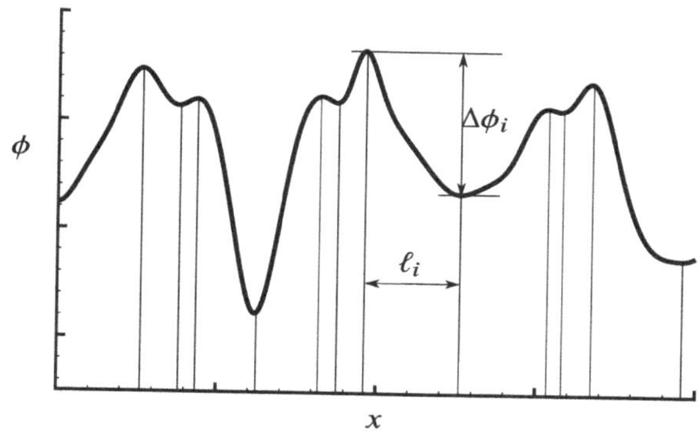

Fig. 2.1 Schematic representation of a DE and its parameters in 1D space. ℓ_i indicates the linear distance and $\Delta\phi_i$ the scalar difference of the ith DE in the scalar profile

judge the spacing between the minima and the maxima. In 1D, DE analysis does the exact same procedure: first, all extremal points in the scalar field are identified. Second, the segments of space linking individual pairs of minima and maxima are then classified as DEs. The DE identified in this way can then be parametrized by its pair of extremal points. The DE length of the ith DE is defined as the separation distance of the minima and maxima in physical space $\ell_i = |x_{\max,i} - x_{\min,i}|$ and the DE scalar difference is defined as the difference of the scalar in these extremal points $\Delta\phi_i = \phi_{\max,i} - \phi_{\min,i}$. These two parameters are henceforth called DE parameters. Additionally, the mean DE scalar gradient is calculated as $g_i = \frac{\Delta\phi_i}{\ell_i}$, which takes the steepness of the entire sub-unit into consideration.

The numerical procedure of DE analysis is trivial for a 1D case, but serves to illustrate the underlying intuitive idea of obtaining scales in this fashion. In higher dimensional spaces, the number of DEs linked to each extremal point is not limited to two. Additionally, the treatment of critical points in the scalar fields gets more complicated as the dimensionality is increased. To link the extremal points in scalar fields in 2D and 3D space, gradient trajectories of ϕ are employed. Starting from any material point, the direction of the scalar gradient $\mathbf{n}_\phi = \frac{\nabla\phi}{|\nabla\phi|}$ is calculated. The gradient trajectory can be determined by tracing along descending $-\mathbf{n}_\phi$ and ascending directions $\mathbf{n}_\phi$. This can be interpreted as moving in the normal direction from each iso-surface of ϕ to the next. The gradient trajectories are traced until they terminate at a maximum in the ascending direction and at a minimum in the descending direction, whereby the definition of extremal points the gradient of ϕ is zero. This procedure in 3D space is illustrated for three exemplary material points and their gradient trajectories in Fig. 2.2. The ensemble of material points linked to the same pair of extremal points in a scalar field is then classified as a DE. It is easy to see that each material point is uniquely assigned to a single DE and since all material points can be assigned, the method provides a space-filling compartmentalization. Therefore, the scalar field itself provides the geometry of the sub-units and the obtained scales are an intrinsic result of the flow.

While the definition of the DE scalar difference $\Delta\phi$ remains the same in higher dimensional fields, the DE length is defined as the Euclidian distance between the

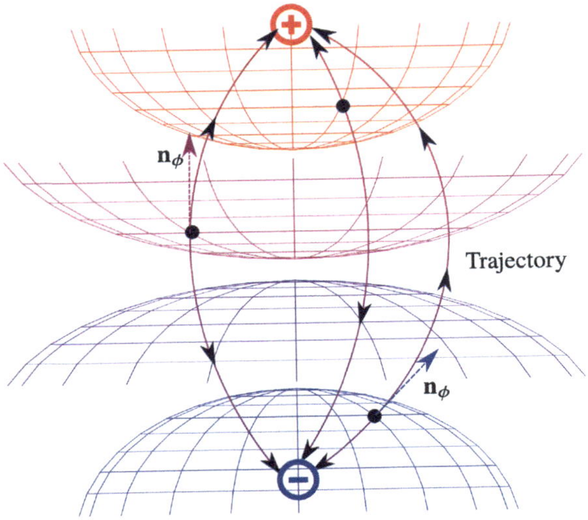

Fig. 2.2 Schematic representation of gradient trajectories connecting the same extremal points in 3D space. The grids represent different iso-surfaces of the scalar ϕ. The black dots indicate material points where gradient trajectories are started in the DE detection process

minimum and the maximum $\ell = |\mathbf{x}_{\max} - \mathbf{x}_{\min}|$. The arc-length s of the gradient trajectories between the extremal points might be a more obvious choice to characterize the length of the DE, but s differs slightly based on the location of the different material points in one DE and is therefore not unambiguous.

A DE obtained from a scalar field from a 3D DNS of a planar temporally evolving non-premixed jet flame is shown in Fig. 2.3. Additionally, a section of an iso-surface of the scalar is shown to further illustrate the tracing of gradient trajectories in the normal direction from iso-surface to iso-surface. In the 3D space, DEs display various shapes and sizes. The overall orientation of each DE, as well as the trajectories used in its detection, proceed approximately in the direction of the straight line connecting the two extremal points. A common characteristic shared by most DEs is the converging of all gradient trajectories as the extremal points are approached. As a multitude of DEs are connected to a single extremal point, these elements form interwoven and entangled structures in 3D space.

While the space-filling parametrization of the turbulent scalar fields by means of the DE already yields a plethora of applications in the analysis of turbulent flows, a lot of physical insight can be gained by using individual DEs as a nonarbitrary and meaningful frame of reference for obtaining spatial statistics. As mentioned above, DEs are regions of monotonic scalar between extremal points. The diffusive flux of ϕ is defined as $J = -D_\phi \nabla \phi$, where D_ϕ is the diffusion coefficient of the scalar.

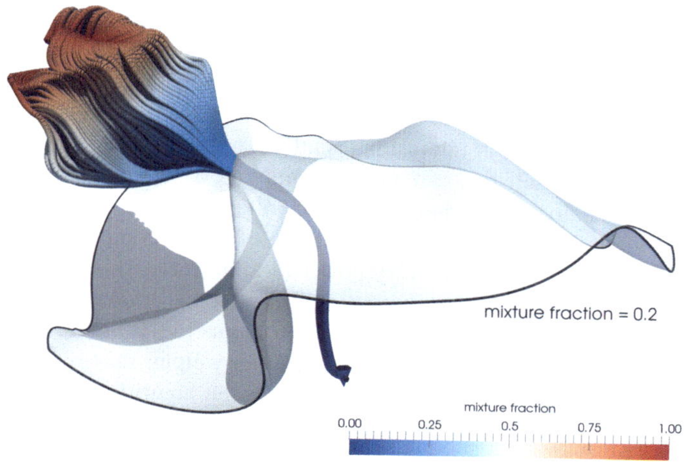

Fig. 2.3 DE in the mixture fraction field Z in a DNS of non-premixed temporally evolving jet flame. The DE is represented by the gradient trajectories used in the detection process. The gradient trajectories are colored by the local value of the passive scalar. The gray iso-surface indicates the position of the stoichiometric mixture fraction

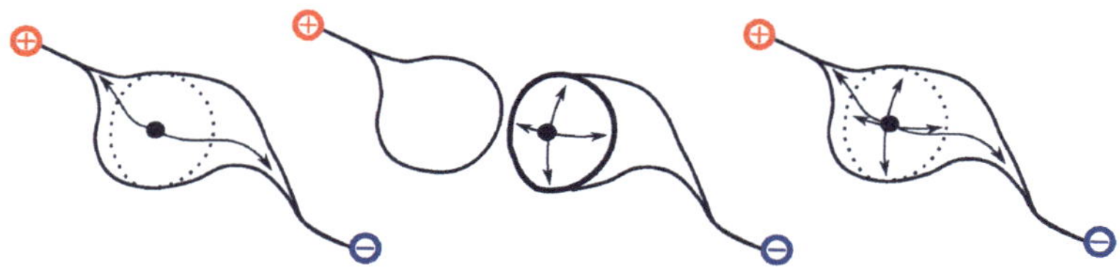

Fig. 2.4 Schematic representation of the potential methods of obtaining spatial statistics in the DE decomposed space. Left: obtaining statistics along the gradient trajectories of the individual DE. Middle: obtaining statistics on an iso-surface (such as mixture fraction or temperature) intersecting the DE. Right: using the DE volume as a frame of reference for statistics

Therefore, it is easy to see that the diffusive flux is interrupted in the extremal points of the scalar field as $J = 0$, since by definition in these points $\nabla\phi = 0$.

Selected methods of using DEs for obtaining spatial statistics are shown in Fig. 2.4. In the cartoon of a DE on the left side, the gradient trajectories are used for obtaining statistics in the normal direction. This provides the ability to obtain structure-function-like statistics along physically meaningful directions and overcome the inherent difficulties in anisotropic flows. Examples and results of normal direction statistics will be given in Sect. 2.2. A second method is to obtain statistics in the tangential direction on the intersection area of an iso-surface and a DE. This is shown for the cartoon of a DE in the middle of Fig. 2.4. Material points on the vertices of the iso-surface of ϕ are chosen as the starting points of the gradient trajectories, uniquely linking sections of the iso-surface to DEs. This is especially useful for scalars which possess physically meaningful absolute values. An example of this was already shown in Fig. 2.3, where the indicated iso-surface is the stoichiometric mixture fraction Z_{st} which indicates the position of the reaction zone in non-premixed flames. Finally, as shown by the DE cartoon on the right side of Fig. 2.4, the entire DE volume can be used to obtain statistics like variances, minimum or maximum values, and other quantities of interest in each DE.

DE analysis lends itself for tracking the temporal evolution of features in time-varying data. Especially when temporal events involving the DEs, like splitting, merging or annihilation, can be meaningfully interpreted in the context of the analyzed scalar and physics, this becomes a very powerful tool. Recently, a robust algorithm was developed which determines the assignment of DEs between two successive time steps by solving two graph optimization problems [3]. It first resolves one-to-one assignments of DEs by computing a maximum-weight, maximum-cardinality matching on a weighted bi-partite graph. Second, the algorithm detects temporal events by creating a graph of potentially conflicting event explanations and finding a weighted, independent set in it.

2.1.2 Physical and Numerical Considerations

The applicability of the DE analysis to a given scalar field depends on several prerequisites, which are of the physical nature of the scalar field and numerical properties of the simulation which provided the scalar field.

Whether a scalar field can be sensibly decomposed into DEs depends on the following characteristics of the scalar [4]:

1. Gradient trajectories must end at definite points without spiraling infinitely.
2. Trajectories and DEs should not be trivial, but display natural and intrinsic characteristics of turbulent fields. For instance, trajectories should neither pass through the entire flow field to be infinitely long, nor be infinitesimally small.
3. The tracing process of trajectories should be relatively stable both analytically and numerically, to resist the perturbation from the inaccuracy of flow fields, or else a unique decomposition of the entire flow field into DEs will not be possible.

Most scalar fields obtained from DNS of turbulent flows satisfy these three conditions if certain numerical prerequisites in the DNS are met. Regarding the first characteristic, it was shown that in a sufficiently smooth scalar field the length of a gradient trajectory will be finite [5]. This implies that since the length of the gradient trajectory is bounded, an infinite spiraling loop should not be possible if the numerical schemes and resolution in the DNS are adequate to provide a sufficiently smooth scalar field.

Satisfying the second condition is linked to a significantly high enough Reynolds number of the flow. If the gradient trajectories can extend over the entire integral scale of the flow, characteristics of the turbulent scales will not be reflected by the DEs due to the lack of scale separation. Transport of most scalars in turbulent flows is governed by the balance of advective and diffusive terms, the former introducing extremal points and the latter removing them. A scalar field dominated by diffusive transport makes a poor choice for DE analysis. Therefore, additional attention should be paid to the Schmidt number of the scalar $Sc_\phi = \nu / D_\phi$, where ν is the kinematic viscosity. Scalars with very low values of Sc_ϕ may not be suitable to estimate turbulent scales accurately even if the Reynolds number is sufficient to result in a well developed turbulent flow field.

The third requirement of the robustness of the traced trajectories against small perturbations is ensured by the previously mentioned diffusion processes in turbulent flows. In the inertial range of turbulence, where viscous effects can be neglected, the random motions of eddies will cause the iso-surfaces of scalars to be highly corrugated. Recalling the analogy between the gradient trajectory tracing and the movement in normal direction from iso-surface to iso-surface, small perturbations caused by inaccuracies will quickly lead to faulty pathing of the gradient trajectories. However, in the viscous range, diffusion smooths the corrugation of the iso-surfaces caused by the turbulent eddies. It follows that the tracing of the gradient trajectories must be performed well below the inertial scale [4]. The consequence for the DNS is the requirement of well resolved viscous scales. This observation implies that DE analysis for scalars obtained from large eddy simulations or experiments that do not resolve the viscous turbulent scales might be problematic.

2.1.2.1 Treatment of Critical Points

In a scalar field, all material points can be classified as critical points and noncritical points. If $\nabla\phi = 0$ is zero at a certain point, this point will be critical. The tracing of the gradient trajectories relies on a well defined direction $\mathbf{n}_\phi$. Due to the previously mentioned physical characteristics of typical scalars in turbulent flows, this is ensured in noncritcal points. Therefore, critical points need to be treated differently. In 3D space, based on the eigenvalues of the Jacobian $\mathbf{J}_\phi = \nabla\nabla\phi$, four types of critical points can be classified for viscous scalar fields: sources, sinks, repelling node saddles, and attracting node saddles [6]. As trajectories converge or diverge in straight lines in all of these critical points, analysis and treatment is fairly simple. Regardless of that, differences between the extremal points and saddle points need to be taken into account.

Saddle points are mathematically not stable and the trajectories in the vicinity of these points can be diverted by a small perturbation, while extremal points are stable and their position can be identified in a robust manner even if small perturbations are present. Due to the fact that the scalar fields used to trace trajectories are diffusion controlled at small scales, the scalar gradient vectors around extremal points will point away from a minimum and in a straight line toward a maximum without rotation.

As soon as the scalar gradient decreases below a certain value compared to the mean gradient in the proximity, an extremal point is expected in the immediate vicinity. Therefore, Wang [4] defines the detection threshold ϵ as the ratio of the local scalar gradient to the mean gradient in the surrounding grid points. The extremal point will be encapsulated by a spherical iso-surface of a scalar value close to the one in the extremal point. The radius of this sphere r depends on the threshold ϵ. For an exemplary threshold of $\epsilon = 2 \cdot 10^{-5}$ (which was used in all of the DE analyses shown in the following sections) the radius of the containing sphere will be $r = 0.02\Delta$, with Δ being the grid spacing between surrounding points. Under the assumption of sufficient numerical resolution and the assumption that the expected critical point is located exactly at the center of the sphere, the local curvature $\kappa = \nabla \cdot \mathbf{n}$ can then be used to identify the type of critical point. For $\kappa = -6$ a maximum is detected, while for $\kappa = 6$ a minimum and $-2 \leq \kappa \leq 2$ a saddle are identified.

In case of a detected extremal point, the gradient trajectory tracing can be stopped for the material point. In case of a saddle point, the gradient trajectory needs to be artificially perturbed, i.e., numerically moved for a small step size in a random direction. However, this procedure is only necessary in the exceedingly rare event that a chosen material point is situated on a saddle point.

From these considerations, it is easily seen that the requirements of the resolution of the DNS exceed the commonly used minimum requirements of $\Delta/\eta \leq 2.1$ [7]. Based on the numerical schemes employed in the DNS, lacking numerical resolution leads to an introduction of false extremal points, if spectral methods are used. Otherwise, extremal points are removed if the numerical diffusivity of the scheme is too high. Clearly, both effects might change the results of the DE analysis. For DNS of isotropic turbulence employing spectral methods, it was found that results of DE analysis do not change for resolutions of $\Delta/\eta \leq 1$ [4]. The same restrictions of $\Delta/\eta \leq 1$ apply to central finite difference schemes and weighted ENO [8] schemes used in the DNS in the following sections.

2.2 DE Parameter Statistics for Inert and Reactive Turbulent Flows

Instead of indirectly obtaining turbulent scales by relating other statistical turbulence quantities to each other, for example expressing the integral length scale as $l_\tau \sim k^{3/2}/\varepsilon$, with the turbulent kinetic energy k and the dissipation rate ε, DE parameters provide scales which are obtained directly from the turbulent flow. The statistics of these parameters allow for some fundamental insight into the governing physical mechanisms and shall be discussed in this section.

The probability density function (PDF) of the DE length $P(\ell)$ displays interesting characteristics as well as helps to understand the physical mechanisms governing the DE formation and annihilation. DE analysis was first applied to isotropic turbulence [9]. It was found that the PDF of the normalized DE length $\ell^* = \ell/\ell_\mathrm{m}$, with the volume average mean DE length ℓ_m, shows a remarkable invariance toward changes in the Reynolds number. This invariance is independent of the scalar field the DE analysis is applied to and carries over to non-isotropic configurations [10], reacting flows [11] and was confirmed in experiments [12]. To demonstrate this, $P(\ell^*)$ is shown for isotropic turbulence as well as other selected cases in Fig. 2.5. These DNS include isotropic turbulence [13], a planar spatially evolving mixing layer [14], a pla-

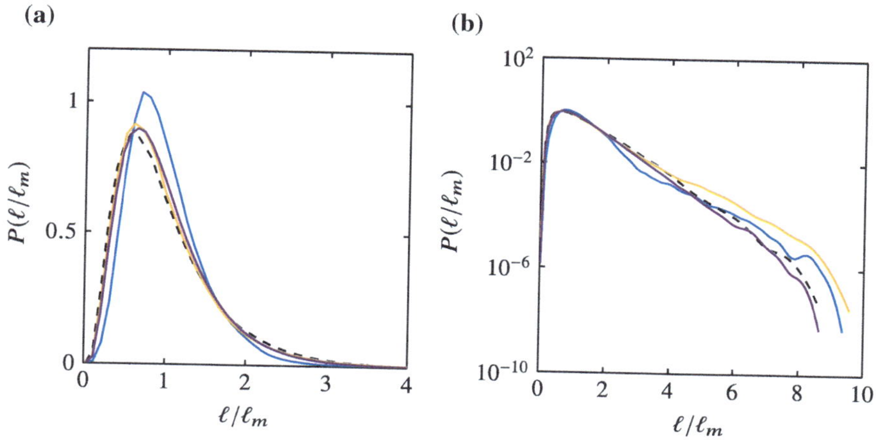

Fig. 2.5 PDF of the normalized DE length ℓ/ℓ_{m} in **a** linear/linear scale and **b** linear/logarithmic scale obtained from DNS of various inert and reacting cases. (—): passive scalar in isotropic turbulence; (—): passive scalar in spatially evolving mixing layer; (—): mixture fraction in temporally evolving diffusion jet flame; and (—): temperature in spatially evolving premixed jet flame

nar temporally evolving non-premixed jet flame [15], and a planar spatially evolving premixed jet flame [11]. Even though the cases are vastly different in terms of the flow configuration and physical properties of the scalar, the PDFs show almost perfect collapse. All PDFs display two distinct features. For small ℓ, a linear increase of the PDF is observed in the linear/linear scaled plot in Fig. 2.5a. This linear increase corresponds to the diffusive drift of the extremal points toward each other and the subsequent annihilation of DEs. The second feature is the random cutting and reconnection of large DEs by turbulent eddies. This manifests itself in the exponential decrease of the PDF for large ℓ, which can be observed in the linear/logarithmic scale plot in Fig. 2.5b.

The scaling of the mean DE length is demonstrated in Fig. 2.6, where the ratio of the mean DE length and the Kolmogorov microscale ℓ_{m}/η is shown for selected flow configurations and Reynolds numbers. A clear Kolmogorov scaling can be observed for all cases. The absolute value of the ratio depends slightly on the scalar field the DE analysis was applied to. Therefore, normalizing ℓ with the η instead of ℓ_{m} yields similar results with regards to the collapse of the PDFs, which was observed in Fig. 2.5.

Adding the second DE parameter to the statistical analyses helps to gain further insight. The joint probability density function (jPDF) of both DE parameters $P(\ell, \Delta\phi)$ is expected to suffice for a complete statistical description of a scalar field [9]. The jPDF of both normalized DE parameters is shown for selected cases in Fig. 2.7. The DE scalar difference was normalized in the same fashion as the DE length, $\Delta\phi^* = \Delta\phi/\Delta\phi_{\mathrm{m}}$, with the volume averaged scalar difference $\Delta\phi_{\mathrm{m}}$. Analogous to the marginal PDF in Fig. 2.5, regions in the jPDF can be attributed to different physical mechanisms. These regions are marked in Fig. 2.7a. The diffusive drift

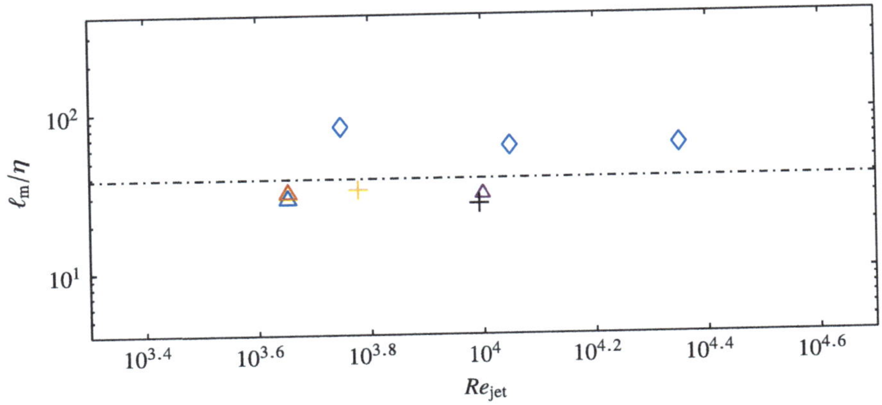

Fig. 2.6 Ratio of the Kolmogorov microscale η to the mean DE length ℓ_m for various inert and reacting cases. Diamonds: spatially evolving premixed jet flames; triangles: temporally evolving diffusion jet flames; and crosses: inert temporally evolving jets. The dash dotted line indicates the average ratio obtained from isotropic turbulence. The line is placed arbitrarily as no jet Reynolds number is defined in homogeneous isotropic turbulence

region is characterized by small ℓ and $\Delta\phi$. The random splitting and reconnection region is indicated for intermediate to large ℓ and $\Delta\phi$. Additionally, the probability density of cliff structures in the scalar fields, the common feature of turbulent scalars [16], can be observed for small ℓ and large $\Delta\phi$. While the scalar difference between the extremal points in Fig. 2.7a–c is solely governed by diffusive and convective transport, for the temperature field in Fig. 2.7a a source term is present in the governing equations. While the normalized spatial distribution of the extremal points remains the same, c.f. Fig. 2.5, the scalar difference is significantly influenced by the source term. The presence of the strongly coherent and robust flame front results in a high probability of cliff structures in the jPDF. The motion of turbulent eddies will introduce extreme points in the temperature field in the vicinity of the flame front, creating smaller scalar differences. However, the inherent difference in the source terms in these newly created points will counteract this. The difference between the fairly universal jPDF of the passive scalar ϕ and the jPDF of a reactive scalar, in this example the temperature T, can be used to gauge the level of scale interaction between turbulence and the physical mechanism governing the source term.

2.3 DE Conditioned Structure Functions

Besides providing turbulent scales, the DEs provide a multitude of ways to analyze turbulent flows or enhance existing methods of analysis. A few examples, which are not limited to inert flows, shall be given in the following section.

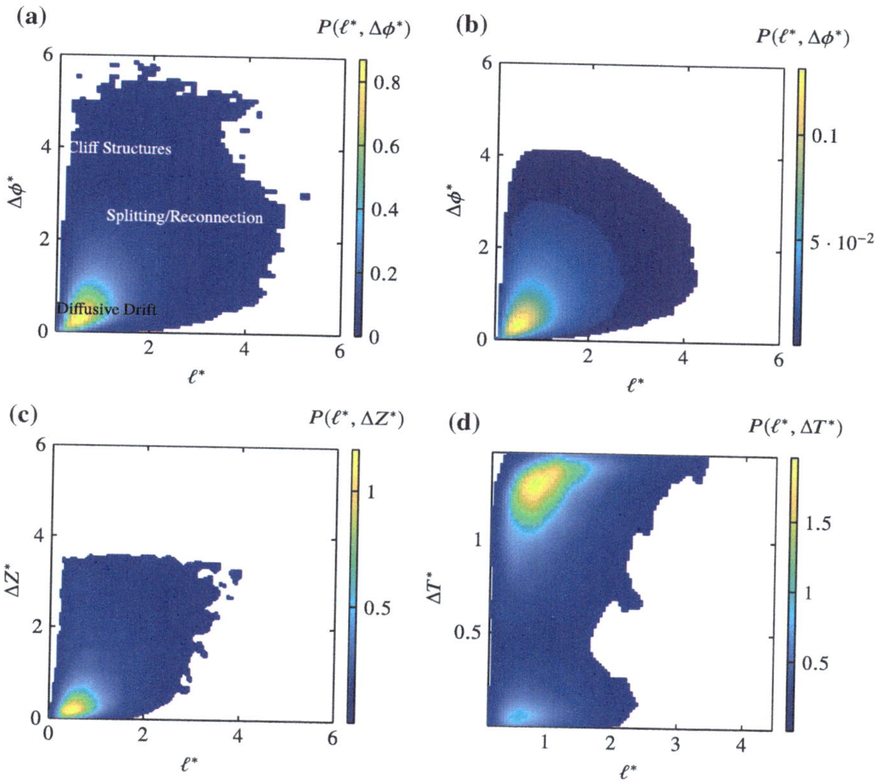

Fig. 2.7 jPDFS of the normalized DE length and DE scalar difference for various fields in different flow configurations. **a** passive scalar in an inert temporally evolving planar jet, **b** passive scalar in an inert spatially evolving mixing layer, **c** mixture fraction in a planar temporally evolving non-premixed jet flame, and **d** temperature in a planar spatially evolving premixed jet flame

It is possible to relate DEs to a different and commonly used method of analysis. Structure functions are a common tool to study the multiscale nature of turbulence to overcome the challenge of non-locality in turbulent flows. Structure functions of order n are defined as

$$S_n(\mathbf{r}) = \langle(\phi(\mathbf{x}+\mathbf{r}) - \phi(\mathbf{x}))^n\rangle\,, \tag{2.1}$$

with $\mathbf{r}$ being the spatial separation between the two points and $\langle\rangle$ indicating an appropriately defined statistical average. Besides the simple case of isotropic turbulence, choosing a physically meaningful $\mathbf{r}$ can prove to be challenging in non-homogeneous or non-isotropic flows. A straightforward way of choosing $\mathbf{r}$ is employing the gradient trajectories of the DEs. The difference of the quantity of interest can be conditioned on the arc length distance s of any point on the gradient trajectory starting in a respective material point as shown by [17]. Even more simplified, only the extremal point pairs of individual DEs can be used to obtain structure functions based on DEs:

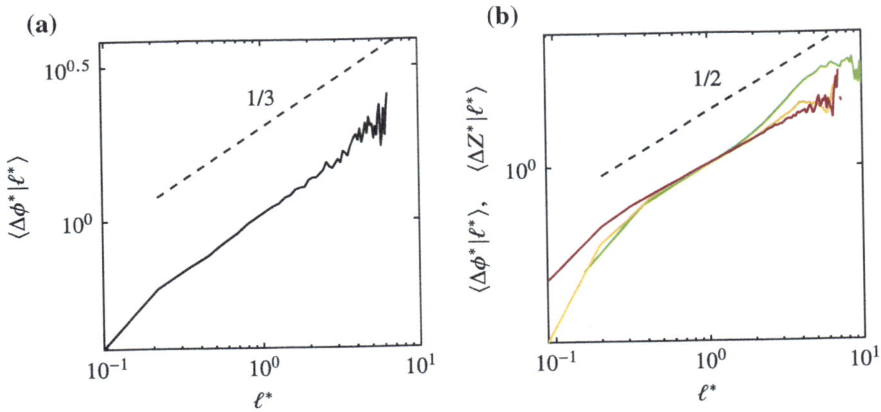

Fig. 2.8 Mean DE scalar difference conditioned on the DE length for various flow configurations. (—): isotropic turbulence; (—): planar temporally evolving non-premixed jet flame; (—): inert temporally evolving planar jet; and (—): inert spatially evolving mixing layer. The dashed black lines indicate in **a** the theoretic scaling and in **b** actual scaling observed in free shear flows

$$S_n \sim \langle \Delta\phi^n | \ell \rangle \,. \tag{2.2}$$

The scalar difference to the nth power conditioned on the DE length $\langle \Delta\phi^n | \ell \rangle$ represents a structure function where the distance $\mathbf{r}$, in this case ℓ, is not arbitrarily chosen, but is determined by the two extremal points of each dissipation element. The statistical average will be done for dissipation elements with the same lengths, instead of using arbitrary points with same separating distances.

The results for the normalized first-order scalar structure function in isotropic turbulence are shown in Fig. 2.8a. The theoretical scaling with this exponent of $1/3$ is recovered in this way of obtaining the structure function [9]. In isotropic turbulence, other theoretical scalings of the structure functions, such as the value of the scaling exponent of $2/3$ for the structure function of the turbulent kinetic energy $\langle (k(\mathbf{x}+\mathbf{r}) - k(\mathbf{x})) \rangle \sim \langle \Delta k | \ell \rangle$, can be recovered as well [17]. However, in free shear flows, the presence of mean shear has an influence on the scaling of the structure functions. This can be investigated accurately with the DE conditioned structure functions as well, as demonstrated for selected shear flow configurations in Fig. 2.8b. Even though the cases are vastly different, the conditional structure functions display a seemingly universal scaling exponent of $1/2$.

Obtaining structure functions in this way is, of course, not limited to the first order, but for the sake of brevity, additional orders are not shown here.

2.4 DE-Based Flow Partitioning in Shear Flows

Another challenge faced in the analysis of free shear flows is the presence of both laminar and turbulent regions. Physical properties of the flow change drastically from one region to the other. Therefore, it is straightforward to see that precisely separating turbulent and laminar parts is of prime importance to obtain meaningful statistics. Turbulent and laminar regions are separated by a thin layer called the turbulent/non-turbulent interface (TNTI). While different and easy to use methods of locating the TNTI exist [18, 19], the flow in the turbulent region just behind the TNTI would not be considered fully turbulent [20]. Simply using all material points behind the TNTI would potentially lead to the inclusion of low Reynolds number effects in the statistics. The region in between the TNTI and the turbulent core was even considered as a third region [21]. It was emphasized that the characteristics of this transition zone might be relevant to the overall behavior if it constitutes a relatively large fraction of the flow.

A nonarbitrary way of partitioning free shear flows and locating the turbulent core regions of shear flows employing DEs was introduced by [22]. This procedure will be explained using the example of a spatially evolving mixing layer. In this flow, two non-rotational or laminar free streams exist, therefore some of the gradient trajectories used in a DE analysis will head toward the outer irrotational streams without reaching an extremal point. These escaping trajectories are used to define the transition regions and to partition the flow. Thus, the general method is as follows: as usual, the gradient trajectory is calculated at each material point located within the TNTI. If this gradient trajectory joins one minimum and one maximum and forms a DE, as represented by trajectories A in the sketch in Fig. 2.9, that point is defined to be inside the turbulent core, which is depicted by the two regions enclosed by the solid lines in that same figure. On the contrary, if the trajectory connects a minimum with the upper outer stream, where the passive scalar is $\phi = 1$, that material point is identified as belonging to the upper turbulence transition region. This case is represented by trajectories B in Fig. 2.9. Correspondingly, if the trajectory moves from a maximum to the lower outer stream, where the passive scalar is $\phi = 0$, the point is inside the lower turbulence transition region. An added benefit of this method is the correct identification of engulfed regions, as shown by trajectory C. Finally, the possibility that a trajectory connects the upper outer region with the lower one without an intermediate extremal point, as in trajectory D, needs to be considered. The material points whose trajectories follow this behavior define internal quasi-laminar diffusion layers. The identification of these regions is especially useful in an early time step in temporally evolving shear flow configurations or in upstream positions in spatially evolving configurations.

The results of this procedure for detecting the turbulent core are shown for the spatially evolving planar mixing layer in a downstream position in Fig. 2.10. It shows that the location of the fully turbulent core in this configuration is by no means trivial. While the majority of the turbulent core regions is situated in the cross-stream center plane, a few isolated regions extend far from this plane. Due to the high

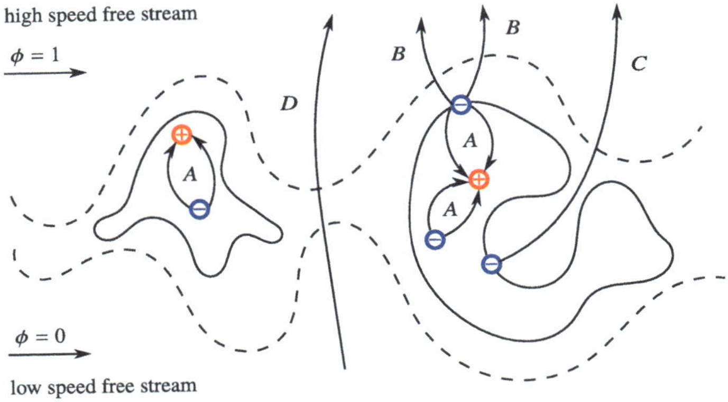

Fig. 2.9 Schematic flow partitioning of a mixing layer-based on DEs: A, trajectories reach from minimum to maximum forming a complete DE indicating a fully turbulent zone; B, trajectories continue from a minimum on the upper part of the mixing layer into irrotational regions; C, trajectory from an engulfed minimum at a lower cross-stream position; D, trajectory passes from one side of the mixing layer to another indicating a quasi-laminar diffusion layer. Solid line separates turbulent zone from turbulence interface. Dashed line indicates the traditional TNTI. Adapted from [22]

Fig. 2.10 Passive scalar in the spanwise-wise centerplane of a spatially evolving mixing layer. Color scale starting at blue ($\phi = 0$) increasing to yellow ($\phi = 1$). Red patches indicate turbulent core regions of the flow which are isolated from the TNTI by means of DE analysis

amount of entrainment of laminar regions into the turbulent flow in this configuration, the turbulent core does not form one coherent region, but displays a fairly patchy appearance. This further demonstrates that simply limiting the investigation to a range of cross-stream coordinates does not completely rule out the inclusion of low Reynolds number effects on the statistics.

The statistics obtained in the fully turbulent regions differ significantly from those obtained in the transition zone [22]. Furthermore, exploiting the universality of DE parameter statistics, demonstrated in Figs. 2.5 and 2.7, this partitioning was used to model a variety of passive scalar statistics in free shear flows using a composite PDF approach for the individual regions [23].

2.5 DE Analysis of Reacting Flows

There are many different physical phenomena in turbulent flows which strongly interact with a certain range of turbulent scales. As the DE analysis provides turbulent scales, a local comparison with the characteristic scales of the additional physical phenomenon is straightforward. The investigation of scale interaction by means of DE analysis is illustrated here in the context of turbulent combustion. This subject lends itself to a DE analysis because in the asymptotic limits of no scale interaction or very limited interaction, turbulent combustion is fairly well understood and can be modeled appropriately, while the intense interaction of flame scales and turbulent scales is still the subject of research.

For premixed combustion, the idea of varying levels of scale interaction leads to the construction of the so-called "Borghi–Peters diagram" [24]. Turbulent scales, such as the Kolmogorov length η are compared to different scales of the flame. Using the DE length ℓ obtained from the temperature field for the local estimation of the turbulent length scales to be compared to the flame thickness l_F, consistent results with regards to the expected combustion regimes were obtained [11].

However, for non-premixed flames this comparison is not as clear as the flame scales strongly vary in space and depend on the local gradient of the mixture fraction Z. The mixture fraction is the most meaningful scalar field in the context of non-premixed flames [24] and is therefore the prime candidate for the DE analysis.

A common configuration to investigate non-premixed laminar flames is the counterflow configuration [25]. An exemplary mixture fraction profile in a counterflow flame is shown in Fig. 2.11a. In the limit of fast, but not infinitely fast chemistry, the so-called "flamelet regime," a corresponding heat release profile in mixture fraction space, can be obtained from the steady-state flamelet solution. This heat release profile is shown in Fig. 2.11b. The decisive parameter linking the flow field and the chemical field in this regime is the scalar dissipation rate at stoichiometric conditions

$$\chi_{\mathrm{st}} = 2D\,(\nabla Z)^2_{\mathrm{st}} \;. \tag{2.3}$$

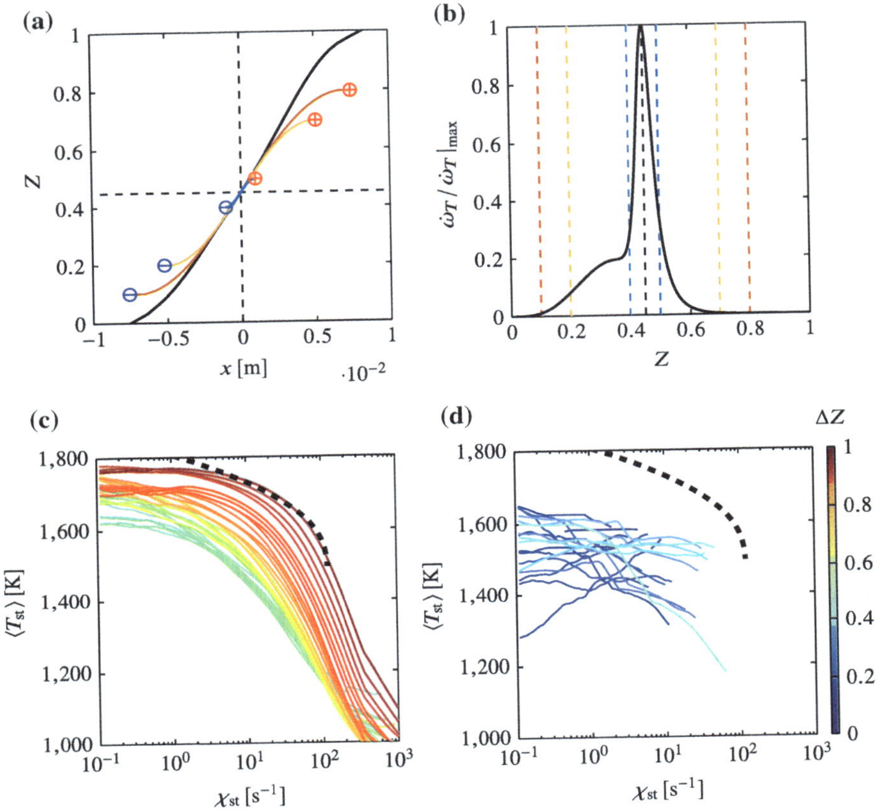

Fig. 2.11 **a** (—): Mixture fraction Z profile in physical space obtained from a counterflow configuration, the black dashed lines indicate the position and the value of the stoichiometric mixture fraction. (—): schematic long DE, (—): schematic intermediate and (—): schematic short DE. **b** (—): Corresponding stationary flamelet solution of the normalized heat release $\dot{\omega}_T$ in mixture fraction space. The colored dashed lines indicate location of the extremal points in mixture fraction space of the respective schematic DEs. **c** and **d** mean stoichiometric temperature $\langle T_{st} \rangle$ in a DNS of planar non-premixed jet flame conditioned on the stoichiometric scalar dissipation rate χ_{st} **c** $\Delta Z \geq \Delta Z_r$ and **d** $\Delta Z \leq \Delta Z_r$). The scalar difference ΔZ is indicated by the color of the lines, ranging from blue ($\Delta Z = 0.01$) over yellow to red ($\Delta Z = 1$). (▪ ▪): stationary flamelet solution in the counterflow configuration

Conveniently, the DE gradient $g = \Delta Z / \ell$ can be used analogously to the scalar dissipation rate. In the limit of small DEs, the following relation is true:

$$\lim_{\ell \to 0} g = |\nabla Z| \ . \tag{2.4}$$

From this follows the proportionality of the scalar dissipation rate χ and the DE gradient g,

$$\chi \sim Dg^2 \,. \tag{2.5}$$

The strong correlation between g and χ was first observed not only in isotropic turbulence [9], but also in free shear flows [10] and in reacting flows [26].

Now let us assume that the mixture fraction field is affected by turbulence of different intensities leading to the introduction of extremal points in the mixture fraction profile in Fig. 2.11a. Consequently, DEs will form around the position of the stoichiometric mixture. Three exemplary conceivable DEs are plotted in Fig. 2.11a; a long DE, an intermediate DE, and a short DE. Let all three DEs have the same mixture fraction gradient at stoichiometric conditions of a laminar flamelet profile so that the scalar dissipation rate χ_{st} is the same. The introduced extremal points signify a forced interruption of the diffusive transport on which the flamelet structure relies. The points of interruption with regards to the heat release for three exemplary DEs are marked in Fig. 2.11b. The long DE interrupts the profile in points of negligible heat release and no effect on the flame structure is expected. However, it is already questionable if the traditional flamelet structure will form for the intermediate DE. In contrast, for the short DE, the formation of an intact flamelet structure can be ruled out. It is straightforward to see that comparing the DE scalar difference ΔZ to the range of Z values in the flamelet in which significant values of the heat release are present is a convenient way to judge whether a conventional flamelet structure is feasible. To this end, a reaction zone thickness ΔZ_{r} in Z-space needs to be defined. Even though the reaction zone thickness varies greatly depending on the fuel and oxidizer composition, ΔZ_{r} can to be easily approximated by fitting the heat release of the flamelet solution for the given composition with a Gaussian:

$$\Delta Z_{\mathrm{r}}' = 2\left(-2\ln(2)\dot{\omega}_{\max} \left.\frac{\partial^2 \dot{\omega}_T}{\partial Z^2}\right|_{Z=Z_{\dot{\omega}_{\max}}}^{-1}\right)^{0.5} , \tag{2.6}$$

with the maximum heat release in the stationary flamelet solution $\dot{\omega}_{\mathrm{T,max}}$ for a given χ_{st} and $Z_{\dot{\omega}_{\max}}$ being the mixture fraction value for which the maximum heat release is observed. In essence, $\Delta Z_{\mathrm{r}}'$ provides a meaningful, nonarbitrary, and easily obtainable flame scale for non-premixed combustion for different values of χ_{st} to be compared to the turbulent scale provided by ΔZ.

This concept is applied to a DNS of a temporally evolving planar non-premixed jet flame. The temperature at stoichiometric conditions is conditioned on the scalar dissipation rate as well as the DE scalar difference ΔZ. This is shown in Fig. 2.11c for large ΔZ. For DEs with $\Delta Z \approx 1$, the stationary flamelet solution is obtained. As the DE scalar difference is reduced, the temperature is lowered as well. The original flamelet profile is compressed along the χ_{st}-axis, but the expected clear correlation with the temperature is still observed. However, for small $\Delta Z < \Delta Z_{\mathrm{r}}$, this correlation ceases as the inner flame structure is interrupted by extremal points. This is shown in Fig. 2.11d, where no discernible relation between χ_{st} and $\langle T_{\mathrm{st}} \rangle$ is observed.

This example shows how the DE scalar difference ΔZ can be used in the context of non-premixed flames to judge the scale interaction between the flame and turbulence.

ΔZ symbolizes the maximum length in Z-space in which a diffusive transport-dominated structure, like a flamelet, can exist.

2.6 Outlook: DE Modeling Potentials

The inherent challenge in simulating turbulent flows in real-world engineering applications is in the high Reynolds numbers and consequently wide separation of turbulent scales. Trying to resolve all of these scales, as is necessitated in DNS, is not currently feasible nor will be in the foreseeable future due to the astronomical computational resource requirements. Therefore, models for the unresolved scales are required. Here, certain features of DE statistics provide a potential for modeling turbulent scales. A potential framework for using DE statistics to model unresolved scales is briefly explained via the example of modeling the probability of flamelet-like combustion in a passenger car direct injection Diesel engine.

The universality of the normalized DE length PDF, which was demonstrated in Fig. 2.5, and to a certain extent the jPDF of both DE parameters, can be exploited to obtain a full statistical description of the turbulent scales in a given scalar field. Additionally, the physical mechanisms governing the evolution of DEs can be used to describe their behavior. Wang [9] derived a semi-empirical stochastic transport equation for the PDF of the normalized DE length $P(\ell^*, t)$, taking the following mechanisms into account:

1. The generation of new extremal points resulting from random eddy turnover of Obukhov–Corrsin eddies. The introduction of extremal points leads to a cutting of DEs.
2. The elimination of extremal points by diffusion and the consecutive joining of dissipation elements.
3. The disappearance of small DEs when ℓ^* approaches zero.

For simplicity, the spatial dependency of the statistics is neglected:

$$\begin{aligned}\frac{\partial P(\ell^*, t)}{\partial t} &+ \frac{\partial}{\partial \ell^*}(v^*(\ell^*)P(\ell^*, t)) = 2\Lambda \int_0^\infty P(\ell^* + z^*, t)\mathrm{d}z \\ &- (\Lambda)\ell^* P(\ell^*, t) + 8 \left.\frac{\partial P(\ell^*, t)}{\partial t}\right|_{\ell^*=0} \Big[\int_0^{\ell^*} \tfrac{z^*}{\ell^*} P(\ell^* - z^*, t)P(z^*, t)\mathrm{d}z^* - P(\ell^*, t)\Big]. \end{aligned} \tag{2.7}$$

Here, the only external parameter is the non-dimensional cutting frequency per unit length $\Lambda = \lambda \ell_{\mathrm{m}}^2/(D_\phi)$, taking the diffusivity of the scalar field D_ϕ as well as the cutting frequency per unit length of the Obukhov–Corrsin eddies λ into account. The non-dimensional diffusive drift velocity is defined as

$$v^*(\ell^*) = -\frac{1}{\ell^*}(1 - c\ell^* \exp(-\ell^*)), \tag{2.8}$$

with the coefficient c to be determined by satisfying the space conservation, i.e., disappearing segments need to be added to longer elements so the overall length of all DEs remains the same.

Considering the same physical mechanisms, the model equation for the normalized jPDF $P(\ell^*, \Delta\phi^*, t)$ was derived using a compensation-defect model for the relationship between ℓ^* and $\Delta\phi^*$ [4]. As the model equation is quite extensive and needs further elaboration, the reader is instead referred to [4].

Both model equations for $P(\ell^*, t)$ and $P(\ell^*, \Delta\phi^*, t)$ yield good results compared to DNS. The exemplary results of the model procedure for $P(\ell^*, t)$ of the mixture fraction field of a temporally evolving diffusion jet flame are shown in Fig. 2.12. Characteristic features, as seen in Figs. 2.5 and 2.7, can be captured and characteristic scalings of all the conditional means, cf. Fig. 2.8, can be reproduced by these models.

In the first step, the model equations for $P(\ell^*, t)$ and $P(\ell^*, \Delta\phi^*, t)$ need to be solved or, exploiting the universality of the normalized statistics, $P(\ell^*, t)$ and $P(\ell^*, \Delta\phi^*, t)$ can be obtained empirically from DNS. As these modeled statistics are normalized by the respective mean quantities, in a second step, ℓ_m and $\Delta\phi_\mathrm{m}$ need to be related to quantities which are readily obtainable in the given CFD simulation. This can be done by exploiting certain ratios or scalings, as seen in Figs. 2.6 and 2.7.

For example, the mean DE parameters are obtained from mean quantities of CFD simulations. The previously observed scaling of the mean DE length with the Kolmogorov microscale η is exploited for the computation of ℓ_m in the CFD simulation:

$$\ell_\mathrm{m}(t, x_i) \sim \left(\overline{\nu}(t, x_i)^3 \widetilde{\varepsilon}(t, x_i)^{-1}\right)^{1/4}, \tag{2.9}$$

with the mean kinematic viscosity $\overline{\nu}$ and the Favre averaged dissipation rate of the turbulent kinetic energy $\widetilde{\varepsilon}$.

Using the theoretical scaling of the first-order scalar structure function with an exponent of 1/3 [9], the mean DE scalar differences is modeled following [27] as

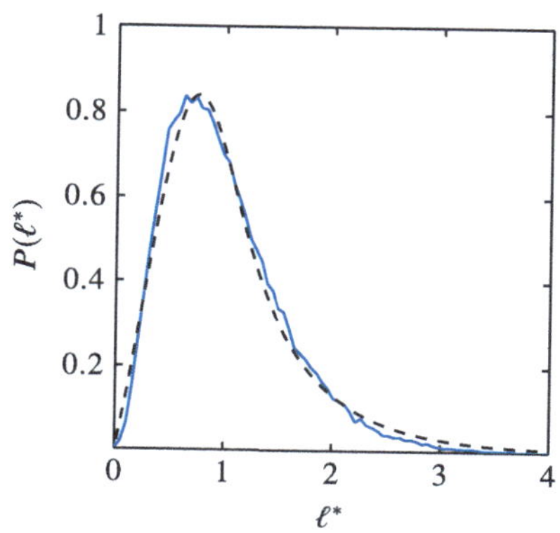

Fig. 2.12 PDF of the normalized DE length ℓ^*. (—): mixture fraction in temporally evolving diffusion jet flame; and (- -): results of the model Eq. (2.7)

$$\Delta Z_{\mathrm{m}}(t, x_i) \sim \ell_{\mathrm{m}}^{1/3} \frac{\widetilde{\chi}(t, x_i)^{1/2}}{\widetilde{\varepsilon}(t, x_i)^{1/6}} \int_0^{\infty} \ell^{*1/3} P(\ell^*, t) \mathrm{d}\ell^* \,. \tag{2.10}$$

The mean DE gradient g_{m} is obtained using the correlation between g and χ analogous to Eq. 2.5

$$g_{\mathrm{m}}(t, x_i) \sim \widetilde{\chi}(t, x_i)^{0.5} \overline{D}^{-0.5} \,. \tag{2.11}$$

In the third step, the correlations between the DE parameters and the physical mechanism of interest, e.g., the combustion behavior, need to be obtained in a non-dimensional form from DNS results using the various methods of obtaining DE-based statistics outlined in Sect. 2.1.1. In this example, the range of DE parameters for which a flamelet-like combustion should be observed in CFD has to be determined from DNS. As shown in Fig. 2.11d, ΔZ_{r} is one of the limits for sustaining a flame structure. Using the approximation for ΔZ_{r} from Eq. 2.6, it is expected to encounter flamelet-like flame structures in the CFD for $\Delta Z > \Delta Z_{\mathrm{r}}$.

A second reason for not encountering burning flamelets, and thus another limit, is flame quenching. For values of $\chi > \chi_{\mathrm{q}}$, with the quenching dissipation rate χ_{q}, the molecular mixing of fuel and oxidizer is insufficient to sustain a burning solution. χ_{q} can be easily obtained from solving the flamelet equations. Exploiting the relationship between χ and g, the quenching DE gradient g_{q} is obtained using the correlation between g and χ from DNS

$$g_{\mathrm{q}}(t) \sim \left(\chi_{\mathrm{q}}/D\right)^{0.5} \,. \tag{2.12}$$

Therefore, the second limit for expecting flamelet structures in CFD, expressed in DE parameters, is $g \leq g_{\mathrm{q}}$.

Finally, the probability of encountering a flamelet-like conditions $\mathcal{P}_{\mathrm{Flamelet}}$ can be computed by denormalizing the DE parameter jPDF with the mean DE parameters $P(g, \Delta Z) = \Delta Z_{\mathrm{m}} g_{\mathrm{m}} P(g^*, \Delta Z^*)$ and integrating the jPDF over the previously determined intervals. The local probability of the "Burning Flamelet" regime is calculated as

$$\mathcal{P}_{\mathrm{Flamelet}} = \int_{Z_{\mathrm{r}}}^{\infty} \int_0^{g_{\mathrm{q}}} P(g, \Delta Z) \mathrm{d}g \mathrm{d}\Delta Z \,. \tag{2.13}$$

The results of this procedure are shown for three exemplary time steps of the simulation of a passenger car diesel engine in Fig. 2.13. While the later time steps are dominated by high values of $\mathcal{P}_{\mathrm{Flamelet}}$, in the early stages of the combustion process, the intense scale interaction prevents the formation of flames characterized by a flamelet behavior. It is expected that regions of low $\mathcal{P}_{\mathrm{Flamelet}}$ might not be adequately predicted by current flamelet-based combustion models. However, this early instant in the combustion process is especially critical to properly match the beginning and the gradient of the pressure rise in the engine, which are of great importance to a

Fig. 2.13 Stoichiometric iso-surface of the mean mixture fraction $\widetilde{Z}_{st}$ for three different crank angle positions in a CFD simulation of a diesel engine. The color of the iso-surface indicates the local probability of the flamelet regime $\mathcal{P}_{\mathrm{Flamelet}}$ from blue ($\mathcal{P}_{\mathrm{Flamelet}} = 0$), over green to yellow ($\mathcal{P}_{\mathrm{Flamelet}} = 1$). The black wireframe indicates the piston bowl geometry and the liner

variety of engine output parameters like pollutant formation and engine noise. This example helps to emphasize the usefulness of DE-based modeling in the context of turbulent combustion as it provides a measure for the validity of model assumptions.

The modeling applications employing DE statistics are of course not limited to this case. In another context, this framework was successfully applied to predict the probability and location of catastrophic detonation events in gasoline engines called “mega knock” [27].

2.7 Conclusions

The concept of DEs was defined and the underlying idea and advantages of estimating turbulent scales in scalar fields in DNS employing DE analysis was motivated. Numerical and physical prerequisites to the scalar fields were explained. Statistics of DE parameters were shown for a wide range of reacting and inert DNS cases and universal features of these statistics demonstrated and physically interpreted. Next, DE parameter statistics were related to more widely used statistics for taken non-locality effects of turbulent flows into consideration. Additionally, the concept of DE-based flow partitioning in free shear flow configurations was outlined. Furthermore, the specific application of DE analysis to reacting flows as a means for the investigation of turbulence/flame interaction was explained. Finally, it was shown how this insight gained from reacting DNS can be applied to CFD of real-world engineering applications, exploiting the universal features of DE statistics in a DE-based modeling framework.

References

1. A.A. Wray, J.C.R. Hunt, Algorithms for classification of turbulent structures, in *Topological Fluid Mechanics*, eds. by H.K. Moffat, A. Tsinober (Cambridge University Press, Cambridge, 1990), pp. 95–104
2. N. Peters, L. Wang, Dissipation element analysis of scalar fields in turbulence. C. R. Mechanique **334**, 493–506 (2006)
3. A. Schnorr, D. Helmrich, D. Denker, T. Kuhlen, B. Hentschel, Feature tracking by two-step optimization. Trans. Vis. Comput, Graph (accepted for publication) (2018)
4. L. Wang, Geometrical description of homogeneous shear turbulence using dissipation element analysis. PhD Thesis, RWTH University, 2008
5. D. D'acunto, K. Kurdyka, Bounds for gradient trajectories and geodesic diameter of real algebraic sets. Bull. Lond. Math. Soc. **38**, 951–968 (2006)
6. L.D. Floriani, M. Spangnuolo (eds.), *Shape Analysis and Structuring* (Springer, Berlin, 2007)
7. S.B. Pope, *Turbulent Flows* (Cambridge University Press, Cambridge, 2000)
8. G.S. Jiang, C.W. Shu, Efficient implementation of weighted eno schemes. J. Comput. Phys. **126**, 202–228 (1996)
9. L. Wang, N. Peters, The length scale distribution function of the distance between extremal points in passive scalar turbulence. J. Fluid Mech. **554**, 457–475 (2006)
10. M. Gampert, P. Schaefer, J. Goebbert, N. Peters, Decomposition of the field of the turbulent kinetic energy into regions of compressive and extensive strain. *Physica Scripta* **155**(014002) (2013)
11. D. Denker, A. Attili, S. Luca, M. Gauding, F. Bisetti, H. Pitsch, Dissipation element analysis of premixed jet flames. Comb. Sci. Tech., volume in press (2019)
12. M. Gampert, P. Schaefer, N. Peters, Experimental investigation of dissipation-element statistics in scalar fields in a jet flow. J. Fluid Mech. **724**, 337–366 (2013)
13. J. Boschung, F. Hennig, D. Denker, H. Pitsch, R.J. Hill, Analysis of structure function equations up to the seventh order. J. Turbul. 1–32 (2017)
14. A. Attili, F. Bisetti, Fluctuations of a passive scalar in a turbulent mixing layer. Phys. Rev. E **3**, 03301 (2013)
15. D. Denker, K. Niemietz, A. Attili, M. Korkmaz, H. Pitsch, Prediction of non-premixed combustion regimes in a di diesel engine in various operation points, in *Proceedings of the 9th European Combustion Meeting*, April 14–17, Lisbon, Portugal (2019)

16. M. Holzer, E.D. Siggia, Turbulent mixing of a passive scalar. Phys. Fluids **6**, 1820–1837 (1994)
17. M. Gampert, J.H. Goebbert, P. Schaefer, M. Gauding, N. Peters, F. Aldudak, M. Oberlack, Extensive strain along gradient trajectories in the turbulent kinetic energy field. New J. Phys. **13**, 043012 (2011)
18. R. Prasad, K.R. Sreenivasan, Scalar interfaces in digital images in turbulent flows. Exp. Fluids **7**, 259–264 (1989)
19. D.K. Bisset, M.M.R.J.C.R. Hunt, The turbulent/non-turbulent interface bounding a far wake. J. Fluid Mech. **451**, 383–410 (2002)
20. M.S. Chong, A.E. Perry, B. Cantwell, The general classification of three dimensional flow fields. Phys. Fluids A **2**, 408–420 (1990)
21. E. Effelsberg, N. Peters, A composite model for the conserved scalar pdf. Combust. Flame **50**, 351–360 (1983)
22. J.P. Mellado, L. Wang, N. Peters, Gradient trajectory analysis of a scalar field with external intermittency. J. Fluid Mech **626**, 333–365 (2009)
23. M. Gampert, P. Schäfer, V. Narayanaswamy, N. Peters, Gradient trajectory analysis in a jet flow for turbulent combustion modeling. J. Turbul. **14**, 147–164 (2013)
24. N. Peters, *Turbulent Combustion* (Cambridge University Press, Cambridge, 2000)
25. H. Pitsch, N. Peters, A consistent flamelet formulation for nonpremixed combustion considering differential diffusion effects. Combust. Flame **114**, 26–40 (1998)
26. M. Gauding, F. Dietzsch, J. Goebbert, D. Thévenin, A. Abdelsamie, C. Hasse, Dissipation element analysis of a turbulent non-premixed jet flame. Phys. Fluids **29**(085103) (2017)
27. N. Peters, B. Kerschgens, G. Paczko, Super-Knock prediction using a refined theory of turbulence. SAE, 2013-01-1109 (2013)

Chapter 3
Computational Singular Perturbation Method and Tangential Stretching Rate Analysis of Large Scale Simulations of Reactive Flows: Feature Tracking, Time Scale Characterization, and Cause/Effect Identification. Part 1, Basic Concepts

M. Valorani, F. Creta, P. P. Ciottoli, R. Malpica Galassi, D. A. Goussis, H. N. Najm, S. Paolucci, H. G. Im, E.-A. Tingas, D. M. Manias, A. Parente, Z. Li, and T. Grenga

Abstract This chapter provides a review of the basic ideas at the core of the Computational Singular Perturbation (CSP) method and the Tangential Stretching Rate (TSR) analysis. It includes a coherent summary of the theoretical foundations of these two methodologies while emphasizing their mutual interconnections. The main

---

M. Valorani (✉) · F. Creta · P. P. Ciottoli · R. Malpica Galassi
Department of Mechanical and Aerospace Engineering,
Sapienza University of Rome, 00184 Rome, Italy
e-mail: mauro.valorani@uniroma1.it

D. A. Goussis
Khalifa University of Science and Technology, Abu Dhabi, UAE

H. N. Najm
Sandia National Laboratories, Albuquerque, CA, USA

S. Paolucci
University of Notre Dame, Notre Dame, IN, USA

H. G. Im
KAUST, Thuwal, Saudi Arabia

E.-A. Tingas
Perth College, University of the Highlands and Islands (UHI), Perth, UK

D. M. Manias
National Technical University of Athens, Athens, Greece

A. Parente · Z. Li
ULB, Bruxelles, Belgium

T. Grenga
Institute for Combustion Technology RWTH Aachen University,
52062 Aachen, Germany

H. Pitsch and A. Attili (eds.), *Data Analysis for Direct Numerical Simulations of Turbulent Combustion*, https://doi.org/10.1007/978-3-030-44718-2_3

theoretical findings are presented in a systematic fashion. Their virtues and limitations will be discussed with reference to auto-ignition systems, laminar and turbulent premixed flames, and non-premixed jet flames. The material presented in the chapter constitutes an effective guideline for further studies.

3.1 Introduction

Advances in computing power of both hardware and software allow today large scale simulations of reacting flows at a relatively low cost. However, together with the ability of generating massively large datasets, it is not clear if we are able to develop post-processing tools capable of extracting valuable information encapsulated in the datasets generated by the simulations.

The most natural selection of observables associated with a simulation of a reacting flow is clearly the set of primitive state variables such as temperature, pressure, velocity, and composition. One can therefore post-process the datasets to extract information about the evolution of these state variables so as to infer sensitivity information of the flow system with respect to the geometry of the control volume, the operating conditions, the history of transients, and so forth. This information can be useful to assess the performance of a specific reactive flow configuration, and possibly to establish a (usually very expensive) design optimization procedure.

It is however difficult to extract directly from the state variables information about qualitative aspects of the flow, such as the identification of regions featuring coherent dynamics (regions of chemically frozen flow, near-equilibrium flow, spontaneous ignition, or deflagration flame fronts), as well as information on the intrinsic flow time scales, and most of all information about cause–effect relationships among the "causing" processes (convective or diffusion transport, chemical reactions) and the "effects" on the state variables or other observables of interest (ignition delay, extinction, etc).

One aspect that makes the post-processing of these large datasets difficult is specifically their size, meaning that the number N_{CFD} of degrees of freedom involved in a DNS or a LES (typically proportional to the product of the number N_{FV} of finite volumes of a 3D mesh times the number N_{SV} of state variables times the number N_{ITS} of integration time steps in an unsteady problem) is a huge number ($N_{CFD} \sim N_{FV} \times N_{SV} \times N_{ITS}$).

Nevertheless, even in systems whose dynamics is described by non-linear PDEs, the effective number $N_{essential}$ of "essential" degrees of freedom is much smaller than N_{CFD}. This is due to the action of the components of the model that generate fast time scales effectively constraining the evolution of the system within an embedding of low-dimensional manifolds, so that the number $N_{essential}$ is substantially smaller than N_{CFD}.

A pictorial illustration of how processes occurring at fast scales are effective in constraining the slow dynamics is provided by Fig. 3.1, which shows a canyon created by the flow of the Colorado River and the various creeks that supply water to the river

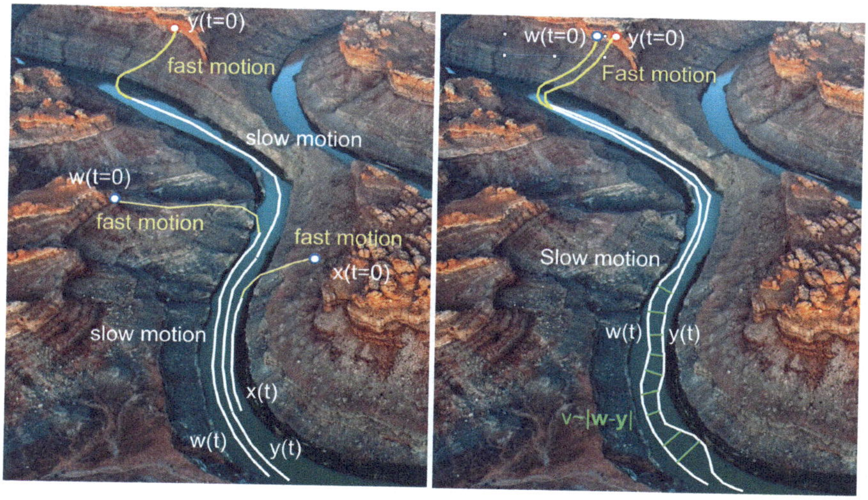

Fig. 3.1 (left) Point dynamics: Water is forced to proceed downhill at a slow pace along the valley. (right) Vector Dynamics: Stream tubes get closer or farther apart from one to another while they both flow downhill

from the surrounding ridges. The mild slope of the valley forces the water to flow downhill at a slow pace, while the large slopes of the ridges cause a fast motion along the creeks. Any stream of water attempting to leave the valley side-wise is quickly contrasted by the action of the ridges that forces the stream to flow back in the valley. Therefore, the presence of ridges reduces the degree of freedom of the flowing water which is not allowed to flow across the full two-dimensional plain surrounding the valley, but is constrained to flow along the one-dimensional valley.

Of course, the actual identification of the valley (a slow invariant manifold) and the reformulation of the original model as a reduced-order model that describes the slow dynamics (the flow of the water in the river)—constrained to evolve within the valley (a slow invariant manifold)—is not an obvious task. The Computational Singular Perturbation (CSP) method has been developed to make this task affordable using only algorithmic tools so that an answer can be readily found even when coping with high dimensional systems.

To fully characterize the slow flow along the valley is however important to consider the fate of two stream tubes emanating from adjacent locations uphill in the river (see Fig. 3.1). The two stream tubes get closer or farther apart from one to another while they both follow the mild downhill slope of the valley. Their mutual distance grows or shrinks as a consequence of the slow flow. The evolution of their mutual distance can be interpreted as a tendency of the flow to expand or contract. The Tangential Stretching Rate (TSR) analysis has been developed to estimate the rate at which the mutual distance grows or shrinks using only algorithmic tools so that the characteristic time scale can easily be determined.

In this chapter, we will review the basic ideas at the core of the CSP method and the TSR analysis. Their virtues and limitations will be discussed with reference to auto-ignition systems, laminar and turbulent premixed flames, and non-premixed jet flames.

The CSP method was first developed by Lam and Goussis [1–6], and extended with new developments and applications by Goussis, Valorani, Paolucci and collaborators [7–15], as well as other investigators [16–22].

The TSR analysis was introduced in the combustion community by Adrover, et al. [23]. Later, Valorani and Paolucci merged the concept of slow/fast decomposition with that of stretching rate in [24, 25]. Applications of the TSR analysis can be found in [26–28].

Consider N scalar fields $\mathbf{z}$ of a reactive system with spatial non-homogeneities. Their time and space evolution can be described by a set of N PDEs where the driving processes are convection and/or diffusion (transport phenomena) as well as chemical reactions. The general form of the governing PDEs is as follows:

$$\frac{\partial \mathbf{z}(\boldsymbol{x}, t)}{\partial t} = \boldsymbol{L}_{\boldsymbol{x}}(\mathbf{z}(\boldsymbol{x}, t)) + \boldsymbol{g}(\mathbf{z}(\boldsymbol{x}, t)), \tag{3.1}$$

where $\boldsymbol{L}_{\boldsymbol{x}}(\mathbf{z})$ is a spatial differential operator (convection and/or diffusion) and $\boldsymbol{g}(\mathbf{z})$ is the chemical source term. If the reactive system is spatially homogeneous then $\boldsymbol{L}_{\boldsymbol{x}}(\mathbf{z})$ is identically zero. In such a case, the independent space variable $\boldsymbol{x}$ plays no role in Eq. (3.1), so that Eq. (3.1) reduces to an autonomous (vector field not dependent explicitly on time) set of ODEs of the form:

$$\frac{d\mathbf{z}(t)}{dt} = \boldsymbol{g}(\mathbf{z}(t)). \tag{3.2}$$

The CSP method and the TSR analysis will be briefly stated in Sects. 3.2 and 3.3, respectively, on the basis of the homogeneous system Eq. (3.2). The application of CSP and TSR to non-homogeneous systems will be discussed in Sects. 3.4.1 and 3.4.2 on the basis of the non-homogeneous system Eq. (3.1).

3.2 CSP Analysis of Homogeneous Systems: Point Dynamics

Let us consider a set of linearly independent column vectors $\boldsymbol{A} = \{\boldsymbol{a}_i\}_{i=1,\dots,N}$ with $\boldsymbol{a}_i \in \boldsymbol{R}^N$ and a set of row vectors $\boldsymbol{B} = \{\boldsymbol{b}^i\}_{i=1,\dots,N}$ with $\boldsymbol{b}^i \in \boldsymbol{R}^N$ orthonormal to A. At each location $\mathbf{z} \in \mathbb{M}$, the column vectors $\boldsymbol{A}$ identify a frame of reference of the tangent space $\mathbb{T}_{\mathbf{z}}$ of the manifold $\mathbb{M}$ within which the system dynamics evolves.

The orthogonality properties provide the following identities:

$$\boldsymbol{I}_N^N = \boldsymbol{A}\boldsymbol{B} \qquad \boldsymbol{B}\boldsymbol{A} = \boldsymbol{I}_N^N. \tag{3.3}$$

The vector field in Eq. (3.2) can be always expressed as the sum of N *modal* contributions as

$$\boldsymbol{g}(\mathbf{z}(t)) = \boldsymbol{A} \cdot (\boldsymbol{B} \cdot \boldsymbol{g}) = \sum_{i=1}^{N} \boldsymbol{a}_i(\mathbf{z}(t))\; f^i(\mathbf{z}(t)), \tag{3.4}$$

where f^i is the *amplitude* of the ith mode given by

$$f^i(\mathbf{z}) = \boldsymbol{b}^i(\mathbf{z}) \cdot \boldsymbol{g}(\mathbf{z}), \quad i = 1, \ldots, N. \tag{3.5}$$

It can be readily shown that the time evolution of f^i satisfies the ODE system

$$\frac{\mathrm{d}f^i(t)}{\mathrm{d}t} = \sum_{k=1}^{N} f^k(\mathbf{z}(t)) \left(\frac{\mathrm{d}\boldsymbol{b}^i(\mathbf{z}(t))}{\mathrm{d}t} + \boldsymbol{b}^i(\mathbf{z}(t)) \cdot \boldsymbol{J}_g(\mathbf{z}(t)) \right) \cdot \boldsymbol{a}_k(\mathbf{z}(t)), \quad i = 1, \ldots, N, \tag{3.6}$$

where $\boldsymbol{J}_g$ is the Jacobian matrix of $\boldsymbol{g}$ with respect to $\mathbf{z}$.

Let us assume that it is possible to associate a time scale τ^i to the ith mode contribution in Eq. (3.4), and to sort all time scales from the smallest to the largest so that:

$$\tau^1 < \cdots < \tau^M \ll \tau^{M+1} < \cdots < \tau^N. \tag{3.7}$$

In this expression, it is assumed that a large time scale gap between τ^M and τ^{M+1} exists and their ratio is a small parameter $\epsilon = |\tau^M/\tau^{M+1}| \ll 1$. We also assume that the fastest M time scales are dissipative, that is, the components of the vector field that generate them tend to lead the system towards equilibrium. This circumstance is very commonly found in chemical kinetics.

Under the above setting the vector field $\boldsymbol{g}$ in Eq. (3.4) can be partitioned in two contributions (one related to the M fast time scales, and the other to the $N - M$ slow time scales) so that Eq. (3.2) reads

$$\frac{d\mathbf{z}(t)}{dt} = \sum_{r=1}^{M} \boldsymbol{a}_r f^r + \sum_{s=M+1}^{N} \boldsymbol{a}_s f^s. \tag{3.8}$$

The occurrence of the time scale gap ϵ allows us to partition the tangent space $\mathbb{T}_\mathbf{z}$ at the location $\mathbf{z}$ in the space of the state variable in two subspaces: the fast subspace spanned by the column vectors $\{\boldsymbol{a}_1 \ldots \boldsymbol{a}_M\}$, and the slow subspace spanned by the column vectors $\{\boldsymbol{a}_{M+1} \ldots \boldsymbol{a}_N\}$.

If the basis vectors $\boldsymbol{A} = \{\boldsymbol{a}_k\}_{k=1,\ldots,N}$ are identified so that the dynamics of the element of the fast mode amplitudes f^r associate only with the M fast time scales and that of the slow mode amplitude f^s associate only with the $(N - M)$ slow time scales, then it holds that after a time period of the order of the fastest of the slow scales τ^{M+1} the following algebraic equations holds:

$$f^r \approx 0 \quad \text{for} \quad r = 1, \ldots, M; \tag{3.9}$$

that is the mode amplitudes of the vector field $\boldsymbol{g}$ in the fast subspace have attained a negligibly small magnitude. The locus of states solution of the algebraic equations (3.9) is a Slow Invariant Manifold (SIM) whose dimension is lower than N. A SIM constitutes a constraint within which the solution of the ODEs in Eq. (3.2) moves driven by the slow processes, according to the reduced order differential-algebraic model:

$$\frac{\mathrm{d}\mathbf{z}(t)}{\mathrm{d}t} \approx \sum_{s=M+1}^{N} \boldsymbol{a}_s(\mathbf{z}(t))\, f^s(\mathbf{z}(t)) \qquad \mathbf{z} \in \boldsymbol{R}^N, \tag{3.10a}$$

$$f^r(\mathbf{z}) \approx 0 \qquad r = 1, \ldots, M. \tag{3.10b}$$

Therefore, the ability to approximate the original dynamics in Eq. (3.2) with the reduced order model of Eqs. (3.10) relies on the ability to identify a set of basis vectors $\boldsymbol{A} = \{\boldsymbol{a}_k\}_{k=1,\ldots,N}$ that allows for fast/slow mode decoupling.

The answer is more or less difficult to find depending on the degree of non-linearity of the vector field $\boldsymbol{g}$: for a linear system, one can achieve full mode decoupling, whereas for non-linear systems, one can at best achieve a decoupling between two subspaces such that only slow time scales prevail in one subspace (therefore referred to as "slow subspace"), and only fast time scales prevail in the other subspace (therefore referred to as "fast subspace").

CSP provides two refinement procedures for the identification of such basis vectors even for non-linear systems.

3.2.1 CSP Basis for Linear Vector Fields

Consider first the case when $\boldsymbol{g} = \boldsymbol{C} \cdot \mathbf{z}$ is a linear mapping, so that the related Jacobian matrix $\boldsymbol{J_g} = \boldsymbol{C}$ is constant. The vectors in $\boldsymbol{A} = \{\boldsymbol{a}_k\}_{k=1,\ldots,N}$ and $\boldsymbol{B} = \{\boldsymbol{b}_k\}_{k=1,\ldots,N}$ reduce to the right and left, respectively, eigenvectors of $\boldsymbol{J_g} = \boldsymbol{C}$. Thus, the time derivative of the left eigenvector $\{\boldsymbol{b}^i(t)\}_{i=1,N}$ is zero in Eq. (3.6) later in this section.

Let $\lambda_1, \ldots, \lambda_N$ be the (possibly complex) eigenvalues of $\boldsymbol{J_g}$ ordered in descending magnitude of their moduli, and $\{\boldsymbol{v}_i\}_{i=1,\ldots,N}$ the corresponding right eigenvectors. Then it is proper to select the CSP vectors as follows:

- for any *real* λ_i of $\boldsymbol{J_g}$, we set $\boldsymbol{a}_i = \boldsymbol{v}_i$;
- for a complex conjugate pair $(\lambda_c, \lambda_{c+1})$ and the corresponding complex conjugate pair right eigenvectors $(\boldsymbol{v}_c, \boldsymbol{v}_{c+1})$, we take as basis vectors $\boldsymbol{a}_c$ and $\boldsymbol{a}_{c+1}$ the

real vectors $\boldsymbol{a}_c := \mathrm{Re}(\boldsymbol{v}_c)$ and $\boldsymbol{a}_{c+1} = \mathrm{Im}(\boldsymbol{v}_{c+1})$. In fact, any linear combination $(\boldsymbol{a}_c, \boldsymbol{a}_{c+1}) = L(\boldsymbol{v}_c, \boldsymbol{v}_{c+1})$, where $(\boldsymbol{a}_c, \boldsymbol{a}_{c+1})$ are real, is satisfactory in delivering a sought-after real basis vector pair that spans the same subspace as that of $(\boldsymbol{v}_c, \boldsymbol{v}_{c+1})$.

Substituting the above choice of the $\boldsymbol{a}_i$ in Eq. (3.6), for each λ_i that is real and simple (for simplicity, we treat eigenvalues with identical algebraic and geometric multiplicity m as m individual simple eigenvalues), its corresponding f^i satisfies

$$\frac{\mathrm{d}f^i(t)}{\mathrm{d}t} = \lambda_i f^i, \tag{3.11}$$

which implies that all modes corresponding to real and simple eigenvalues are decoupled from each other with descending magnitude of exponential rate of changes. When there are real and repeated eigenvalues with algebraic and geometric multiplicity different, e.g. $\lambda_l = \lambda_{l+1} = \cdots = \lambda_{l+r}$ where $r+1$ is the algebraic multiplicity, their corresponding modes then satisfy

$$\frac{\mathrm{d}}{\mathrm{d}t}\begin{pmatrix} f^l \\ \vdots \\ f^{l+r} \end{pmatrix} = \Lambda_l \begin{pmatrix} f^l \\ \vdots \\ f^{l+r} \end{pmatrix}, \tag{3.12}$$

where Λ_l is an $(r+1) \times (r+1)$ upper diagonal Jordan block. Thus the time changes of all eigenmodes correspond to the eigenvalue λ_l are decoupled from modes with $i \neq l, \ldots, l+r$.

When there are complex conjugate eigenvalues, λ_c and λ_{c+1}, then f^c and f^{c+1} satisfy the two-dimensional subsystem

$$\frac{\mathrm{d}}{\mathrm{d}t}\begin{pmatrix} f^c \\ f^{c+1} \end{pmatrix} = \begin{bmatrix} \mathrm{Re}(\lambda_c) & \mathrm{Im}(\lambda_c) \\ -\mathrm{Im}(\lambda_c) & \mathrm{Re}(\lambda_c) \end{bmatrix}\begin{pmatrix} f^c \\ f^{c+1} \end{pmatrix}, \tag{3.13}$$

and hence are also decoupled from all other f^i with $i \neq c, c+1$.

In summary, when $\boldsymbol{g}$ is linear, the changes in mode amplitudes corresponding to eigenvalues with different real parts are all decoupled while the CSP basis vectors are used to span $\boldsymbol{g}$. Thus, the above choice of the $\boldsymbol{a}_i$ is the ideal choice of basis vectors for the linear system.

The closed form solution of Eq. (3.6) in the linear case reads:

$$\boldsymbol{f}(t) = e^{\Lambda t}\boldsymbol{f}(0), \tag{3.14}$$

and thus for $t > O(\lambda_M^{-1})$ we have $f^r(t) \approx 0,\ r = 1, \ldots, M$, that is Eq. (3.9).

3.2.2 CSP Basis for Non-linear Vector Fields

When $\boldsymbol{g}$ is non-linear, the ODEs in Eq. (3.6) are generally coupled and a two-step CSP refinement procedure [1, 6, 11] can be applied recursively to weaken the coupling among different modes, in particular, the coupling between fast modes and slow modes. A "good" set of basis vectors $\{\boldsymbol{a}_i\}$ in the CSP context after refinement, that provides leading order accuracy and a reduced model free of the fast time scales, gives

$$\frac{\mathrm{d}}{\mathrm{d}t}\begin{pmatrix} f^s \\ f^r \end{pmatrix} = \begin{bmatrix} \Lambda^r_r & \Lambda^s_r \sim \mathrm{O}(\epsilon) \\ \Lambda^r_s \sim \mathrm{O}(\epsilon) & \Lambda^s_s \end{bmatrix} \cdot \begin{pmatrix} f^s \\ f^r \end{pmatrix}, \tag{3.15}$$

where

$$\boldsymbol{\Lambda}^i_j = \left(\frac{\mathrm{d}\boldsymbol{b}^i}{\mathrm{d}t} + \boldsymbol{b}^i \cdot J_g\right) \cdot \boldsymbol{a}_j \qquad i = r, s \qquad j = r, s. \tag{3.16}$$

Note that $\boldsymbol{\Lambda}^r_r \in \boldsymbol{R}^{M\times M}$ and $\boldsymbol{\Lambda}^s_s \in \boldsymbol{R}^{(N-M)\times(N-M)}$ characterize the fast and slow dynamics of modes, respectively. More precisely, the small magnitudes of the off-diagonal blocks $\boldsymbol{\Lambda}^s_r$ and $\boldsymbol{\Lambda}^r_s$ imply the separation of fast and slow dynamics.

The CSP refinements can provide information about non-linear vector fields. Each application of the first step decreases the magnitude of the upper-right matrix block $\boldsymbol{\Lambda}^s_r$ by an order ϵ. Each application of the second step decreases the magnitude of the lower-left matrix block $\boldsymbol{\Lambda}^r_s$ by an order ϵ.

Note however that the first CSP refinement requires the evaluation of the Jacobian of the vector field $\boldsymbol{g}$, the second refinement requires the evaluation of its Hessian and so on. As discussed in [29], the proper expansion of the SIM might involve multiple small parameters when the system manifests an explosive behaviour.

Interested in leading order accuracy, the fast–slow decomposition can be based on the eigensystem of the local Jacobian of the non-linear vector field $\boldsymbol{g}$. This approximation is usually a good compromise between accuracy and computational cost.

3.2.3 Identification of the Fast and Slow Subspace Dimensions

In the CSP method, the dimension of the fast subspace M is defined by the number of the fastest modes, which, within the limit of accuracy specified by the user, provide a negligible contribution to the change of the state vector on a time period of the order of the fastest of the active time scale τ_{M+1}.

Operatively, one can define M as the largest integer lying between 0 and $N-1$ which satisfies the following inequality for each component of the N-dimensional state vector $\mathbf{z}$:

$$\left\{\max\{M \in (0, \ldots, N-1)\} : \delta z^i_{\text{fast}} \approx \tau^{M+1} \sum_r |a^i_r f^r| < \delta z^i_{\text{error}}\right\} \tag{3.17}$$

with $r = 1, \dots, M$, and where the error vector is defined as

$$\delta z^i_{\text{error}} = \text{rtol}^i |z^i| + \text{atol}^i. \tag{3.18}$$

This criterion[1] defines the fast subspace as the subspace of the tangent space spanned by modes that contribute an amount δz^i_{fast} in the time period τ^{M+1} that is considered negligible according to a user-defined error threshold $\delta z^i_{\text{error}}$.

Implicit, in the summation in the criterion (3.17), is the approximation of keeping constant $\left|\sum_r \boldsymbol{a}_r f^r\right|$ over a time period of the order of τ^{M+1}. In contrast, both the mode direction $\boldsymbol{a}_r$ and its amplitude f^r vary with time.

The variation of $\boldsymbol{A}_r$ does not affect the magnitude of the contribution of the fast modes, while it does the magnitude of $\boldsymbol{f}^r$ whose time evolution can be approximated, at any time t^n, by the closed-form solution of the linear system Eq. (3.14) by

$$\boldsymbol{f}^r(t = t^n + \tau) = e^{\tau \Lambda_r} \boldsymbol{f}^r_0(t^n). \tag{3.19}$$

This estimate shows that the approximation of keeping the value of the fast mode amplitudes constant over a time period of the order of τ^{M+1} in Eq. (3.17) overestimates the contribution of the M fastest modes. As a consequence, the criterion of Eq. (3.17) might provide a too conservative estimate of the fast subspace dimension, which might also lead to numerical noise since the fast mode amplitudes are typically small but not consistently so with time.

A more accurate estimate of the fast mode contribution can be obtained by first recasting the summation in Eq. (3.17) as an integral where we retain the functional dependence of $\boldsymbol{f}^r$ with time, but we neglect a possible rotation of $\boldsymbol{A}_r$, and then replacing $\boldsymbol{f}^r(t)$ with $e^{\tau \Lambda_r} \boldsymbol{f}^r(t^n)$ according to Eq. (3.19), where $\boldsymbol{f}^r(t^n)$ is the value of the fast mode amplitude at the beginning of a time interval of duration τ^{M+1}, so that

$$\delta z^i_{\text{fast}} \approx \int_0^{\tau^{M+1}} \sum_r a^i_r f^r(t) dt = \int_0^{\tau^{M+1}} \sum_r a^i_r f^r(t^n) e^{\lambda_r t} dt \tag{3.20}$$

with $r = 1, \dots, M$. The closed form of the last integral provides the sought-after form of the criterion [30]:

$$\left\{ \max\{M \in (0, \dots, N-1)\} : \delta z^i_{\text{fast}} = \sum_r \left| a^i_r f^r(t^n) \frac{e^{\lambda^r \tau^{M+1}} - 1}{\lambda^r} \right| < \delta z^i_{\text{error}} \right\}. \tag{3.21}$$

[1] In the application of the criterion defined by Eq. (3.17), special care is required to manage the contribution of complex conjugate eigenvalues and eigenvectors.

3.3 TSR Analysis of Homogeneous Systems: Vector Dynamics

The characterization of the dynamical features of a spatially homogeneous system whose dynamics is described by Eq. (3.2) can be carried out by viewing the dynamical system as a generalized *advection machine* of all the possible geometric entities that can be defined on the state space. The simplest measure element constructed in the tangent space $\mathcal{T}_{\mathbf{z}_0}$ at $\mathbf{z}_0$ is the vector $\boldsymbol{v}$ defined as follows:

$$\boldsymbol{v}(t) := \lim_{|\boldsymbol{\epsilon}|\to 0} \frac{\mathbf{z}_2(t) - \mathbf{z}_1(t)}{|\boldsymbol{\epsilon}|}, \qquad (3.22)$$

where $\mathbf{z}_1(t)$ and $\mathbf{z}_2(t)$ are two nearby trajectories of Eq. (3.2) originating by the initial conditions $\mathbf{z}_1(0) = \mathbf{z}_0$ and $\mathbf{z}_2(0) = \mathbf{z}_0 + \boldsymbol{\epsilon}$, where $\boldsymbol{\epsilon}$ is a small perturbation vector.

By construction, the vector $\boldsymbol{v}(t)$ belongs to the tangent bundle along $\mathbf{z}_1(t)$, and $\boldsymbol{v}(t)$ is a scaled measure, at time t, of the difference between the two trajectories.

As detailed in [24], subtracting the *point dynamics* Eq. (3.2) as applied to the two nearby initial conditions $\mathbf{z}_1(0)$ and $\mathbf{z}_2(0)$ yields:

$$\frac{d\,(\mathbf{z}_2 - \mathbf{z}_1)}{dt} = \boldsymbol{g}(\mathbf{z}_2) - \boldsymbol{g}(\mathbf{z}_1), \qquad \mathbf{z}_2(0) - \mathbf{z}_1(0) = \boldsymbol{\epsilon}. \qquad (3.23)$$

Expanding in Taylor series $\boldsymbol{g}(\mathbf{z}_2)$ about $\mathbf{z}_1$ yields the evolution equation for the *vector dynamics* of the line element $\boldsymbol{v}$:

$$\frac{d\boldsymbol{v}}{dt} = \boldsymbol{J_g}(\mathbf{z}_1) \cdot \boldsymbol{v}, \qquad \boldsymbol{v}(0) = \mathbf{1}, \qquad (3.24)$$

where $\boldsymbol{J_g}(\mathbf{z}_1)$ is the Jacobian of the vector field $\boldsymbol{g}$ evaluated along the reference trajectory, and $\mathbf{1}$ is a unit vector at $\mathbf{z}_0$ taken along any direction.

The norm v of the vector $\boldsymbol{v}$ is defined as $v = \sqrt{\boldsymbol{v} \cdot \boldsymbol{v}}$ and its time evolution is readily found by taking the scalar product of the left- and right-hand side of Eq. (3.24) with $\boldsymbol{v}$, and reads

$$\frac{dv}{dt} = \left(\frac{\boldsymbol{v} \cdot \boldsymbol{J_g} \cdot \boldsymbol{v}}{v^2}\right) v, \qquad v(0) = 1. \qquad (3.25)$$

The rate at which v changes (grows/shrinks) with time is governed by the quadratic form enclosed by the parentheses on the right-hand side of Eq. (3.25). It is thus proper to name this coefficient "the (local) rate of stretching of the dynamics", evaluated along the direction identified by the unit vector $\tilde{\boldsymbol{u}} = \boldsymbol{v}/v$ and defined as

$$\omega_{\tilde{u}} := \tilde{\boldsymbol{u}} \cdot \boldsymbol{J_g} \cdot \tilde{\boldsymbol{u}}. \qquad (3.26)$$

The (local) stretching rate $\omega_{\tilde{u}}$ takes positive/negative values when the dynamics acts so as to stretch/shrink the initial unit vector.

As a first example, if we choose a Jacobian eigenvector as the direction of stretch, we obtain the associated eigenvalue. However, more interesting results may be obtained along with other *special* directions, as follows.

If the unit vector $\tilde{\boldsymbol{u}}$ is taken in the direction of the vector field $\boldsymbol{g}$, it is natural to name the associated stretching rate as *tangential* $\tilde{\boldsymbol{\tau}}$ (to the vector field).

Thus, $\tilde{\boldsymbol{\tau}}$ is defined as:

$$\tilde{\boldsymbol{\tau}} = \frac{\boldsymbol{g}(\mathbf{z})}{|\boldsymbol{g}(\mathbf{z})|}, \tag{3.27}$$

which can be rewritten after projecting the vector field over the right eigenvector basis as:

$$\tilde{\boldsymbol{\tau}} = \frac{1}{g} \sum_{i=1}^{N} \boldsymbol{a}_i f^i . \tag{3.28}$$

with $g = |\boldsymbol{g}|$.

Given that $\boldsymbol{J_g} = \boldsymbol{A} \cdot \Lambda \cdot \boldsymbol{B}$, we obtain the Tangential Stretching Rate (TSR) definition:

$$\omega_{\tilde{\tau}} = \tilde{\boldsymbol{\tau}} \cdot \boldsymbol{J_g} \cdot \tilde{\boldsymbol{\tau}} = \frac{1}{g^2} \left(\boldsymbol{g} \cdot \boldsymbol{A} \cdot \Lambda \cdot \boldsymbol{B} \cdot \boldsymbol{g} \right) . \tag{3.29}$$

Expanding $\boldsymbol{g}$ in terms of eigenmodes, as detailed in [24], one obtains:

$$\omega_{\tilde{\tau}} = \sum_{i=1}^{N} W_i \lambda_i , \tag{3.30}$$

where the weights W_i are defined as:

$$W_i = \frac{f^i}{g} \left(\tilde{\boldsymbol{\tau}} \cdot \boldsymbol{a}_i \right) = \frac{f^i}{g} \sum_{k=1}^{N} \frac{f^k}{g} \left(\boldsymbol{a}_k \cdot \boldsymbol{a}_i \right) . \tag{3.31}$$

The definition of the TSR, Eq. (3.30), together with the weights defined as in Eq. (3.31), combines three aspects of the dynamics, that turn out to be equally important in the generation of the most energetic scale:

1. the eigenvalues, λ_i, of the Jacobian matrix, $\boldsymbol{J_g}$, of the local vector field, $\boldsymbol{g}$, of the system;
2. the mode amplitude, $f^i = \boldsymbol{b}^i \cdot \boldsymbol{g}$, of each eigenmode $\boldsymbol{a}_i$;
3. the degree of co-linearity, $\boldsymbol{g} \cdot \boldsymbol{a}_i$, of each eigenvector $\boldsymbol{a}_i$ with respect to the vector field $\boldsymbol{g}$.

The TSR is a weighted average of all eigenvalues in 1) with weights that depend on both 2) and 3). Each mode can in principle contribute to the TSR providing its mode amplitude is large (high energy), and its direction sufficiently aligned with that of the vector field.

The analysis presented indicates that the proper characteristic chemical time scale in problems involving multi-step chemical kinetic mechanisms is $\tau_{chem} := |1/\omega_{\bar{\tau}}|$. This estimate can be adopted to define a Damköhler number, Da, in conjunction with a characteristic fluid dynamical scale τ_f to yield $Da := \tau_f/\tau_{chem}$. This conclusion can be drawn by observing that τ_{chem} is (i) the most relevant time scale during both the explosive and relaxation regimes, (ii) is intrinsic to the dynamics, and, as such, can be identified without the need of any ad-hoc assumption.

3.3.1 Tangent Space Decomposition Induced by TSR

After ordering the terms in the sum which defines $\omega_{\bar{\tau}}$ by the descending value of the modulus of the eigenvalue, and adopting the value $\boldsymbol{g} \cdot \boldsymbol{a}_i = f^i$ for all modes, which implies $\boldsymbol{g}$ co-linear with $\boldsymbol{a}_i$, we have

$$\omega_{\bar{\tau}} = \sum_{r=1}^{L} \left(\frac{f^r}{g}\right)^2 \lambda_r + \sum_{a=L+1}^{K-1} \left(\frac{f^a}{g}\right)^2 \lambda_a + \sum_{s=K}^{N} \left(\frac{f^s}{g}\right)^2 \lambda_s, \tag{3.32}$$

where the labels "r", "a", "s" denote fast, active, slow modes, respectively.

Now consider a typical situation for which, at some point in the phase space, there exists a number of fast and slow modes with a vanishing amplitude, that is $\left\{f^r = \boldsymbol{b}^r \cdot \boldsymbol{g} \approx 0\right\}_{r=1,\dots,L}$ and $\left\{f^s = \boldsymbol{b}^s \cdot \boldsymbol{g} \approx 0\right\}_{s=K,\dots,N}$, with $L < K$. This situation can develop when there exist two spectral gaps between the slow and active subspaces, $\varepsilon_s = |\lambda_L/\lambda_{L+1}|$, and between the active and fast subspaces, $\varepsilon_r = |\lambda_{K-1}/\lambda_K|$, both (much) smaller than one. In such a case, it happens that

$$\omega_{\bar{\tau}} \approx \sum_{a=L+1}^{K-1} \left(\frac{f^a}{g}\right)^2 \lambda_a, \tag{3.33}$$

which implies that only the active scales contribute to $\omega_{\bar{\tau}}$. If an active eigenvalue is positive, say λ_{a+}, it is likely that f^{a+} will be the largest of all active mode amplitudes. In this case $\omega_{\bar{\tau}}$ will be mostly affected by λ_{a+}.

From Eq. (3.32), three scales can be identified: *(i)* one associated with $\omega_{\bar{\tau}}$, which is the most energy containing scale; *(ii)* two others associated with λ_{K-1} and λ_{L+1}, which are the fastest and slowest scales, respectively, contributing to $\omega_{\bar{\tau}}$, and these bracket the range of active (energy containing) scales. The integers K and L underline a decomposition of the tangent space $\mathcal{T}_x$ in three subspaces: the fast subspace $\mathcal{F}_{TSR}$ ($r = 1, \dots, L$), the active $\mathcal{A}_{TSR}$ ($a = L+1, \dots, K-1$), and the slow $\mathcal{S}_{TSR}$ ($s = K, \dots, N$), such that

$$\mathcal{T}_x = \mathcal{F}_{TSR} \bigoplus \mathcal{A}_{TSR} \bigoplus \mathcal{S}_{TSR}. \tag{3.34}$$

The decomposition of the tangent space in three subspaces (plus one subspace related to invariant properties of the system) is analogous to that employed in the G-Scheme framework proposed in [31]. Note that the classic decomposition in the CSP method involves only the slow $\mathcal{S}_{CSP}$ and fast $\mathcal{F}_{CSP}$ subspaces. The following equivalence among subspaces hold:

$$\begin{aligned} \mathcal{S}_{CSP} &= \mathcal{A}_{TSR} \bigoplus \mathcal{S}_{TSR}, \\ \mathcal{F}_{CSP} &= \mathcal{F}_{TSR}. \end{aligned} \tag{3.35}$$

In turn, the three-fold decomposition in the G-Scheme and TSR accounts for both the singularly perturbed *slow system* ($\mathcal{T}_x = \mathcal{F}_{TSR} \bigoplus \mathcal{A}_{TSR}$) and the regularly perturbed *fast system* ($\mathcal{T}_x = \mathcal{A}_{TSR} \bigoplus \mathcal{S}_{TSR}$) options for model reduction, whereas CSP accounts for the *slow system* ($\mathcal{T}_x = \mathcal{F}_{CSP} \bigoplus \mathcal{S}_{CSP}$) option only. Here, we employed the definition of *slow* and *fast* systems according to Fenichel [32].

3.4 Extension to Non-homogeneous Systems

Consider a reactive flow system whose dynamics is described by the set of PDEs of the form Eq. (3.1), that is one involving spatial non-homogeneities that trigger transport processes (convection, diffusion, ...). The most natural extension of the CSP method, originally developed for systems of ODEs, to systems of PDEs involves casting the set of PDEs into a set of ODEs by discretization of the right-hand side (method of lines) involving infinite dimensional space operators as well as finite dimensional vector fields (chemical source term). However the dimension of the space increases proportionally to the number of cells K used in the space discretization, that is the state vector lives in an extended space of dimensions $N \times K$. Moreover, the discrete eigensystem is related to the combined kinetic plus transport nature of the system, and the discrete eigenvectors are approximation of the eigenfunctions of the spatial operators. All these concepts are theoretically sound, but practically difficult to manage. Moreover their information has a global (spatially speaking) nature and does not provide local information about the flow.

Therefore, an alternative, albeit less rigorous, approach is to accept to work in the original N-dimensional space defined by the kinetics only. Such a formulation is based on the assumption that the fastest time scales are generated by the kinetics term. In this context, transport affects only the amplitude of the kinetic eigenmodes but not the structure of the mode themselves. More details on how to relate reactive with reactive-diffusive time scales can be found in [33].

3.4.1 CSP (local) Analysis of Non-homogeneous Reactive Systems

As discussed in [6, 9, 34, 35], one can recast Eq. (3.1) as an expansion over the kinetic modes as follows:

$$\frac{\partial \mathbf{z}}{\partial t} = \boldsymbol{L}(\mathbf{z}) + \boldsymbol{g}(\mathbf{z}) = \sum_{i=1}^{N} \boldsymbol{a}_i(\mathbf{z}) h^i(\mathbf{z}), \tag{3.36}$$

where the modal amplitudes h^i are defined as:

$$h^i = \boldsymbol{b}^i \cdot (\boldsymbol{L}(\mathbf{z}) + \boldsymbol{g}(\mathbf{z})) \tag{3.37a}$$
$$= (\boldsymbol{b}^i \cdot \boldsymbol{e}_1) L^1 + \cdots + (\boldsymbol{b}^i \cdot \boldsymbol{e}_N) L^N + (\boldsymbol{b}^i \cdot \boldsymbol{S}_1) R^1 + \cdots + (\boldsymbol{b}^i \cdot \boldsymbol{S}_{2N_r}) R^{2N_r}. \tag{3.37b}$$

and the CSP basis is evaluated with respect to the finite-dimensional vector field $\boldsymbol{g}(\mathbf{z})$. The N CSP-modes in Eq. (3.36) are ordered according as in Eq. (3.7).

At any time epoch and space location, where the fast time scales are generated by the chemical kinetics term, one can partition the tangent space into the slow and fast subspaces on the basis of the criterion Eq. (3.21) where the mode amplitude is evaluated in terms of h^i and not f^i. When the amplitudes of the M fastest modes have vanished, i.e. when the following relations hold,

$$h^i \approx 0 \qquad i = 1, \ldots, M, \tag{3.38}$$

we declare *exhausted* the chemical time scales τ^i_{chem} $(i = 1, \ldots, M)$ and *fast* the corresponding modes, while the remaining $N - M$ time scales and modes are declared as being *active* and *slow*.

Equation (3.38) expresses the equilibration of the different terms in the expression of each h^i, as in Eq. (3.37b), the cancellations among the elementary rates being the most important [6].

It might also occur that the amplitude of any mode h^i is zero or very small not because of the action of the different participating processes balancing one another along the direction of the mode, but because each term in the expression h^i, as in Eq. (3.37b), is either very small or zero. In this case, we will refer to the mode, either fast or slow, as a *dormant* (in case of ODEs) or *frozen* (in case of PDEs) mode [5].

A mode is declared *dormant/frozen* if the following inequality is satisfied:

$$\tilde{h}^i = |\,\boldsymbol{b}^i \cdot \boldsymbol{L}\,| + |(\boldsymbol{b}^i \cdot \boldsymbol{S}_1) R^1| + |(\boldsymbol{b}^i \cdot \boldsymbol{S}_2) R^2| + \cdots + |(\boldsymbol{b}^i \cdot \boldsymbol{S}_{2N_r}) R^{2N_r}| < h_{\text{tolerance}} \tag{3.39}$$

for $i = 1, \ldots, N$, with $h_{\text{tolerance}}$ being a user defined error threshold.

Unlike exhausted fast modes, the dormant/frozen modes do not produce equations of state. However, like an exhausted fast mode, the presence of a dormant/frozen mode indicates a decrease by one of the dimensions of the subdomain in which the solution can evolve, by not allowing the spreading of the manifold along their directions.

A dormant mode can originate when the chemistry is frozen with respect to convection and diffusion, that is when the characteristic chemical time scales are much slower than the characteristic time scales of convection and diffusion. Note that a dormant mode can relate to either the M exhausted time scales or the $N - M$ slow time scales.

Therefore, at each point in time and space, the role of the reactions and the other physical processes in the problem (convection and diffusion) which are responsible for the emergence of fast exhausted and/or dormant/frozen modes is in a sense passive, when viewed from the fast directions point of view. In particular they tend to diminish the domain in which the solution is allowed to move; the exhausted fast modes by constraining the evolution of the trajectory within the manifold and the dormant/frozen fast modes by not allowing the trajectory to evolve along their directions, both by virtue of the constraints expressed in Eq. (3.38). On the other hand, the role of the processes responsible for the emergence of the non-exhausted, non-dormant modes is active, that is they determine the trajectory speed according to the set of simplified (non-stiff) PDE's:

$$\frac{\partial \mathbf{z}}{\partial t} = \sum_{s=1}^{M+1,N} \boldsymbol{a}_s(\mathbf{z}) h^s(\mathbf{z}), \tag{3.40}$$

in which slow active modes participate. Note here that a physical process, either reaction or transport, can participate simultaneously in the fast exhausted or dormant/frozen modes and the slow active or dormant/frozen ones.

A dormant mode can be safely dropped in the modal expansion of the right-hand side of Eq. (3.40) without any loss in accuracy. This also implies that when a number N_d of dormant modes is present at a given space and time location, then the actual dimension of the system locally reduces to $N - N_d$ corresponding to the total number of degrees of freedom available to the system space vector in the phase space.

Some of the dormant modes will be found among the M exhausted time scales, say N_d^{fast}. In this case, the number of algebraic relations describing the manifold reduces to $M - N_d^{fast}$. The number of active modes describing the slow dynamics of the system is therefore: $N - (N_d - N_d^{fast}) - M$. This number turns out to be the dimension of the manifold as well.

It should be noted that in the case of zero transport, i.e. $\boldsymbol{L}_x = \mathbf{0}$, h^i translates to f^i and all the previously introduced relations are consistent with the CSP approach for homogeneous systems of ODEs. Also note that $h^i \approx 0$ might express a balance among reactions or a balance between reactions and transport, along the direction $\boldsymbol{a}_i$ of the it ith mode, which is established on a time scale defined by the ith eigenvalue. When $|\boldsymbol{b}^i \cdot \boldsymbol{L}(\mathbf{z})|$ contributes little to the cancellations occurring in $h^i \approx 0$, the ith exhausted mode denotes equilibrium among reactions, while when $|\boldsymbol{b}^i \cdot \boldsymbol{L}(\mathbf{z})|$ contributes significantly the mode denotes equilibrium among reactions and transport. However, when $h^i \approx 0$ and $|\boldsymbol{b}^i \cdot \boldsymbol{L}(\mathbf{z})|$ is not negligible, then kinetics cannot reach equilibrium because transport keeps it off chemical equilibrium, and $h^i \approx 0$ represents a stationary condition where transport balances non-equilibrium chemical kinetics.

It should be pointed out that in an ODE system involving only kinetics, a zero (kinetic) eigenvalue, say $\lambda^e = 0$, corresponds to a mode $\boldsymbol{a}_e$ which defines a (linear) conservation law (possibly an elemental mass conservation) according to the expression: $\boldsymbol{b}^e \cdot \boldsymbol{g}(\mathbf{z}) = 0$. However, in a PDE system involving diffusion, it might occur that diffusion could have a non-negligible projection over the same mode, say $\boldsymbol{b}^e \cdot \boldsymbol{L}(\mathbf{z}) \neq 0$, so that in this circumstance the conservation law does not hold anymore ($\boldsymbol{b}^e \cdot (\boldsymbol{L}(\mathbf{z}) + \boldsymbol{g}(\mathbf{z})) \neq 0$).

3.4.2 TSR (local) Analysis of Non-homogeneous Reactive Systems

Having recalled how CSP has been applied to systems of PDEs, it is rather natural to extend the definition of the TSR to systems of PDEs by simply replacing the definition of $\tilde{\boldsymbol{\tau}}$ in (3.27), which is based on kinetics only, with a $\tilde{\boldsymbol{\tau}}_{pde}$ based on the algebraic sum of kinetics and transport as follows:

$$\tilde{\boldsymbol{\tau}}_{pde} =: \frac{\boldsymbol{L}(\mathbf{z}) + \boldsymbol{g}(\mathbf{z})}{|\boldsymbol{L}(\mathbf{z}) + \boldsymbol{g}(\mathbf{z})|}, \tag{3.41}$$

which can be rewritten, after projecting the vector field over the right (kinetic) eigenvector basis, as

$$\tilde{\boldsymbol{\tau}}_{pde} = \frac{1}{|\boldsymbol{L}(\mathbf{z}) + \boldsymbol{g}(\mathbf{z})|} \sum_{i=1}^{N} \boldsymbol{a}_i h^i, \tag{3.42}$$

with

$$h^i = \boldsymbol{b}^i \cdot (\boldsymbol{L}(\mathbf{z}) + \boldsymbol{g}(\mathbf{z})).$$

After having introduced $\tilde{\boldsymbol{\tau}}_{pde}$, it is straightforward to arrive at the extended definition of the TSR, $\omega_{\tilde{\boldsymbol{\tau}}_{pde}}$, which reads:

$$\omega_{\tilde{\boldsymbol{\tau}}_{pde}} = \tilde{\boldsymbol{\tau}}_{pde} \cdot \boldsymbol{J_g} \cdot \tilde{\boldsymbol{\tau}}_{pde} = \sum_{i=1}^{N} W_{i,pde} \lambda_i \tag{3.43}$$

where the weights $W_{i,pde}$ are defined as follows

$$W_{i,pde} = \frac{h^i}{\mathrm{L} + \mathrm{g}} \left(\tilde{\boldsymbol{\tau}}_{pde} \cdot \boldsymbol{a}_i \right) = \frac{h^i}{\mathrm{L} + \mathrm{g}} \sum_{k=1}^{N} \frac{h^k}{\mathrm{L} + \mathrm{g}} (\boldsymbol{a}_k \cdot \boldsymbol{a}_i), \tag{3.44}$$

and L is the norm of $\boldsymbol{L}$.

The explosive/dissipative nature of the flow will still be identified by a positive/negative value of the TSR, $\omega_{\tilde{\tau}_{pde}}$. Also, it can happen that a purely kinetic propensity to explosion, as marked by a large and positive value of $\omega_{\tilde{\tau}}$, can be mitigated by an adverse effect of transport, that is with a lower (albeit still positive) value of $\omega_{\tilde{\tau}_{pde}}$. In the limit, $\omega_{\tilde{\tau}_{pde}}$ can become zero or negative when transport overwhelms kinetics. This is the case when ignition cannot occur at low Damköhler numbers.

3.5 Cause/Effect Identification

In the general framework of the CSP theory, various algorithmic mathematical tools have been developed and their validity has been widely tested.

These tools provide physical interpretation about the role of each driving process to specific aspects of the dynamics, such as time scales (Time scale Participation Index, TPI), mode amplitudes (Amplitude Participation Index, API), time rates of change of specific state variables (Slow and Fast Importance Indices), and the most energetic time scale (TSR Participation Index, TSR-PI), as outlined before.

We also introduced in [36] an index that addresses the process most affecting specific subspace. We referred to this index as G-Scheme Participation Index, because it has been introduced in conjunction with the development of the G-Scheme solver.

Finally, it is possible to utilize the information on entropy production associated with reactions within each subspace to define an *Entropy Participation Index* (EPI) as done in [37]. This index identifies the reactions (and all those species associated with them) which are responsible for the largest entropy production.

All these indices exploit the common feature that these observables are defined as sums of modal contributions, and/or sums of contributions from specific driving processes.

3.5.1 Time Scale Participation Index

The eigenvalue of the ith mode is defined as $\lambda_i = \boldsymbol{b}^i \cdot \boldsymbol{J}_g \cdot \boldsymbol{a}_i$, where $\boldsymbol{a}_i$ and $\boldsymbol{b}^i$ are the ith right (column) and left (row), respectively, eigenvectors of the Jacobian $\boldsymbol{J}_g$. Taking into account the fact that the vector field $\boldsymbol{g}(\mathbf{z}) = \sum_{k=1}^{2K} \hat{\mathbf{S}}_k R^k$ is the sum of $2K$ terms, with e.g. K reversible reactions or $2K$ irreversible reactions, the nth eigenvalue can be expressed as:

$$\lambda_n = \boldsymbol{b}^n \cdot \boldsymbol{J}_g \cdot \boldsymbol{a}_n = \boldsymbol{b}^n \cdot \sum_{k=1}^{2K} \frac{\partial}{\partial \mathbf{z}} \left(\hat{\mathbf{S}}_k R^k \right) \cdot \boldsymbol{a}_n = c_1^n + \cdots + c_{2K}^n \tag{3.45}$$

since $J_g = \frac{\partial}{\partial \mathbf{z}}(\hat{S}_1 R^1) + \cdots + \frac{\partial}{\partial \mathbf{z}}(\hat{S}_{2K} R^{2K})$ [38, 39]. Considering real eigenvalues (the extension to the case where some of the eigenvalues are complex conjugate pairs is straightforward [39]), the magnitude of the terms c_k^n is indicative of the contribution of the kth reaction to the nth eigenvalue. The c_k^n terms can be either positive or negative; when positive, it implies that the kth reaction contributes to an explosive character of the nth time scale τ_n, while when negative, it implies that the kth reaction contributes to a dissipative character.

The contribution of each of the $2K$ reactions to the time scale τ_n can be assessed through the related term in Eq. (3.45), by employing the Time scale Participation Index (TPI) [38–42]:

$$J_k^n = \frac{c_k^n}{|c_1^n| + \cdots + |c_{2K}^n|}. \tag{3.46}$$

J_k^n measures the relative contribution of the kth reaction to the nth eigenvalue λ_n and, therefore, to the time scale τ_n. By definition $|J_1^n| + \cdots + |J_{2K}^n| = 1$.

3.5.2 Amplitude Participation Index

Considering that the amplitude h^n of the nth mode can be expanded as

$$h^n = \boldsymbol{b}^n \cdot (\boldsymbol{L}_x(z) + \boldsymbol{g}(z)) = \sum_{j=1}^{N} b_j^n L_{\mathbf{z},j} + \sum_{k=1}^{2K} (\boldsymbol{b}^n \cdot \hat{\boldsymbol{S}}_k) R^k = \sum_{l=1}^{2K+N} \beta_l^n, \tag{3.47}$$

allows us to introduce an *Amplitude Participation Index* (API), P_k^n, that measures the relative contribution of the kth process (reaction or transport term) to the amplitude of the nth mode [5, 43–46]:

$$P_k^n = \frac{\beta_k^n}{\sum_{j=1}^{2K+N} |\beta_j^n|}. \tag{3.48}$$

In the case of an exhausted mode $h^r = 0\,(r = 1, M)$. Therefore, relatively large values of P_k^r identify the processes (reaction or transport) contributing to the equilibrium stated by the relation $h^r = 0$. By definition $|P_1^n| + \cdots + |P_{2K+N}^n| = 1$.

3.5.3 Slow and Fast Importance Indices

The time rates of change of a specific state variable can be first partitioned in slow and fast contributions as $dz^i/dt = g^i = g_s^i + g_r^i$. Next, each slow and fast contribution can be expanded to isolate the contribution of a specific driving process [13, 43] to yield:

$$
\begin{aligned}
g_r^i &= \sum_{r=1}^{M} a_r^i \boldsymbol{b}^r \cdot (\boldsymbol{L}_{\boldsymbol{x}}(z) + \boldsymbol{g}(z)) \\
&= \sum_{j=1}^{N} \sum_{r=1}^{M} a_r^i (\boldsymbol{b}_j^r \cdot \boldsymbol{L}_{\mathbf{z}}^j) + \sum_{k=1}^{2K} \sum_{r=1}^{M} a_r^i (\boldsymbol{b}_j^r \cdot \hat{\mathbf{S}}_k)\, R^k = \sum_{l=1}^{2K+N} \psi_{r,l}^i
\end{aligned}, \tag{3.49}
$$

$$
\begin{aligned}
g_s^i &= \sum_{s=M+1}^{N} a_s^i \boldsymbol{b}^s \cdot (\boldsymbol{L}_{\boldsymbol{x}}(z) + \boldsymbol{g}(z)) \\
&= \sum_{j=1}^{N} \sum_{s=M+1}^{N} a_s^i (\boldsymbol{b}_j^s \cdot \boldsymbol{L}_{\mathbf{z}}^j) + \sum_{k=1}^{2K} \sum_{s=M+1}^{N} a_s^i (\boldsymbol{b}_j^s \cdot \hat{\mathbf{S}}_k)\, R^k = \sum_{l=1}^{2K+N} \psi_{s,l}^i .
\end{aligned} \tag{3.50}
$$

We now introduce the *Slow/Fast Importance Index* (S/FII), $I_{s,k}^i$ and $I_{r,k}^i$, that measure the relative contributions of the kth process (reaction or transport term) to the time rates of change of a specific state variables in the slow and fast subspaces

$$
I_{s,k}^i = \frac{\psi_{s,k}^i}{\sum_{j=1}^{2K+N} |\psi_{s,j}^i|} \quad \text{no sum on k,} \tag{3.51}
$$

$$
I_{r,k}^i = \frac{\psi_{r,k}^i}{\sum_{j=1}^{2K+N} |\psi_{r,j}^i|} \quad \text{no sum on k.} \tag{3.52}
$$

The index k in Eqs. (3.51) and (3.52) indicates either a forward ($k \leq N_r$) or a reverse ($N_r + N < k \leq 2N_r + N$) chemical reaction, or a transport term. When considering a transport process, such importance indices will therefore measure the relative influence of the convective or diffusive transport process of the kth species to the time rate of change of the ith species.

3.5.4 Tangential Stretching Rate Participation Index

The contribution of the ith mode to $\omega_{\tilde{\tau}}$ or to $\omega_{\tilde{\tau}_{pde}}$ can be assessed through the TSR-PI, $V_i^{\omega_{\tilde{\tau}_{pde}}}$, defined as in [25]:

$$
V_i^{\omega_{\tilde{\tau}_{pde}}} = \text{sign}(Re(\lambda_i)) \frac{W_{i,pde}|\lambda_i|}{\sum_{s=1}^{N} |W_{s,pde}|\lambda_s||} . \tag{3.53}
$$

Based on this expression, modes with a large and positive/negative $V_i^{\omega_{\bar{\tau}_{pde}}}$ are those that contribute the most to the growth/decay of the most energetic time scale.

The contribution of the kth reaction to $\omega_{\bar{\tau}}$, $H_k^{\omega_{\bar{\tau}}}$, can be assessed for systems involving kinetics only (that is in the absence of transport) by combining the API, P_k^n, which identifies reactions important to the nth mode, to the TSR-PI, $V_n^{\omega_{\bar{\tau}}}$, which identifies modes important to $\omega_{\bar{\tau}}$, as follows: $H_k^{\omega_{\bar{\tau}}} = V_n^{\omega_{\bar{\tau}}} \times P_k^n$.

Similarly, for systems involving kinetics and transport, the contribution of the kth process (either a reaction or transport term) to $\omega_{\bar{\tau}_{pde}}$, $H_k^{\omega_{\bar{\tau}_{pde}}}$, can be assessed by combining the API, P_k^n, which identifies processes important to the nth mode, to the TSR-PI, $V_n^{\omega_{\bar{\tau}_{pde}}}$, which identifies modes important to $\omega_{\bar{\tau}_{pde}}$, as follows: $H_k^{\omega_{\bar{\tau}_{pde}}} = V_n^{\omega_{\bar{\tau}_{pde}}} \times P_k^n$.

Therefore, the sum of the contributions of all the reactions ($H_{Chem}^{\omega_{\bar{\tau}_{pde}}}$) can provide a clear understanding about the role of chemical kinetics in the system. Likewise, the summation of the contributions of all the convective/diffusive terms ($H_{Conv}^{\omega_{\bar{\tau}_{pde}}}/H_{Diff}^{\omega_{\bar{\tau}_{pde}}}$) can clarify the role of convection/diffusion in the system and their sum ($H_{Transport}^{\omega_{\bar{\tau}_{pde}}}$) can be used to determine the regions where transport prevails over kinetics.

It is noted that a similar approach has been proposed in the past in the context of CSP [6, 44], where the summation of the individual terms is performed over the determined dominant mode.

For a more convenient evaluation of the order of magnitude of both the positive and negative TSR, we introduce three ancillary parameters defined as $\Omega = \text{Sign}(\omega_{\bar{\tau}}) \, \text{Log}_{10}|max(1, \omega_{\tau})|$, $\Omega_{pde} = \text{Sign}(\omega_{\bar{\tau}_{pde}}) \, \text{Log}_{10}|max(1, \omega_{\tau_{pde}})|$ and $\Lambda_i = \text{Sign}(\lambda_i) \, \text{Log}_{10}|max(1, \lambda)|$.

Thus, both $\omega_{\bar{\tau}}$ and $\omega_{\bar{\tau}_{pde}}$ are linear combinations of the chemical system eigenvalues and represent the stretch rate induced on the state vector by the chemical source term. Physically, $\omega_{\bar{\tau}_{pde}}$ represents the stretching rate along the unit vector aligned to the state evolution direction due to the reactive-transport system, while the $\omega_{\bar{\tau}}$ is the stretching rate along the unit vector oriented in the direction of the chemical source term, i.e. the direction that the state vector evolution would follow if the transport term was abruptly set to zero. In other words, $\omega_{\bar{\tau}}$ can be seen as the time scale at which the chemical source term would evolve to recover its local chemical equilibrium, if transport would be suddenly set to zero.

3.5.5 *Feature Tracking*

The ultimate goal of the mathematical machinery illustrated in the preceding sections is to extract useful information from detailed flow simulations. However, the post-processing of the flow datasets produces a large number of derived parameters (APIs, TPIs, TSR-PIs, etc.). It is therefore mandatory to provide a rational procedure which allows one to cope with the vast amount of data. There is no unique answer to this quest, and thus we simply propose our favorite recipe.

When dealing with a multi-dimensional reactive flow fileld, the first thing is to identify the space-time flow regions featuring the same number M of exhausted modes. As already noted, a certain value of exhausted modes can correspond to modes either in *near-equilibrium*, *frozen*, or *non-equilibrium* state. A certain number M of exhausted modes might correspond to the dimension of a SIM featuring a qualitative persistence of properties in term of cause–effect relationships. Thus, it makes sense to carry out statistics of CSP/TSR indices conditioned on the number of *near-equilibrium*, *frozen*, or *non-equilibrium* modes.

Flow regions associated with the development of deflagrative fronts or spontaneous ignition fronts can be detected by comparing the values of the TSR (Eq. (3.30)) and the extended TSR (Eq. (3.43)). Later, we will discuss how and why large differences of these two parameters are found where diffusive transport is significant, as occurs in deflagrative fronts, whereas small differences of these two parameters are found where diffusive transport is negligible as occurs in spontaneous ignition fronts.

3.6 Epilogue

In this chapter, we summarized the concepts behind the CSP method and the TSR analysis. In Chap. 4, we will illustrate and discuss a few applications using these techniques.

References

1. S.H. Lam, D.A. Goussis, Proc. Combust. Inst. **22**, 931 (1988)
2. S. Lam, D. Goussis, in *Modeling and Simulation of Systems*, ed. by P.B. et al. (Baltzer Scientific Publishing Co, 1989), pp. 303–306
3. S. Lam, D. Goussis, in *Reduced Kinetic Mechanisms and Asymptotic Approximations for Methane-Air Flames*, Springer Lecture Notes, ed. by M. Smooke (Springer, Berlin, 1991), pp. 227–242
4. S.H. Lam, Combust. Sci. Technol. **89**, 375 (1993)
5. S.H. Lam, D.A. Goussis, Int. J. Chem. Kinet. **26**, 461 (1994)
6. M. Hadjinicolaou, D.A. Goussis, SIAM J. Sci. Comput. **20**(3), 781 (1998)
7. M. Valorani, D.A. Goussis, H.N. Najm, Using CSP to analyze computed reactive flows, in *8th SIAM International Conference on Numerical Combustion*, Amelia Island, FL (2000)
8. M. Valorani, D. Goussis, H.N. Najm, in *Ninth International Conference on Numerical Combustion*, Sorrento, Italy (2002)
9. M. Valorani, H.N. Najm, D. Goussis, Combust. Flame **134**(1–2), 35 (2003)
10. D.A. Goussis, M. Valorani, F. Creta, H.N. Najm, in *Computational Fluid and Solid Mechanics*, vol. 2, ed. by K. Bathe (Elsevier Science, Cambridge, 2003), pp. 1951–1954
11. M. Valorani, D.A. Goussis, F. Creta, H.N. Najm, J. Comput. Phys. **209**, 754 (2005)
12. M. Valorani, F. Creta, D.A. Goussis, J.C. Lee, H.N. Najm, in *Computational Fluid and Solid Mechanics*, ed. by K. Bathe (Elsevier Science, Amsterdam, 2005), pp. 900–904
13. M. Valorani, F. Creta, D.A. Goussis, J.C. Lee, H.N. Najm, Combust. Flame **146**, 29 (2006)

14. M. Valorani, F. Creta, F. Donato, H.N. Najm, D.A. Goussis, Proc. Combust. Inst. **31**, 483 (2007)
15. M. Valorani, F. Creta, F. Donato, H.N. Najm, D.A. Goussis, in *ECCOMAS CFD 2006*, Delft, Holland (2006)
16. T.F. Lu, Y.G. Ju, C.K. Law, Combust. Flame **126**, 1445 (2001)
17. T.F. Lu, C.K. Law, Combust. Flame **154**, 761 (2008)
18. A. Zagaris, H.G. Kaper, T.J. Kaper, in *SIAM Conference on Applications of Dynamical Systems*, Snowbird, Utah (2003). Accessed 27–31 May 2003
19. A. Zagaris, H.G. Kaper, T.J. Kaper, Nonlinear Sci. **14**, 59 (2004)
20. M. Salloum, A. Alexanderian, O. Le Maître, H.N. Najm, O. Knio, Comput. Methods Appl. Mech. Eng. **217–220**, 121 (2012)
21. J.C. Lee, H.N. Najm, S. Lefantzi, J. Ray, M. Frenklach, M. Valorani, D.A. Goussis, Combust. Theory Model. **11**(1), 73 (2007)
22. J. Prager, H.N. Najm, M. Valorani, D.A. Goussis, Proc. Combust. Inst. **32**(1), 509 (2008)
23. A. Adrover, F. Creta, M. Giona, M. Valorani, J. Comput. Phys. **225**(2), 1442 (2007)
24. M. Valorani, S. Paolucci, E. Martelli, T. Grenga, P.P. Ciottoli, Combust. Flame **162**(8), 2963 (2015)
25. M. Valorani, P.P. Ciottoli, R.M. Galassi, Proc. Combust. Inst. **36**(1), 1357 (2017)
26. P. Pal, M. Valorani, P.G. Arias, H.G. Im, M.S. Wooldridge, P.P. Ciottoli, R. Malpica Galassi, Proc. Combust. Inst. **36**(3), 3705 (2017)
27. D.M. Manias, E.A. Tingas, F.E.H. Pérez, R.M. Galassi, P.P. Ciottoli, M. Valorani, H.G. Im, Combust. Flame **200**, 155 (2019)
28. Z. Li, R.M. Galassi, P.P. Ciottoli, A. Parente, M. Valorani, Combust. Flame **208**, 281 (2019)
29. D.A. Goussis, Phys. D: Nonlinear Phenom. **248**, 16 (2013)
30. M. Valorani, P. Ciottoli, R. Malpica Galassi, S. Paolucci, T. Grenga, E. Martelli, Flow, Turbul. Combust. **101**(4), 1023 (2018)
31. M. Valorani, S. Paolucci, J. Comput. Phys. **228**, 4665 (2009)
32. N. Fenichel, J. Differ. Equ. **31**, 53 (1979)
33. D.A. Goussis, M. Valorani, F. Creta, H.N. Najm, Prog. Comput. Fluid Dyn. Int. J. **5**(6), 316 (2005)
34. U. Maas, S. Pope, in *Symposium (International) on Combustion*, vol. 25,1 (1994), pp. 1349–1356
35. S. Lam, Combust. Sci. Technol. **179**(4), 767 (2007)
36. T. Grenga, S. Paolucci, M. Valorani, Combust. Flame **189C**, 275 (2018)
37. M. Valorani, S. Paolucci, P.P. Ciottoli, R.M. Galassi, Combust. Theory Model. **21**(1), 137 (2017)
38. D.A. Goussis, G. Skevis, in *Computational Fluid and Solid Mechanics*, vol. 1, ed. by K. Bathe (Elsevier, Amsterdam, 2005), pp. 650–653
39. D.A. Goussis, H.N. Najm, Multiscale Model. Simul. **5**(4), 1297 (2006)
40. E.A. Tingas, D.C. Kyritsis, D.A. Goussis, Fuel **183**, 90 (2016)
41. E.A. Tingas, D.C. Kyritsis, D.A. Goussis, Combust. Theory Model. **21**(1), 93 (2017)
42. D.J. Diamantis, D.C. Kyritsis, D.A. Goussis, Proc. Combust. Inst. p. in press (2014)
43. D. Goussis, S. Lam, in *Symposium (International) on Combustion*, vol. 24 (1992), pp. 113–120
44. J. Prager, H.N. Najm, M. Valorani, D. Goussis, Combust. Flame **158**(11), 2128 (2011)
45. D.M. Manias, E.A. Tingas, C.E. Frouzakis, K. Boulouchos, D.A. Goussis, Combust. Flame **164**, 111 (2016)
46. D.J. Diamantis, E. Mastorakos, D.A. Goussis, Combust. Theory Model. **19**(3), 382 (2015)

Chapter 4
Computational Singular Perturbation Method and Tangential Stretching Rate Analysis of Large Scale Simulations of Reactive Flows: Feature Tracking, Time Scale Characterization, and Cause/Effect Identification. Part 2, Analyses of Ignition Systems, Laminar and Turbulent Flames

M. Valorani, F. Creta, P. P. Ciottoli, R. Malpica Galassi, D. A. Goussis, H. N. Najm, S. Paolucci, H. G. Im, E.-A. Tingas, D. M. Manias, A. Parente, Z. Li, and T. Grenga

M. Valorani (✉) · F. Creta · P. P. Ciottoli · R. Malpica Galassi
Department of Mechanical and Aerospace Engineering, Sapienza University of Rome, 00184 Rome, Italy
e-mail: mauro.valorani@uniroma1.it

D. A. Goussis
Khalifa University of Science and Technology, Abu Dhabi, UAE

H. N. Najm
Sandia National Laboratories, Albuquerque, USA

S. Paolucci
Aerospace and Mechanical Engineering, University of Notre Dame, Notre Dame, USA

H. G. Im
KAUST, Thuwal, Saudi Arabia

E.-A. Tingas
Perth College, University of the Highlands and Islands (UHI), Perth, UK

D. M. Manias
National Technical University of Athens, Athens, Greece

A. Parente · Z. Li
ULB, Brussels, Belgium

T. Grenga
Institute for Combustion Technology RWTH Aachen University, 52062 Aachen, Germany

H. Pitsch and A. Attili (eds.), *Data Analysis for Direct Numerical Simulations of Turbulent Combustion*, https://doi.org/10.1007/978-3-030-44718-2_4

4.1 Introduction

Chapter 3 summarized the concepts behind the CSP method and the TSR analysis. In this chapter, we discuss applications of these techniques. It must be noted that the CSP method and the G-Scheme extension provided the theoretical framework for designing solvers for deterministic stiff problems (the CSP solver [1] and the G-Scheme solver [2, 3]), for stiff uncertain problems [4], and for stiff stochastic problems [5, 6]. Moreover, CSP and TSR have been useful in the design of effective and user-friendly algorithms for chemical kinetic simplification [7, 8] and reduction [9–11]. CSP has been employed for studying problems outside the combustion community, namely in biology and medical sciences [12–14]. All these computational tools have been coded into a suite of libraries referred to as *CSP Tool kit (CSPTk)* [15].

4.2 CSP Analysis on Auto-ignition Systems

The dynamics of auto-igniting systems are typically characterized by timescales spanning a wide range. Two representative examples are shown in Fig. 4.1 where the timescales of CH_4/air (T(0) = 800 K, p(0) = 3 MPa, $\phi = 0.8$) and dimethyl-ether (DME)/air (T(0) = 1,100 K, p(0) = 5 MPa, $\phi = 1.0$) mixtures are displayed along with the temperature evolution [16, 17]. As indicated in the figure, almost all timescales at any given instance are dissipative (indicated by solid curves) and very few are explosive (indicated by dotted curves). This is a common feature of the homogeneous auto-ignition process that has been demonstrated extensively; e.g., [16–29]. The few explosive timescales appear from the start of the process until the point where the temperature undergoes a steep increase. The period of the auto-ignition process in which explosive timescales are present was introduced as the "*explosive stage*" in [21]. In addition, the fastest of the explosive timescales is denoted by $\tau_{e,f}$ and the second fastest is denoted by $\tau_{e,s}$; the corresponding amplitudes are then be denoted as $f^{e,f}$ and $f^{e,s}$, respectively. Of all timescales that develop in a system, the characteristic one is usually $\tau_{e,f}$. In that case, $\tau_{e,f}$ sets the time frame of the action of the CSP mode $\alpha_{e,f} f^{e,f}$ in Eq. (3.2) and generally controls the system's time evolution. However, the examination of $\tau_{e,f}$ as the characteristic timescale implies that the following two conditions are met. First, $f^{e,f}$ is the largest of all modes and second, the $\tau_{e,f}$ is among the fastest of the slow timescales, a feature which is typically manifested when a timescale gap is developed and $\tau_{e,f}$ is just above this gap, placed among the fastest of the slow (non-exhausted) timescales. In that case, the gap among the timescales suggests that a valid reduced model can be constructed, while the placement of the $\tau_{e,f}$ suggests that this is the characteristic of the slow system that governs the process when the M fastest timescales become exhausted. In both cases shown in Fig. 4.1 both conditions are met, i.e., the examination of the system's amplitudes reveals that $f^{e,f}$ is always the largest of all amplitudes during the explosive stage while the sufficiently large timescale gap that is generated soon after the start of the process shows that $\tau_{e,f}$ is among the fastest of the slow timescales, thus, the characteristic one.

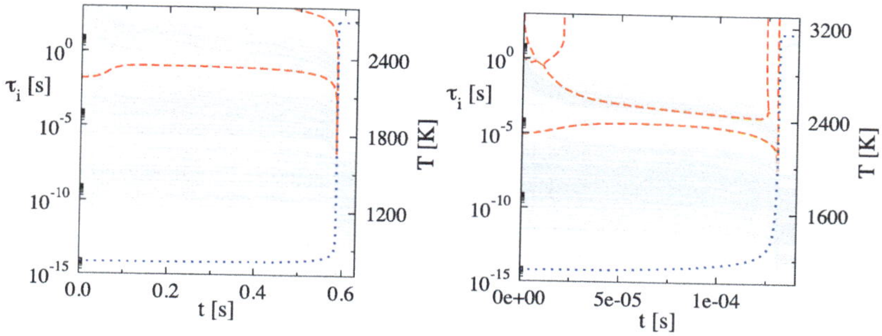

Fig. 4.1 The developing timescales during the constant volume auto-ignition of CH_4/air, T(0) = 800 K, p(0) = 3 MPa, $\phi = 0.8$ (left) and DME/air, T(0) = 1,100 K, p(0) = 5 MPa, $\phi = 1.0$ (right) mixtures [16, 17]. Solid (grey) and dashed (red) lines represent dissipative and explosive modes, respectively. The dotted (blue) line denotes the temperature evolution

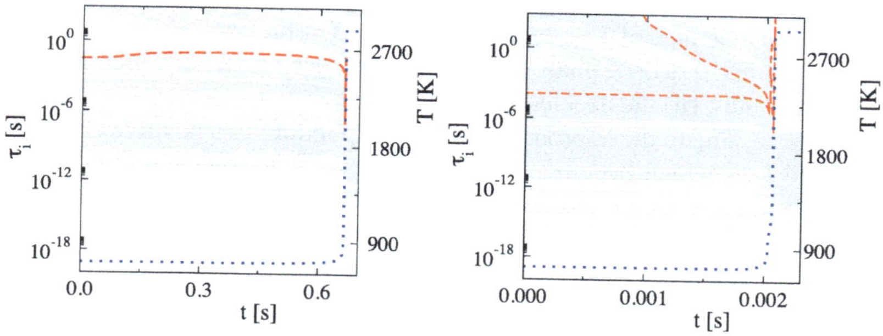

Fig. 4.2 The developing timescales during the constant volume auto-ignition of EtOH/air, T(0) = 700 K, p(0) = 3 MPa, $\phi = 1.0$ (left) and nC_7H_{16}/air, T(0) = 720 K, p(0) = 6 MPa, $\phi = 1.0$ (right) mixtures [19, 29]. Solid (grey) and dashed (red) lines represent dissipative and explosive modes, respectively. The dotted (blue) line denotes the temperature evolution

It is noted though that the examination of $\tau_{e,f}$ as the system's characteristic timescale, can still be legitimate even if the second condition is not met (i.e., the existence of a timescale gap and the placement of $\tau_{e,f}$ among the fastest of the slow timescales), on condition that the dynamics of the system are quasi-linear [19, 28, 29]. For instance, as shown in Fig. 4.2 for the auto-ignition of ethanol (EtOH)/air (T(0) = 700 K, p(0) = 3 MPa, $\phi = 1.0$), $\tau_{e,f}$ lies at the top of a dense pack of dissipative timescales while in the case of n-heptane (nC_6H_{16})/air (T(0) = 720 K, p(0) = 6 MPa, $\phi = 1.0$) $\tau_{e,f}$ lies in the middle of a dense pack of dissipative timescales and no timescale gap exists. However, due to the dominance of $f^{e,f}$ against all other amplitudes and the quasi-linear character of the dynamics (as manifested by the constancy of all fastest timescales including $\tau_{e,f}$), the analysis of $\tau_{e,f}$ as the main driving mode remains valid in both cases.

Typically, $\tau_{e,f}$ develops throughout or for the most part of the explosive stage while $\tau_{e,s}$ may exist either throughout or partly during the explosive stage, as is the case for DME/air (Fig. 4.1) and $EtOH_4$/air (Fig. 4.2), respectively. However, in all

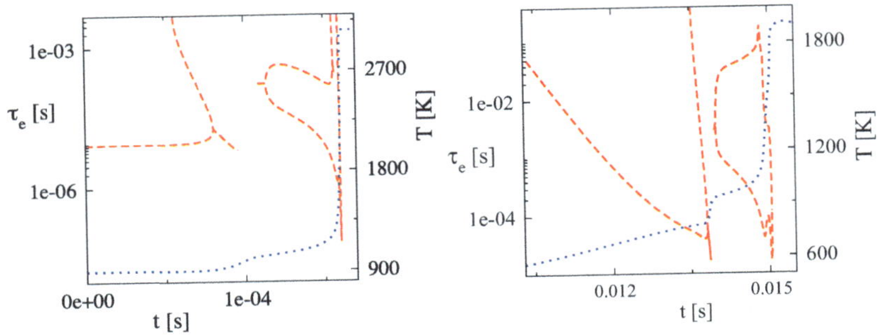

Fig. 4.3 The evolution of the explosive timescales (dashed-red lines) and temperature (dotted blue lines) during the constant volume auto-ignition of DME/air, T(0) = 900 K, p(0) = 5 MPa, $\phi = 1.0$ (left) and the HCCI of nC_6H_{14}/air, T(0) = 336 K, p(0) = 0.1 MPa, $\phi = 0.4$ (right) [19, 28]

cases, the disappearance of $\tau_{e,f}$ during or at the end of the explosive stage bears the following features: (i) a brief phase where $\tau_{e,f}$ firstly accelerates and then decelerates or only decelerates, (ii) the deceleration of $\tau_{e,f}$ ends when $\tau_{e,f}$ meets or approaches significantly $\tau_{e,s}$ due to the approach of the corresponding two positive eigenvalues, say $\lambda_{e,f}$ and $\lambda_{e,s}$, which become a complex pair with positive real part, then they evolve into a complex pair with a negative real part and finally to two negative real eigenvalues, and (iii) the time instant of the disappearance of $\tau_{e,f}$ is well correlated with a local or global maximum temperature gradient.

Figure 4.3 displays the temporal evolution of the explosive timescales along with temperature for the auto-ignition of DME/air (T(0) = 900 K, p(0) = 5 MPa, $\phi = 1.0$) and the HCCI of n-hexane (nC_6H_{14})/air (T(0) = 336 K, p(0) = 0.1 MPa, $\phi = 0.4$) [19, 28]. The temperature profiles indicate that both cases exhibit two-stage ignition and share the following common features: (i) there is a period in the explosive stage in which $\tau_{e,f}$ disappears, (ii) the point where $\tau_{e,f}$ disappears in the explosive stage marks the development of first stage ignition, (iii) the point where $\tau_{e,f}$ at the end of the explosive stage correlates with the steep temperature increase during the second stage ignition, (iv) for the most part of the second ignition stage $\tau_{e,f}$ accelerates as opposed to the first ignition stage where it can either accelerate or remain practically constant, (v) before the disappearance of $\tau_{e,f}$ at the second ignition stage, a brief period precedes where $\tau_{e,f}$ rapidly accelerates and then decelerates and (vi) the disappearance of $\tau_{e,f}$ occurs after meeting with $\tau_{e,s}$.

Non-typical findings regarding the evolution of $\tau_{e,f}$ were reported in the case of the n-heptane 3-stage ignition [29]. Figure 4.4 shows that the $\tau_{e,f}$ profile during the first ignition stage, resembles a lot the typical first stage of a two-stage ignition, since $\tau_{e,f}$ is constant for the most part and only in the end accelerates and then decelerates meeting $\tau_{e,s}$. In the second ignition stage $\tau_{e,f}$ gradually accelerates and in the end starts decelerating rapidly but it eventually does not meet $\tau_{e,s}$ and disappears. Instead, its deceleration stops signaling the end of the second ignition stage, and slowly accelerates again indicating, thus, the introduction of the third ignition stage. Notably,

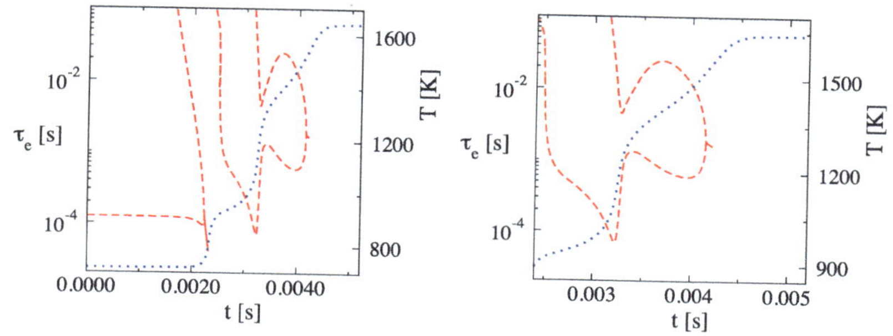

Fig. 4.4 The developing explosive timescales (red-dashed lines) overlaid with the temperature profile (blue-dotted line) for the constant volume auto-ignition of a lean ($\phi = 0.3$) n-heptane/air mixture at T(0) = 720 K, p(0) 40 atm. The right figure is a zoomed-in view of the left [29]

during the third ignition stage $\tau_{e,f}$ is considerably slower compared to the previous two ignition stages, thus, suggesting that the mixture's reactivity is decreased.

In order to assess the dependence of the system's characteristic timescale ($\tau_{e,f}$) on temperature, the explosive timescales can be calculated, using a truncated Jacobian ($\mathbf{J}_C$), where the row and column of the full Jacobian ($\mathbf{J}$) relating to temperature are eliminated, a method first introduced in [21]. The explosive timescales computed from J_C are compared with those computed from J. Clearly, if the system's explosive timescales are practically independent of (i) the dynamics introduced by the temperature equation, or (ii) by the chemical/thermal coupling, then the explosive timescales of J_C will closely approximate those of J. Therefore, if $\tau_{e,f}$ based on J_C approximates well the $\tau_{e,f}$ of J, this suggests that $\tau_{e,f}$ of J is weakly dependent on temperature, hence, the system is in the chemical runaway regime, while if the opposite stands (i.e., if $\tau_{e,f}$ based on J_C poorly approximates the $\tau_{e,f}$ of J), then the system lies in the thermal runaway regime.

The aforementioned approach can provide valuable physical insight into both simple and complex systems. For instance, Fig. 4.5 displays the explosive timescales of the full and the truncated Jacobians for a hydrogen/air mixture with initial conditions around the third explosion limit; p(0) = 1 atm is closely below the third limit, while

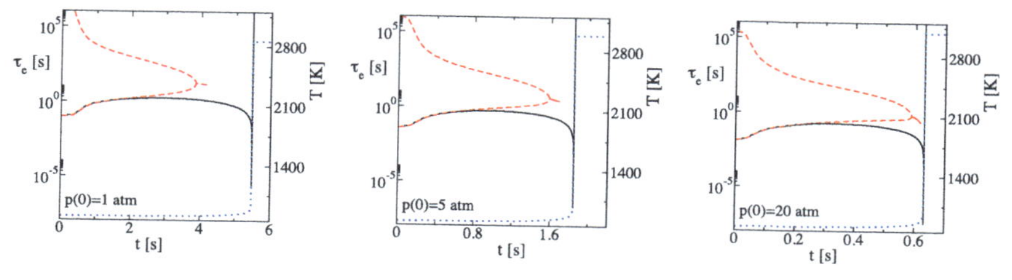

Fig. 4.5 The developing explosive timescales of the full (solid-black lines) and the truncated (dashed-red lines) Jacobians across hydrogen's third explosive limit; T(0) = 800 K, $\phi = 1.0$ [23]. Temperature has been overlaid with dotted-blue lines

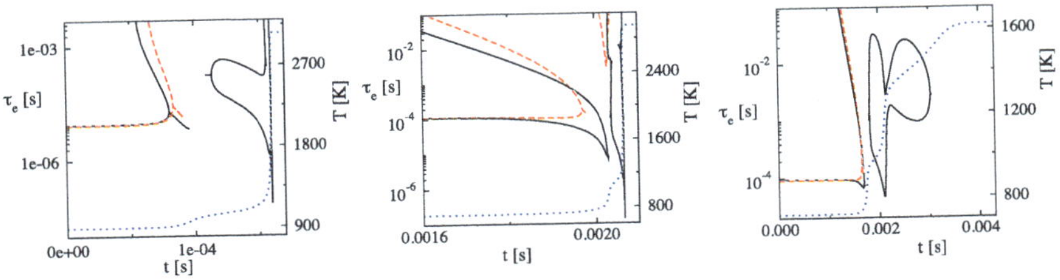

Fig. 4.6 The developing explosive timescales of the full (solid-black lines) and the truncated (dashed-red lines) Jacobians during the constant volume auto-ignition of DME/air on the left (T(0) = 900 K, p(0) = 5 MPa, ϕ = 1.0), nC_7H_16/air in the middle (T(0) = 720 K, p(0) = 4 MPa, ϕ = 0.3) and nC_6H_{14}/air on the right (T(0) = 720 K, p(0) = 4 MPa, ϕ = 1.0) [19, 29]. The temperature has been overlaid with dotted-blue lines

p(0) = 5 and 20 atm are closely and well above the third limit, respectively. While the initial pressure decreases, $\tau_{e,f}$ becomes more and more dependent on temperature, since $\tau_{e,f}$ of J_C approximates poorly the $\tau_{e,f}$ of J, thus, indicating that the thermal runaway plays a more important role in the system's evolution, at the expense of the chemical runaway. This is in agreement with the stipulation in Ref. [30] that the hydrogen system around the third explosion limit is driven to auto-ignition through a thermal explosion of the Frank-Kamanetski type. The relative importance of the two mechanisms, of course, can change depending on initial conditions, e.g., one would expect more intense chain branching at higher initial temperatures [23].

Regarding multi-stage igniting systems, the first stage typically relates to the chemical runaway while the second/third stage is part of the thermal runaway regime. This is clearly displayed in Fig. 4.6 where three representative cases of multi-stage ignition are depicted; two-stage ignition of mixtures of DME/air (T(0) = 900 K, p(0) = 5 MPa, ϕ = 1.0) and nC_7H_16/air (T(0) = 720 K, p(0) = 4 MPa, ϕ = 0.3) and a three-stage ignition of nC_6H_{14}/air (T(0) = 720 K, p(0) = 4 MPa, ϕ = 1.0). In particular, it is shown that in all three cases, $\tau_{e,f}$ of J_C approximates very well the $\tau_{e,f}$ of J during the first ignition stage, thus suggesting a very weak dependence of $\tau_{e,f}$ on temperature, an indication that the system there is in the chemical runaway regime. On the other hand, during the second/third ignition stage, $\tau_{e,f}$ of J_C approximates very poorly the $\tau_{e,f}$ of J. In fact, during the second/third ignition stage, $\tau_{e,f}$ of J_C is practically absent. These findings suggest that the system there is in the thermal runaway regime.

4.3 TSR Analysis of Three-Stage Ignition

TSR addresses the identification of the most energy containing modes within the active subspace. When the active subspace attains a large dimensionality, the identification of the processes most affecting the modes that contain the most energy

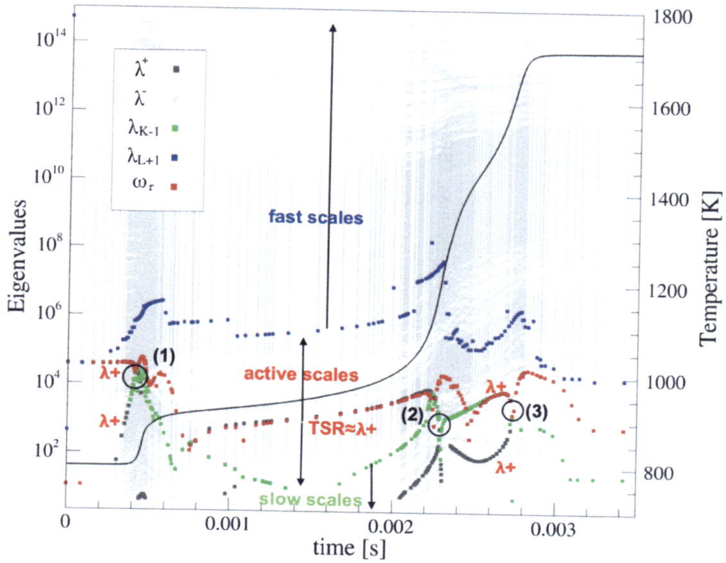

Fig. 4.7 Temperature (black line), absolute value of negative eigenvalues (light grey), positive eigenvalues (dark grey), λ_{L+1} (blue), λ_{K-1} (green), and TSR (red). The test case is n-heptane/air mixture $\phi = 0.3$, p = 30 bar, and T = 800 K

becomes tedious and time-consuming. The TSR and the TSR participation indices offer a strategy to sort out the important cause/effect relationships within the active subspace.

Here, we consider the three-stage ignition of a homogeneous n-heptane/air mixture at the equivalence ratio of 0.3, pressure of 30 bar, and initial temperature of 800 K. The chemical source term is described by a 1268 species and 10668 reactions kinetic scheme [31].

The whole spectrum of scales ($\tau^i = |\lambda_i^{-1}|$) ranges from 10^{-15} s to ~1 s, as shown in Fig. 4.7 by the system's eigenvalues in light grey. The time variation of the magnitude of each eigenvalue with time is caused by the action of the system's nonlinearities (for a linear system, the eigenvalues would be constant with time). The range of active scales, bracketed between λ_{L+1} (blue) and λ_{K-1} (green) according to the tangent space decomposition of Sect. 3.3.1, keeps changing throughout the ignition as can be noted in the figure. The positive eigenvalues are depicted in dark grey. Two pairs of positive eigenvalues merge at each of the three stages of the ignition process (marked as point (1), (2), and (3) in the figure). The TSR is depicted in red and lies within the active subspace. The TSR may happen to coincide with the largest in magnitude of the two positive eigenvalues, and especially so between the first and second ignition stage. Note that the explosive regime is a slow active process!

The characterization of the dynamics involves (i) the identification of a driving timescale, among the 1269 present in the spectrum at every time instant, and (ii)

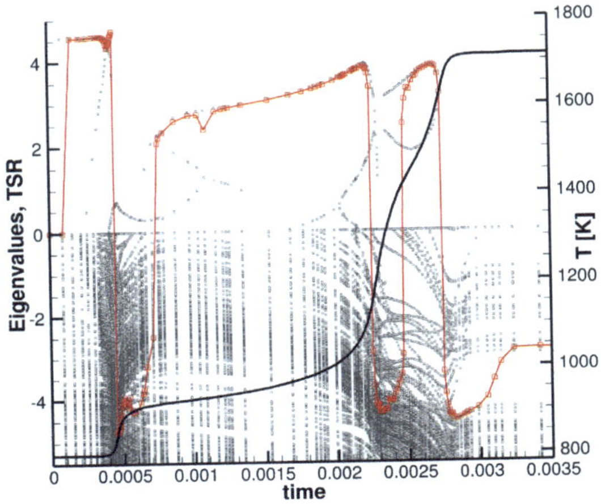

Fig. 4.8 Temperature (black), eigenvalues (grey) and TSR (red) time evolution. Both are expressed as sign(λ) Log10(1+|λ|). The test case is n-heptane/air mixture $\phi = 0.3$, p = 30 bar, and T = 800 K

the classification of the combustion regime, based on the magnitude and the sign of the TSR, respectively. Figure 4.8 shows the time evolution of the system eigenvalues and the TSR, represented in the form sign(λ) Log10(1+|λ|) to better appreciate the sign transition. Multiple positive eigenvalues develop throughout the whole ignition process. The TSR is able to select at all time instants the currently active timescale, which is not necessarily coincident with (the reciprocal of) an eigenvalue with a positive real part, this making the TSR a suitable marker for staged-ignition behaviors. In fact, each of the three stages of the ignition is characterized by the currently active timescale that first becomes positive/explosive and later negative/contractive.

4.4 TSR Analysis of Reactive-Diffusive Systems

A numerical dataset is generated employing the unsteady laminar flamelet model introduced in [32], to analyze the ignition processes in a non-premixed system resorting to the TSR index.

The flamelet model [32] describes the competition between chemical kinetics and molecular diffusion, under a unity Lewis number assumption.

The comparison of the flamelet model [32] with Eq. (3.1) allows us to establish that the N-dimensional state vector is equal to $\mathbf{z} := (T, Y_\alpha)$, so that $N = N_s + 1$. Moreover, the space operator $\boldsymbol{L}(\mathbf{z})$ can be defined as $\boldsymbol{L}(T, Y_\alpha) := \left(\frac{1}{2}\chi\left(\frac{\partial^2 T}{\partial \xi^2} + \frac{1}{c_p}\frac{\partial c_p}{\partial \xi}\frac{\partial T}{\partial \xi}\right), \frac{1}{2}\chi\frac{\partial^2 Y_\alpha}{\partial \xi^2}\right)$, and the source term $\boldsymbol{g}(\mathbf{z})$ can be defined as $\boldsymbol{g}(T, Y_\alpha) := \left(\frac{\dot{\omega}_T}{c_p \rho}, \frac{\dot{\omega}_\alpha}{\rho}\right)$.

We obtained time-accurate solutions of the flamelet model [32] by means of the CSPTk package, which integrates in time using CVODE and where the thermo–kinetic databases are parsed and handled using the TChem package [33].

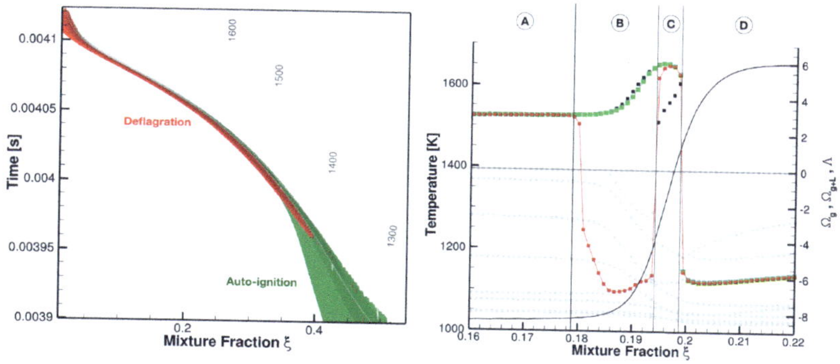

Fig. 4.9 Left: Case $\chi = 10\ s^{-1}$. Flame front identification and characterization: auto-ignition versus deflagration. Temperature isocontours in grey. Right: Slice at $t = 4.059 \times 10^{-3}$, zoom on the flame front. Ω_g (red) versus $\Omega_g + \mathbf{L}$ (green). Positive eigenvalues (black), negative eigenvalues (gray). Temperature (black line)

The reactive mixture considered is syngas/air, resorting to the same detailed mechanism employed in the auto-ignition case, with 12 species and 33 chemical reactions [34]. The fuel is syngas (H2:CO molar ratio of 0.7:1), and the oxidizer is oxygen diluted with three times the amount of nitrogen present in the air (i.e., a molar ratio of N2:O2 = 11.28). This reactive mixture is relevant to gas turbine operations and has been extensively studied in [35, 36]. The domain of integration relates to mixture fraction, and thus it ranges from $\xi = 0$ to 1. The mixture fraction space discretization involves 1024 cells of equal size. The boundary conditions considered are pure oxidizer at $\xi = 0$ and pure fuel at $\xi = 1$, both set at all times at the temperature of 1000 K, while the pressure is kept constant in the whole domain and at all times and equal to 20 atm. The initial condition at time $t = 0$ is temperature of 1000 K in the whole domain, and the mass fractions of fuel and oxidizer along the ξ domain are obtained with a linear distribution between the left and right boundaries. The scalar dissipation used is $\chi = 10\ s^{-1}$. As discussed in [37], after ignition takes place at the most reactive mixture fraction of $\xi \approx 0.5$, diffusion steps in as soon as kinetics creates spatial non uniformities, in such a way as to propagate heat and mass outward with respect to the most ignitable mixture fraction value. The reaction-diffusion wave weakens while traveling toward the rich mixture, while it becomes stronger (faster) moving toward the leaner mixtures. The steep temperature front generated by the reaction-diffusion wave moves from right to left.

Figure 4.9 (left) shows the essential representation and characterization of the critical features of the flame front: the green region is the region where ω_{g+L} attains a peak ($\omega_{g+L} > 10^4$), tracking the largest positive eigenvalue. This region may be identified as the flame front, i.e., where the dynamics of the reactive-diffusive system is explosive and sufficiently fast. We refer to this region as an autoigniting region, since ω_g and ω_{g+L} coincide, meaning that the role of transport processes here is not pivotal. The red region corresponds to the region where ω_g and ω_{g+L} differ, i.e., the

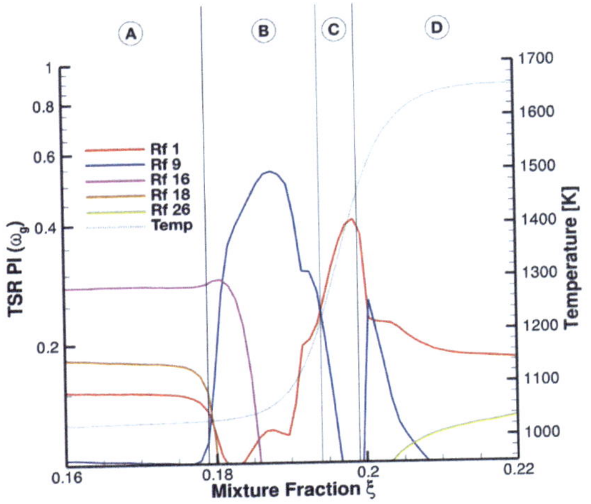

Fig. 4.10 Participation Indices to ω_g at $t = 4.059 \times 10^{-3}$ s in the flame front region. Legend of reactions: R-1: $H + O_2 \Leftrightarrow O + OH$, R-9: $H + O_2(+M) \Leftrightarrow HO_2(+M)$, R-16: $H_2O_2(+M) \Leftrightarrow 2OH(+M)$, R-18: $H_2O_2 + H \Leftrightarrow HO_2 + H_2$, R-26: $HCO + M \Leftrightarrow H + CO + M$

deflagrative, diffusion-contributed, part of the flame front, which is also depicted in Fig. 4.9 (right) as the red symbols valley in the region marked as (B). This figure shows a time slice at $t = 4.059 \times 10^{-3}$ s, with the deflagrative wave traveling from right to left. Analyzing this figure from left to right, we observe four distinct regions: (i) region (A), where $\omega_\mathbf{g}$ and $\omega_\mathbf{g+L}$ coincide and are small and positive; (ii) region (B), where $\omega_\mathbf{g+L}$ remains close to the positive eigenvalue, indicating the explosive nature of the reaction/diffusion system, while $\omega_\mathbf{g}$ becomes negative, pointing out the contractive nature of the system dynamics due to the chemical source term; (iii) region (C), where $\omega_\mathbf{g}$ becomes positive and coincident with $\omega_\mathbf{g+L}$; and (iv) region (D), where the diffusive processes become negligible, and the system exhibits a contractive nature with both $\omega_\mathbf{g}$ and $\omega_\mathbf{g+L}$ being negative and coincident.

It is straightforward to identify the processes, i.e., reactions and diffusion of species, participating to the modes mostly contributing to $\omega_\mathbf{g}$ and $\omega_\mathbf{g+L}$, by resorting to the participation indices. It is worth to investigate the participation indices to both ω_g and $\omega_\mathbf{g+L}$, shown in Figs. 4.10 and 4.11 for the time slice $t = 4.059 \times 10^{-3}$ s.

We observe that in region (A) the behavior of the participation indices to $\omega_\mathbf{g}$ in the mixture fraction space, moving toward richer mixtures (from left to right), resembles the evolution in time of the indices computed for the below-crossover auto-ignition problem. If diffusion were not present, i.e., $\chi = 0$, this time slice would have shown the state of several batches of the mixture, with different equivalence ratio, evolving independently in time, and displayed at $t = 4.059 \times 10^{-3}$ s. Since the reactive-diffusion wave has not yet reached this region, the left-to-right evolution is approximately an initial composition effect that translates into different ignition delays (supposing that the chemical pathways do not vary much around stoichio-

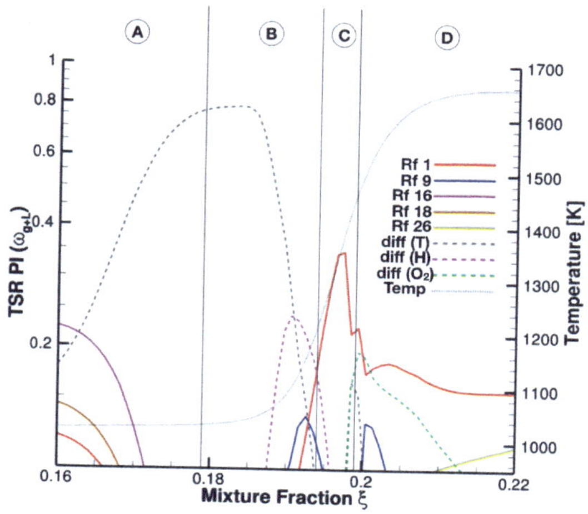

Fig. 4.11 Participation Indices to ω_{g+L} at $t = 4.059 \times 10^{-3}$ s in the flame front region. Legend of reactions same as in Fig. 4.10

metric conditions). In particular, Rb-18 and Rf-16 are dominant, while Rf-9 and Rf-1 show a growing importance. Note that, although the mixture is syngas, the ignition process is started by the hydrogen sub-mechanism. Note also that the temperature/pressure condition of this flamelet falls in the below-crossover regime of the hydrogen/air kinetics [38–40]. In region B, the reaction-diffusion wave travels from right to left. Here, the diffusion of temperature is the dominant process of the reactive-diffusive system right in front of the flame (wave) front. This is the region where ω_g and ω_{g+L} differ, and where also the respective PIs differ the most. Here, ω_g is negative and mostly contributed by Rf-9. The effect of the reaction-diffusion wave is to enhance ignition (deflagrative effect), and chemistry alone would react in a dissipative way with the chain terminating reaction Rf-9. This explains the Rf-9 peak in Fig. 4.10, which differs from the auto-ignition case. In region C, the chain branching reaction Rf-1 is the largest contributor to heat release. Finally, in region D, thus behind the flame front, the carbon chemistry takes place as well, mostly through reaction Rb-26.

In summary, the deflagrative wave accelerates the ignition process of the mixture to the left of the front, and the driving processes are mostly the diffusion of temperature and, to a lesser extent, diffusion of H, which in turn trigger Rf-1 earlier than the sequence Rb-18/Rf-16 would have done in absence of transport.

4.5 CSP/TSR Analysis of Turbulent Premixed Combustion at High Karlovitz Numbers

Toward higher efficiencies and lower emissions, modern combustion devices are designed to operate at unconventional and often extreme conditions such as higher pressure and turbulence intensities, and highly diluted and preheated modes known

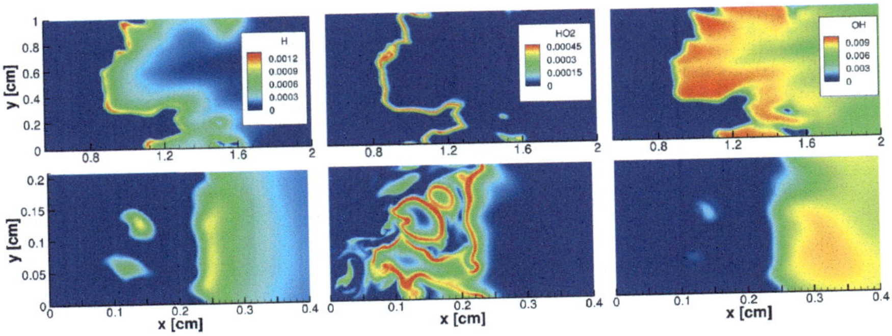

Fig. 4.12 The spatial distribution of mass fracton of H (left), HO_2 (middle) and OH (right) at Ka =14.4 (top) and Ka = 126 (bottom). Reprinted from [45], Copyright (2019), with permission from Elsevier

as the moderate or intense low-oxygen dilution (MILD) combustion [41]. From a fundamental standpoint, these conditions often increase the characteristic Karlovitz numbers (Ka), such that the turbulence-flame interaction manifests the combustion behavior that is distinct from that of the laminar flamelet regime [42]. Turbulent flames in such "reaction sheet" or "distributed" regimes exhibit highly complex topologies that are difficult to characterize based on conventional approaches such as the progress variable representation based on scalar iso-surfaces. To this end, there is a need to unambiguously identify specific regions of importance that are relevant to key global quantities of interest, such as heat release or pollutant formation rates, for a systematic analysis of the bulk combustion performance. The CSP/TSR approach offers an automated algorithmic tool for feature detection and characterization of the flame dynamics, especially for large scale 3D DNS datasets.

As the turbulence intensity increases in premixed combustion, smaller turbulent eddies penetrate inside the flame, making the structure significantly deviate from that of the one-dimensional laminar flame. Figure 4.12 shows cross-sectional views from three-dimensional DNS of hydrogen/air premixed flames at Ka = 14.4 and 126 [43, 44]. Isocontours of HO_2, H, and OH represent the preheat, main reaction, and downstream regions, respectively. As is particularly evident in the HO_2 profiles, the flame structure becomes highly disrupted at higher Karlovitz numbers, clearly losing the flamelet-like characteristics. In these conditions, the traditional way to identify the flame structure based on the progress variable, regardless of the choice of the specific scalar variable, leads to large ambiguities. Therefore, an alternative metric using the CSP/TSR diagnostics is proposed as an objective way to demarcate important regions in highly complex turbulent combustion systems.

The explosive mode, $\Lambda_{e,f}$, usually represents an important dynamical feature such as an ignition event, and is thus commonly used as the basic diagnostics of explosive behavior in reacting flows. When the transport effects become large, however, as in highly turbulent flows, the explosive mode alone may not consistently identify the important regions of interest where strong interactions between kinetics and transport

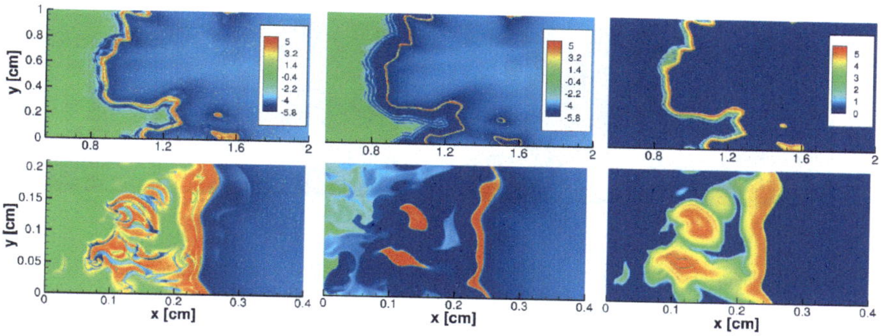

Fig. 4.13 The spatial distribution of Ω_{pde} (left), Ω (middle) and $\Lambda_{e,f}$ (right) at Ka = 14.4 (top) and Ka = 126 (bottom) regimes. Reprinted from [45], Copyright (2019), with permission from Elsevier

exist. This is because the timescale of the explosive mode is not necessarily the fastest of the important dynamical modes, or its mode amplitude may be too small to have any significant effect on the system dynamics. In this case, the TSR metric serves as an integrated diagnostic tool to identify the system characteristics as a net result combining both chemistry and transport effects. To illustrate this, Fig. 4.13 shows the spatial distributions of $\Lambda_{e,f}$, along with Ω_{pde} and Ω, for the cases shown in Fig. 4.12. Comparing $\Lambda_{e,f}$ to Ω_{pde} (left) and Ω (middle), it is evident that $\Lambda_{e,f}$ exists in a significantly larger region compared to the region of positive Ω_{pde} and Ω, which are more localized in the region of major heat release as demarcated by the HO_2 isocontours in Fig. 4.12.

To assess which metric provides with the most accurate identification of the flame characteristics, the correlation of each quantity with the heat release rate may be compared. For the cases considered, it was found that the extended TSR, Ω_{pde}, best correlates with the heat release rate. Figure 4.14 shows the heat release rate isocontours (color) overlaid with $\Omega_{pde} = 0$ (white) and $\Lambda_{e,f} = 0$ (black) isocontour lines. It is seen that the explosive regions based on $\Omega_{pde} > 0$ follows the heat release rate more closely than those of $\Lambda_{e,f} > 0$ which is much more widespread throughout the preheat transport zone. This indicates that a positive eigenvalue alone does not necessarily represent the actual explosive events in the system, and may be misleading in characterizing turbulent flame dynamics [45]. The alternative TSR metric properly identifies the most energetic eigenvalues within both the explosive and dissipative regimes, since it accounts for the combination of all eigenvalues weighted with their amplitudes. This way TSR ensures that the explosive dynamics is properly captured.

Furthermore, the comparison between Ω_{pde} and $\Lambda_{e,f}$ reveals that the differences between the two quantities are found mainly in the preheat zone, to varying degrees depending on the level of turbulence intensity. The main cause of the difference is that the explosive modes ($\Lambda_{e,f} > 0$) exist but the net system behavior is dissipative. As a result, for the higher Ka case, $\Lambda_{e,f}$ overpredicts the explosiveness of the system without properly capturing the additional dissipation associated with turbu-

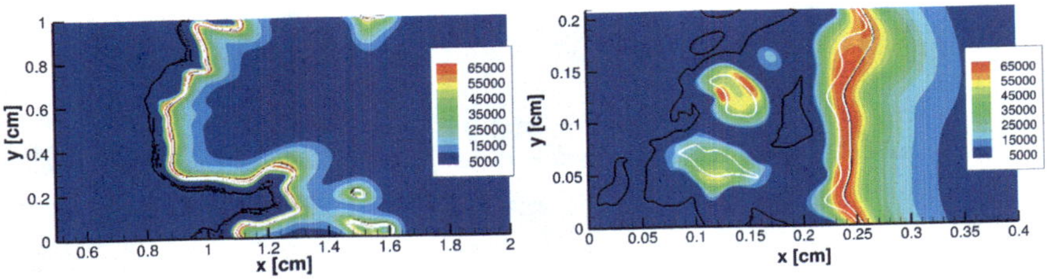

Fig. 4.14 The spatial distribution of the heat release rate, overlaid with the $\Omega_{pde} = 0$ (white solid lines) and $\Lambda_{e,f} = 0$ contour lines (black solid lines), at Ka = 14.4 (left) and Ka = 126 (right). Reprinted from [45], Copyright (2019), with permission from Elsevier

lent transport. It is important to note that transport-induced explosive behavior may occur as seen in the higher Ka case (right), which is adequately identified by the TSR diagnostics.

These results demonstrate the challenges in characterizing the highly turbulent flames based solely on the chemical behavior. At high Ka, the chemical explosive mode analysis may incorrectly represent the large transport zone as the reaction zone broadening. In contrast, the combined CSP/TSR approach not only captures the explosive dynamics correctly, but also allows the identification of the cause of the explosiveness to be either transport or reaction processes.

4.6 TSR Dynamics in MILD Combustion

Toward higher efficiencies and lower emissions, the Moderate or Intense Low-oxygen Dilution (MILD) combustion is utilized in various industrial combustion systems [46, 47]. Under MILD conditions, a reactant mixture exhibits intense preheating and high dilution in order to achieve stable combustion and relatively low flame temperature. As a result, MILD combustion exhibits highly complex morphology with nearly invisible flames and distributed temperature field [48–50]. Indeed, there is growing evidence to believe that the theoretical distributed combustion regime is most likely relevant in highly turbulent MILD combustion applications. Therefore, it is anticipated that the TSR diagnostics serves as a valuable tool to analyze such flames.

To investigate this, recent DNS data for methane-air flames at MILD conditions diluted with exhaust gas recirculation (EGR) [52–54] was analyzed. Figure 4.15 shows the distribution of temperature (left), and the HCO (middle) and OH (right) mass fractions using an instantaneous solution field, where HCO and OH are often used as a flame marker [55, 56]. Highly distributed structures in the upstream region are evident, confirming that the combustion occurs in a distributed manner, such that none of the variables shown here can serve as a flame marker based on their isosurfaces. Furthermore, since the preheated reactant mixture is more reactive, the characteristics of transport-chemistry interaction are significantly different from those at normal flame conditions.

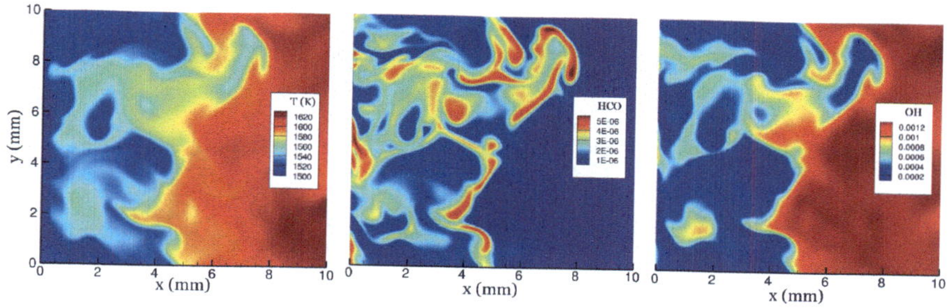

Fig. 4.15 The mid-z plane spatial distribution of temperature (left) and the mass fractions of HCO (middle) and OH (right) of methane MILD combustion. Reprinted from [51], Copyright (2019), with permission from Elsevier

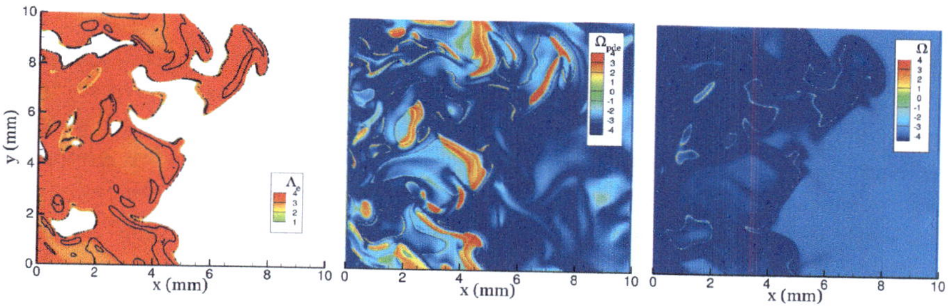

Fig. 4.16 The mid-z plane spatial distributions of $\Lambda_{e,f}$, overlaid with the $\Omega_{pde} = 0$ isoline (left), along with the Ω_{pde} (middle) and Ω (right) of methane MILD combustion. Reprinted from [51], Copyright (2019), with permission from Elsevier

For the solution field shown in Fig. 4.15, the three quantities, $\Lambda_{e,f}$ (left), Ω_{pde} (middle), and Ω (right) are shown in Fig. 4.16. Compared to the high Ka flames in the previous section, a much more significant contrast is evident. In the present MILD case, the chemically explosive modes are largely suppressed by the dissipative action of turbulent transport. As a result, Ω is negative nearly everywhere, indicating that the net effect of all CSP modes is dissipative. When transport is taken into account, however, there appear regions where $\Omega_{pde} > 0$, such that the explosive dynamics emerge from turbulent mixing. In this region, one or more explosive modes were found to be enhanced by transport, leading to positive Ω_{pde} [51], contrary to its general dissipative nature. In addition, similar to the high Ka regular flames, the heat release rate is well correlated with Ω_{pde}, but not with to $\Lambda_{e,f}$. As such, it is confirmed that Ω_{pde} serves as the appropriate quantity to identify the regions of significant dynamical activities for complex turbulent combustion phenomena.

To further examine the underlying processes that contribute to the explosive dynamics, the contributions of reaction, convection, and diffusion to the explosive dynamics of Ω_{pde} are determined by $H_{Chem}^{\omega_{\tilde{\tau}_{pde}}}$, $H_{Conv}^{\omega_{\tilde{\tau}_{pde}}}$, and $H_{Diff}^{\omega_{\tilde{\tau}_{pde}}}$, respectively, and these quantities are compared in Fig. 4.17. The net explosive regions, $\Omega_{pde} > 0$, are marked by the black lines. Although chemistry undoubtedly has the largest con-

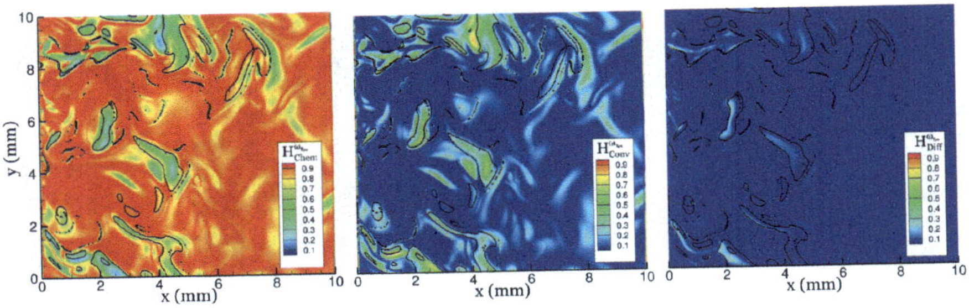

Fig. 4.17 The mid-z plane spatial distributions of the $H_{Chem}^{\omega_{\tilde{\tau}_{pde}}}$ (left), $H_{Conv}^{\omega_{\tilde{\tau}_{pde}}}$ (middle) and $H_{Diff}^{\omega_{\tilde{\tau}_{pde}}}$ (right) of methane MILD combustion, overlaid with the $\Omega_{pde} = 0$ isoline. Reprinted from [51], Copyright (2019), with permission from Elsevier

tribution to the explosive dynamics, the turbulent convective transport is also seen as a significant contributor to drive the explosive nature of the system. In contrast, the molecular diffusion shows little effect on either explosive or dissipative dynamics. Therefore, a typical turbulent MILD combustion is characterized as important reaction zones driven by the combination of chemical reactions and turbulent advection, while there is a wide region with explosive chemical modes which becomes dissipative due to strong turbulent transport effects.

Having identified the most significant CSP modes that drive the system's slow dynamics, the timescale and amplitude participation indices (TPI/API) can provide further information about the role of individual modes to the system's explosive/dissipative character. For example, for the cases considered, the TPI/API diagnostics reveals that the dissipative mode is favored mainly by HCO chemistry, while the explosive one is favored by hydrogen-related reactions and CH_4 convection. Consequently, the competition between these two different sets of processes is becoming the main source of the competition to determine the net explosive or dissipative characteristics.

Similarly, the TSR-PI identifies the contributions from the chemistry or transport to the positive Ω_{pde}. The TSR-PI from chemistry and transport are denoted as $H_{Chem}^{\omega_{\tilde{\tau}_{pde}}}$ and $H_{Transport}^{\omega_{\tilde{\tau}_{pde}}}$, respectively. The TSR-PI diagnostics was applied to three different instantaneous solution fields of a MILD combustion case. Within the explosive regions denoted by $\Omega_{pde} > 0$ the relative contributions between the chemistry and transport are quantified and the region is marked in red if the explosive behavior is driven mainly by chemistry ($H_{Chem}^{\omega_{\tilde{\tau}_{pde}}} > 0.5$) and in blue if it is by transport ($H_{Transport}^{\omega_{\tilde{\tau}_{pde}}} > 0.5$). This criterion in effect determines whether the propagating front is auto-ignitive or deflagrative, respectively. Figure 4.18 shows the results. It is found that auto-ignition prevails for all three instantaneous solution fields, as expected for a MILD condition. However, the results also show that a significant level of deflagration fronts also exists, indicating that MILD combustion in general should be characterized with a varying degree of mixed-mode combustion. In summary,

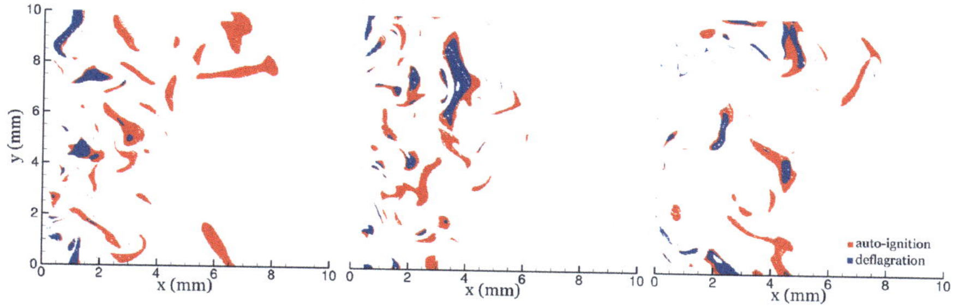

Fig. 4.18 Auto-ignition versus deflagration regions within the region of positive Ω_{pde} of methane MILD combustion at three timesteps: $t = 0.7$ ms (left), $t = 0.8$ ms (middle) and $t = 0.9$ ms (right). Reprinted from [51], Copyright (2019), with permission from Elsevier

CSP/TSR analysis is found to be a valuable unified approach to characterize in detail the dynamical nature of highly turbulent or MILD combustion systems which are likely to exhibit complex mixed-mode combustion phenomena.

4.7 CSP/TSR Analysis of Jet-in-hot-coflow Burners

Although MILD combustion and similar technologies have been implemented in some practical devices, the complex features of low-emission regimes have proven difficult to predict making it difficult to transpose solutions from one industrial configuration to another. To alleviate the dependency on the burner geometry and allow for high-fidelity experiments, model flames like the jet-in-hot-coflow (JHC) [57–59] were built to emulate MILD conditions. The JHC burner has an insulated and cooled central jet with an inner diameter of 4.25 mm providing an equimolar mixture of CH_4 and H_2 [57]. A secondary burner mounted upstream of the exit plane has an inner diameter of 82 mm and provides the hot combustion products. The combustion products are mixed with air and nitrogen, to control the oxygen levels to 3, 6, and 9%. The 3% case reaches MILD conditions, while the 6 and 9% cases are transitional between MILD and conventional regimes. Therefore, it provides a good comparison between the MILD and non-MILD combustion regimes under the same burner configuration and fuel component. The experimental data (mean and root-mean-square) of temperature and mass fractions of species (CH_4, H_2, H_2O, CO_2, N_2, O_2, NO, CO, and OH) along the centerline as well as at various axial positions (30/60/120/200 mm) are available for validation [57]. The available high-fidelity experimental data was used to validate Large Eddy Simulation (LES) and, in particular, to assess the accuracy of reactor-based models in the context of MILD combustion.

The numerical simulation domain starts from the jet exit location and extends 180 mm axially. The radial direction expands 90 mm from the centerline. The whole domain is discretized with a 3D cylinder structured mesh having $\sim$1.5 million cells; 52 cells are used to discretize the jet diameter. The LEMOS [60] inflow generator for velocity field is used on all the three streams—fuel, coflow, and air. The measured mean species mass fraction, temperature, and velocity taken 4 mm downstream of the burner exit are used as boundary condition for the simulation. A reduced skeletal mechanism KEE58 [61] with 17 species and 58 reactions is used in combination with the Laminar Finite Rate (LFR) combustion model [62, 63]. In LFR model, the filtered source term comes directly from the chemical term, without inclusion of the turbulent effect. Compared to the reactor-based models like Partially Stirred Reactor (PaSR), the LFR model emulates the condition that the factor κ which represents the fraction of the reactive zone in the cell is set to 1.0—meaning that the whole cell is either reacting or non-reacting at all [63].

Satisfactory agreement was obtained by comparing the LES results of the mean and rms temperature and species mass fraction (H_2O, CO, and OH) with the experimental data [63]. MILD combustion is characterized by a strong overlap between chemistry and fluid dynamics timescales, leading to Damköhler numbers of order one. As a result, turbulence-chemistry interactions becomes of paramount importance. In order to study turbulence-chemistry interaction and the relationship between auto-ignition and flame propagation under MILD conditions, advanced post-processing tools are required. The computational singular perturbation (CSP) method [64, 65] enables the definition of rigorous mathematical tools capable of extracting local information from chemically reactive flow fields. The CSP analysis and the concept of TSR were thus used to shed light on the intrinsic characteristics of MILD combustion.

The CSP/TSR analysis was applied to the instantaneous LES resolved fields, following the same procedure employed in a DNS context. In general, the validity of this approach relies on the inclusion of the sub-grid fluctuations in the local chemical source term $\boldsymbol{g}$ and transport term $\boldsymbol{L}$ employed in the construction of the CSP decomposition. In our specific case, chemical space scales are found to be fully resolved by the LES grid, allowing one to compute the chemical source term as a function of the filtered composition, using the Laminar Finite Rate (LFR) model [63]. The employed diffusive terms, on the other hand, embed the sub-grid scale effects, being the diffusion coefficients modified to model the sub-grid scale turbulent motions. Hence, the effects of transport on the chemical modes are to be intended as both resolved and sub-grid parts. More details regarding the analysis can be found in Li et al. [66].

The chemical TSR (ω_{g}) field along with heat release rate contour lines on the jet axi-symmetric plane for the cases with coflow oxygen levels of 3 and 9% are presented in Figs. 4.19a and 4.20a. Most of the domain in both cases show negative ω_{g} levels, which indicate the dissipative/non-explosive nature of the active chemical processes. Positive values of ω_{g} are captured in the coflow area and close to the jet exit. The latter is only captured for the 3% case, indicating that the ignition of the 3% case is purely controlled by chemical reaction.

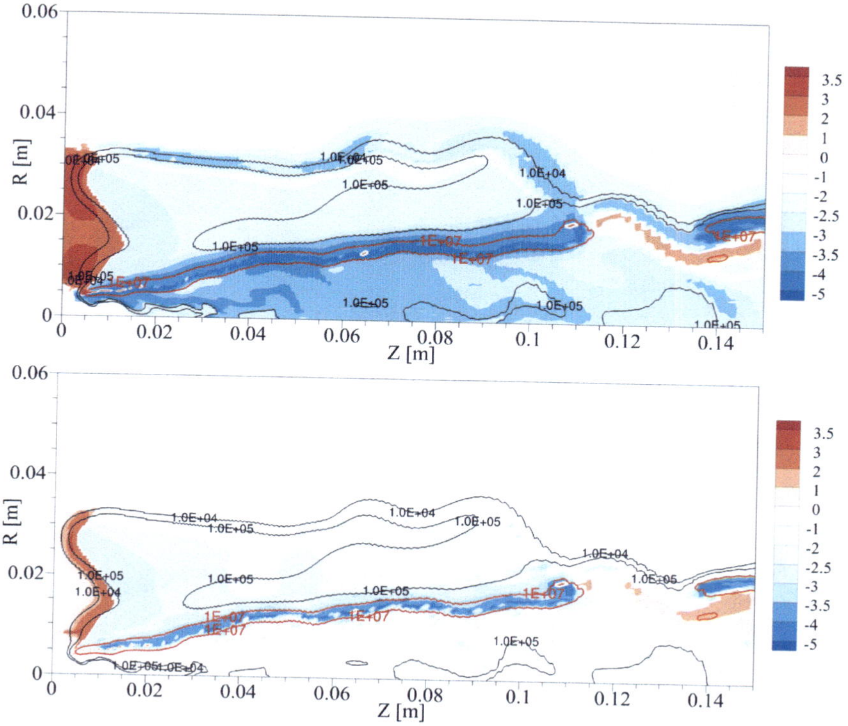

Fig. 4.19 Contour plots of the chemical TSR and extended TSR for the 3% coflow oxygen level [66]. Heat release rate values are shown with contour lines. The TSR values are shown in logarithmic form, as: sign(TSR) × $\log_{10}$ | TSR |. Zero (white) regions refer to values smaller than $10^0 = 1\ \mathrm{s}^{-1}$, hence timescales larger than 1 s

Close to the hot coflow exit location upstream, positive values for the chemical TSR ($\omega_\mathbf{g}$) can be observed, pointing out a chemically explosive area. Since the boundary condition provided in the coflow stream is not a strictly equilibrium mixture and there is no pre-inlet pipe used for the current LES, the components (CO, O_2, H_2O and CO_2) in the coflow stream show a propensity to evolve on a rather slow explosive timescale of the order of 10^{-3} s, and then to reach their equilibrium state. The equilibrium state is partially reached at around Z = 0.01 m, before the entrainment of air.

The most negative $\omega_\mathbf{g}$ region overlaps with the high heat release rate. The magnitude of $\omega_\mathbf{g}$ indicates that the timescale associated with the active chemical processes in this region is of the order of $10^{(-5)-(-6)}$ s, being the fastest chemical processes in the field. The inner structure of the region with high heat release rate shows a distinctive two-branch behavior, which might resemble the lean (upper) and rich (lower) branches of a triple-flame structure [67]. In addition to the interaction zone between the fuel stream and the coflow stream, a fast chemical contractive region can be seen

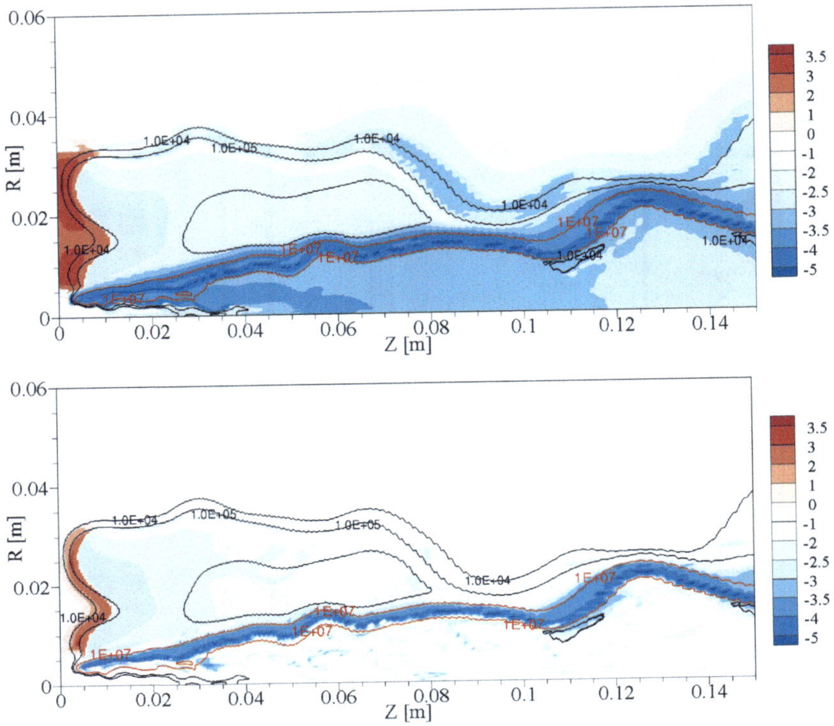

Fig. 4.20 Contour plots of the chemical TSR and extended TSR for the 9% coflow oxygen level [66]. Heat release rate values are shown with contour lines. The TSR values are shown in logarithmic form, as: sign(TSR) × $\log_{10}$ | TSR |. Zero (white) regions refer to values smaller than $10^0 = 1\ s^{-1}$, hence timescales larger than 1 s

where the mixing between the coflow stream and air stream takes place. However, the chemical activity develops on slower timescales with respect to the region with high heat release rate.

For the 3% case, positive ω_g values are observed downstream of the jet, starting from around Z = 0.112 m. This indicates local extinction and explosive re-ignition of the flame around those locations. A break between the contour lines of heat release rate is also found in this region. The re-ignition comes from the entrainment of air into the heated fuel stream. The entrainment of air for the 3% case happens at around Z = 0.12 m, which is much later than the 9% case, at around Z = 0.09 m. The lack of O_2 in the coflow stream and late entrainment of air lead to local extinction for the 3% case. Because of slightly increased oxygen levels, local extinction is not present in the 9% case.

Figures 4.19b and 4.20b show the field of extended TSR for the 3% and 9% cases, respectively. Comparing the extended TSR with the chemical TSR, we observe that: (i) fast chemistry is mitigated by transport, which lowers the magnitude of the driving scale in most of the field, and (ii) there are no purely diffusion-driven fast explosive regions. More specifically, the coflow injection appears to be influenced by transport: diffusion mitigates the chemical ignition propensity. This is substantiated by a reduction of both the magnitude and the spatial extension of the positive extended TSR region with respect to the chemical TSR. In the mixing layer with diluted air, fast chemistry (chemical scale $\sim 10^{-4}$ s) is strongly balanced by transport: the extended TSR approaches zero. Even though this behavior might resemble the condition in which the reactive-diffusive system approaches equilibrium ($\boldsymbol{L}(\mathbf{z}) + \boldsymbol{g}(\mathbf{z}) \approx 0$), this is not the case. Actually, in this area, the extended TSR tracks the positive eigenvalue λ_e, which however is of order $10^0 = 1\ s^{-1}$, yielding a very slow explosive dynamics (not appreciable in the logarithmic scale of the figures).

The same equilibrating effect is found in the region with high heat release rate, specifically in the rich branch of the flame, where λ_e is again tracked by the extended TSR. These two regions, even though characterized by extremely slow timescales, represent diffusion-driven explosive zones. It appears that transport is shifting the system's dynamics toward slower scales: the major chemically explosive regions, as pointed out by the chemical TSR, appear less explosive and reduced in size, due to the activation of other contractive and slow modes. The chain-shaped explosive area at the leading edge of the flame is tracked by the extended TSR as well, supporting the auto-igniting nature of the regime at that location, with diffusion that plays an opposing role.

A number of positive eigenvalue spots along the high HRR layer, which are not tracked by the chemical TSR, are partially tracked by the extended TSR, however other contractive modes maintain equal importance and the overall value of the extended TSR remains slightly negative.

In summary, the current section presents a numerical study of a jet-in-hot-coflow (JHC) burner which emulates Moderate or Intense Low-oxygen Dilution (MILD) conditions. Such combustion regime offers reduction in pollutant emissions and improvements in efficiency. The fundamental mechanism of turbulence-chemistry interactions in such a regime is not fully understood, therefore, the advanced post-processing tools based on the theories of computational singular perturbation (CSP) and tangential stretching rate (TSR) are adopted to investigate the large eddy simulation (LES) results of the JHC burner with different coflow oxygen levels.

The analysis of the reactive layers with TSR suggests that the flame under MILD condition (the 3% case before axial direction of 100 mm) is initiated by auto-ignition. However, the extended TSR with diffusive term shows the importance of both auto-ignition and diffusion-controlled ignition under MILD conditions. This might substantiate findings from previous research works [68] stating that the initiated ignition kernels from auto-ignition are transported further downstream. At the location close to the flame ignition point, the 3% case is more chemical reaction driven while the 9% case is transport driven.

Acknowledgements The authors wish to dedicate this contribution to the memory of Prof. S.H. Lam who passed away on Oct. 29, 2018.

HGI and MV acknowledge the support of a competitive research funding from King Abdullah University of Science and Technology (KAUST).

MV acknowledges the support of the Italian Ministry of University and Research (MIUR).

HNN acknowledges the support of the US Department of Energy (DOE), Office of Basic Energy Sciences (BES), Division of Chemical Sciences, Geosciences, and Biosciences. Sandia National Laboratories is a multimission laboratory managed and operated by the National Technology and Engineering Solutions of Sandia, LLC., a wholly owned subsidiary of Honeywell International, Inc., for the U.S. Department of Energy's National Nuclear Security Administration under Contract No. DE-NA-0003525.

AP and ZL have received funding from the European Research Council (ERC) under the European Union's Horizon 2020 research and innovation programme under grant agreement No 714605.

References

1. M. Valorani, D.A. Goussis, J. Comput. Phys. **169**, 44 (2001)
2. M. Valorani, P. Ciottoli, R. Malpica Galassi, S. Paolucci, T. Grenga, E. Martelli, Flow Turbul. Combust. **101**(4), 1023 (2018)
3. M. Valorani, S. Paolucci, J. Comput. Phys. **228**, 4665 (2009)
4. H.N. Najm, M. Valorani, J. Comput. Phys. **270**, 544 (2014)
5. L. Wang, X. Han, Y. Cao, H.N. Najm, J. Comput. Phys. **335**, 404 (2017)
6. X. Han, M. Valorani, H.N. Najm, J. Chem. Phys. **150**(19), 194101 (2019)
7. M. Valorani, F. Creta, D.A. Goussis, J.C. Lee, H.N. Najm, in *Computational Fluid and Solid Mechanics 2005*, ed. by K. Bathe (Elsevier Science, 2005), pp. 900–904
8. R.M. Galassi, P.P. Ciottoli, S.M. Sarathy, H.G. Im, S. Paolucci, M. Valorani, Combust. Flame **197**, 439 (2018)
9. A. Massias, D. Diamantis, E. Mastorakos, D. Goussis, Combust. Flame **117**(6), 685–708 (1999)
10. A. Massias, D. Diamantis, E. Mastorakos, D. Goussis, Combust. Theory Model. **3**(6), 233–257 (1999)
11. E.A. Tingas, D.J. Diamantis, D.A. Goussis, Combust. Theory Model. **22**(6), 1049 (2018)
12. D.A. Goussis, H.N. Najm, Multiscale Model. Simul. **5**(4), 1297 (2006)
13. D.G. Patsatzis, D.A. Goussis, Math. Biosci. **315**, 108220 (2019)
14. D.A. Michalaki, I. Lida Goussis, J. Math. Biol. **77**(3), 821 (2018)
15. M. Valorani, P. Ciottoli, R. Malpica Galassi, CSPTk - A software toolkit for the CSP and TSR analysis of kinetic models and the simplification and reduction of chemical kinetics mechanisms. The software can be obtained upon request to M.Valorani (mauro.valorani@uniroma1.it) (2015)
16. E.A. Tingas, D.C. Kyritsis, D.A. Goussis, Combust. Flame **162**(9), 3263 (2015)
17. E.A. Tingas, H.G. Im, D.C. Kyritsis, D.A. Goussis, Fuel **211**, 898 (2018)
18. E.A. Tingas, D.C. Kyritsis, D.A. Goussis, Fuel **183**, 90 (2016)
19. E.A. Tingas, D.C. Kyritsis, D.A. Goussis, Combust. Theory Model. **21**(1), 93 (2017)
20. D.M. Manias, E.A. Tingas, C.E. Frouzakis, K. Boulouchos, D.A. Goussis, Combust. Flame **164**, 111 (2016)
21. D.J. Diamantis, E. Mastorakos, D.A. Goussis, Combust. Theory Model. **19**(3), 382 (2015)
22. A. Kazakov, M. Chaos, Z. Zhao, F.L. Dryer, J. Phys. Chem. A **110**(21), 7003 (2006)
23. E.A. Tingas, D.C. Kyritsis, D.A. Goussis, J. Energ. Eng. **145**(1), 04018074 (2018)
24. Y. Li, A. Alfazazi, B. Mohan, E.A. Tingas, J. Badra, H.G. Im, S.M. Sarathy, Fuel **247**, 164 (2019)
25. W. Song, E.A. Tingas, H.G. Im, Combust. Flame **195**, 84 (2018)

26. E.A. Tingas, D.C. Kyritsis, D.A. Goussis, Fuel **169**, 15 (2016)
27. E.A. Tingas, D.M. Manias, S.M. Sarathy, D.A. Goussis, Fuel **223**, 74 (2018)
28. E.A. Tingas, Z. Wang, S.M. Sarathy, H.G. Im, D.A. Goussis, Combust. Flame **188**, 28 (2018)
29. S.M. Sarathy, E.A. Tingas, E.F. Nasir, A. Detogni, Z. Wang, A. Farooq, H. Im, *Proc* (Combust, Inst, 2018)
30. A.L. Sánchez, F.A. Williams, Prog. Energy Combust. Sci. **41**, 1 (2014)
31. K. Zhang, C. Banyon, J. Bugler, H.J. Curran, A. Rodriguez, O. Herbinet, F. Battin-Leclerc, C. B'Chir, K.A. Heufer, Combustion and Flame **172**, 116 (2016)
32. N. Peters, *Turbulent Combustion* (Cambridge University Press, UK, Cambridge, UK, 2000)
33. C. Safta, H.N. Najm, O. Knio, Sandia Report **SAND2011-3282** (2011). http://www.sandia.gov/tchem
34. J. Li, Z. Zhao, A. Kazakov, M. Chaos, F.L. Dryer, J.J.J. Scire, International Journal of Chemical Kinetics **39**, 109 (2007)
35. P. Pal, M. Valorani, P.G. Arias, H.G. Im, M.S. Wooldridge, P.P. Ciottoli, R. Malpica Galassi, Proc. Combust. Inst. **36**(3), 3705 (2017)
36. P. Pal, A.B. Mansfield, M.S. Wooldridge, H.G. Im, Combustion Theory and Modelling **66**(5), 1 (2015)
37. M. Valorani, P.P. Ciottoli, R.M. Galassi, Proc. Combust. Inst. **36**(1), 1357 (2017)
38. C. Trevino, Progress in Astronautics and Aeronautics. AIAA **131**, 19 (1991)
39. P. Boivin, C. Jimenez, A.L. Sànchez, F.A. Williams, Proceedings of the Combustion Institute **33**(1), 517 (2011)
40. P. Boivin, A.L. Sànchez, F.A. Williams, Combustion and Flame **159**(2), 748 (2012)
41. A. Cavaliere, M. de Joannon, Progress in Energy and Combustion science **30**(4), 329 (2004)
42. H.G. Im, in *Modeling and Simulation of Turbulent Combustion* (Springer, 2018), pp. 99–132
43. H.G. Im, P.G. Arias, S. Chaudhuri, H.A. Uranakara, Combustion Science and Technology **188**(8), 1182 (2016)
44. P. Arias, H. Uranakar, S. Chaudhuri, H. Im, in *APS Meeting Abstracts* (2015)
45. D.M. Manias, E.A. Tingas, F.E.H. Pérez, R.M. Galassi, P.P. Ciottoli, M. Valorani, H.G. Im, Combust. Flame **200**, 155 (2019)
46. J.A. Wünning, J.G. Wünning, Progress in Energy and Combustion Science **23**, 81 (1997)
47. A. Cavaliere, M. de Joannon, Progress in Energy and Combustion Science **30**, 329 (2004)
48. B.B. Dally, E. Riesmeier, N. Peters, Combustion and Flame **137**, 418 (2004)
49. M. de Joannon, G. Sorrentino, A. Cavaliere, Combustion and Flame **159**, 1832 (2012)
50. X. Gao, F. Duan, S.C. Lim, M.S. Yip, Energy **59**, 559 (2013)
51. D.M. Manias, E.A. Tingas, Y. Minamoto, H.G. Im, Combustion and Flame **208**, 86 (2019)
52. Y. Minamoto, N. Swaminathan, R.S. Cant, T. Leung, Combustion Science and Technology **186**(8), 1075 (2014)
53. Y. Minamoto, N. Swaminathan, Combustion and Flame **161**(4), 1063 (2014)
54. Y. Minamoto, N. Swaminathan, S.R. Cant, T. Leung, Combustion and Flame **161**(11), 2801 (2014)
55. Z.M. Nikolaou, N. Swaminathan, Combustion and Flame **161**(12), 3073 (2014)
56. P.R. Medwell, P.A. Kalt, B.B. Dally, Combustion and Flame **148**(1–2), 48 (2007)
57. B.B. Dally, A.N. Karpetis, R.S. Barlow, Proceedings of the Combustion Institute **29**, 1147 (2002)
58. E. Oldenhof, M.J. Tummers, E. van Veen, D. Roekaerts, Combustion and Flame **157**, 1167 (2010)
59. E. Oldenhof, M.J. Tummers, E.H. van Veen, D.J.E.M. Roekaerts, Combustion and Flame **158**(8), 1553 (2011)
60. N. Kornev, H. Kröger, E. Hassel, Communications in Numerial Methods in Engineering **24**, 875 (2008)
61. R.W. Bilger, S.H. Starner, R.J. Kee, Combustion and Flame **80**(2), 135 (1990)
62. C. Duwig, P. Iudiciani, Fuel **123**, 256 (2014)
63. Z. Li, A. Cuoci, A. Parente, Proceedings of the Combustion Institute (2018)
64. S.H. Lam, D.A. Goussis, Proc. Comb. Inst. **22**, 931 (1988)

65. S.H. Lam, Combust. Sci. Technol. **89**, 375 (1993)
66. Z. Li, R.M. Galassi, P.P. Ciottoli, A. Parente, M. Valorani, Combust. Flame **208**, 281 (2019)
67. H.N. Najm, M. Valorani, D.A. Goussis, J. Prager, Combustion Theory and Modelling **14**(2), 257 (2010)
68. N.A.K. Doan, N. Swaminathan, Combustion and Flame **201**, 234 (2019)

Chapter 5
Chemical Explosive Mode Analysis for Diagnostics of Direct Numerical Simulations

Chun Sang Yoo, Tianfeng Lu, and Jacqueline H. Chen

Abstract Direct numerical simulation (DNS) has become an important tool to predict and understand complex structures and behaviors of turbulent flames over the last two decades, enabled by the rapid growth of supercomputer power and development of more efficient and accurate Navier–Stokes equation solvers [1]. To predict the strongly nonlinear chemical kinetic processes and their interactions with the flow, detailed chemistry is typically employed in DNS while the computational cost is high even after aggressive mechanism reduction [2]. DNS on today's supercomputer is capable to generate massive datasets, say tens or hundreds of terabytes, even in cleaned forms, such that systematic computational diagnostic tools need to be developed to extract salient information from the massive raw data. Canonical diagnostic methods based on individual scalars, such as temperature or a species concentration and their combinations (e.g., progress variable and mixture fraction) have been widely employed in previous studies. However, the use of such scalars typically requires semi-empirical criteria that need to be adjusted for different flame types and conditions, rendering them difficult to be automated for the processing of large flame data. Tools universally applicable to different flames and suitable for DNS data diagnostics are scarce and need to be developed. To address this need, a method of chemical explosive mode analysis (CEMA) was recently developed to systematically detect critical flame features for general reacting flows, particularly when local ignition, extinction, and premixed flame fronts are involved [3–6]. CEMA has been demonstrated in elementary reactors, laminar flames and a variety of turbulent flames [3–9]. It was found that CEMA-based criteria are rather robust and reliable for

C. S. Yoo
Department of Mechanical Engineering, Ulsan National Institute of Science and Technology, Ulsan 44919, Republic of Korea
e-mail: csyoo@unist.ac.kr

T. Lu (✉)
Department of Mechanical Engineering, University of Connecticut, Storrs, CT 06269-3139, USA
e-mail: tianfeng.lu@uconn.edu

J. H. Chen
Combustion Research Facility, Sandia National Laboratories, Livermore, CA 94550, USA
e-mail: jhchen@sandia.gov

H. Pitsch and A. Attili (eds.), *Data Analysis for Direct Numerical Simulations of Turbulent Combustion*, https://doi.org/10.1007/978-3-030-44718-2_5

limit phenomena detection for both premixed and partially premixed flames, and the use of CEMA in computational diagnostics of different types of flames is discussed in the present chapter.

5.1 Overview of CEMA

A typical reacting flow is governed by the following equations in the Lagrangian system:

$$\frac{D\boldsymbol{y}}{Dt} = \boldsymbol{s} + \boldsymbol{\omega}, \tag{5.1}$$

where $\boldsymbol{y}$ is the vector of local thermodynamic variables including temperature and species mole concentrations, $\boldsymbol{s}$ is the nonchemical source terms, such as molecular diffusion or any compression effects, and $\boldsymbol{\omega}$ is the chemical source term. Note that $\boldsymbol{s}$ may directly involve quantities in the entire domain, while $\boldsymbol{\omega}$ only depends on the local thermodynamic state, i.e., $\boldsymbol{\omega} = \boldsymbol{\omega}(\boldsymbol{y})$. By applying the chain rule the dynamics of $\boldsymbol{\omega}$ can be expressed as

$$\frac{D\boldsymbol{\omega}(\boldsymbol{y})}{Dt} = \mathbf{J}\frac{D\boldsymbol{y}}{Dt} = \mathbf{J}(\boldsymbol{s} + \boldsymbol{\omega}), \quad \mathbf{J} = \frac{\partial \boldsymbol{\omega}}{\partial \boldsymbol{y}}. \tag{5.2}$$

Equation (5.2) can be projected in any direction $\boldsymbol{b}$, which can be time-dependent, as

$$\boldsymbol{b} \cdot \frac{D\boldsymbol{\omega}}{Dt} = \boldsymbol{b} \cdot \mathbf{J}(\boldsymbol{s} + \boldsymbol{\omega}), \tag{5.3}$$

and if $\boldsymbol{b}$ is chosen to be a left eigenvector of the Jacobian, that is $\boldsymbol{b} \cdot \mathbf{J} = \lambda \boldsymbol{b}$, where λ is the eigenvalue, Eq. (5.4) becomes

$$\boldsymbol{b} \cdot \frac{D\boldsymbol{\omega}}{Dt} = \lambda \boldsymbol{b} \cdot (\boldsymbol{s} + \boldsymbol{\omega}), \tag{5.4}$$

which can be further transformed to obtain the dynamics of a chemical mode, $f \equiv \boldsymbol{b} \cdot \boldsymbol{\omega}$

$$\frac{Df}{Dt} = \lambda f + \lambda(\boldsymbol{b} \cdot \boldsymbol{s}) - \frac{D\boldsymbol{b}}{Dt} \cdot \boldsymbol{\omega}. \tag{5.5}$$

The first source term in Eq. (5.5) is the self-driving force, the second term indicates the forcing by diffusion, and the third term is attributed to the rotation of the eigenvector in a nonlinear system. Note that in the Eulerian system, Eq. (5.1) can be expressed as

$$\frac{\partial \boldsymbol{y}}{\partial t} = \boldsymbol{c} + \boldsymbol{s} + \boldsymbol{\omega}, \tag{5.6}$$

where $\boldsymbol{c}$ is the convection source term, and the dynamics of the chemical mode is governed by

$$\frac{\partial f}{\partial t} = \lambda f + \lambda(\boldsymbol{b} \cdot \boldsymbol{c}) + \lambda(\boldsymbol{b} \cdot \boldsymbol{s}) - \frac{\partial \boldsymbol{b}}{\partial t} \cdot \boldsymbol{\omega}. \tag{5.7}$$

Nevertheless, if only the behavior of the chemical explosive mode (CEM) is of interest, it makes no significant difference whether Eq. (5.5) or Eq. (5.7) is used.

Due to the conservation of elements and energy, the chemical Jacobian **J** typically has M+1 zero eigenvalues, where M is the number of elements involved in the species. Since an element conservation mode is associated with a constant eigenvector and a zero-eigenvalue, the RHS of Eq. (5.5) is trivial for an element conservation mode in a flow, and thus it is typically straightforward to computationally identify element conservation modes in any flow field by simply examining the magnitude of the eigenvalues. The energy conservation mode is however more involved to identify because the thermodynamic properties, such as heat capacity, can be a nonlinear function of the dependent variables. In such cases, the energy conservation mode is typically not trivial due to the nonlinear effect of eigenvector rotation, and it may be computationally challenging to distinguish the energy conservation mode from other slow chemical modes. As an effective work-around, for cases where the dynamics of the energy conservation mode is not of interest, the time dependence of heat capacity, etc. can be neglected such that the energy conservation mode becomes trivial and associated a zero-eigenvalue [3].

A chemical mode is nonexplosive if it is associated with an eigenvalue with a negative real part. Chemical nonexplosive modes typically prevail in near-equilibrium products, for which the chemical reaction rates can be linearized as:

$$\boldsymbol{\omega}(\boldsymbol{y}) \approx \boldsymbol{\omega}(\boldsymbol{y}_\infty) + \mathbf{J}_\infty(\boldsymbol{y} - \boldsymbol{y}_\infty) = \mathbf{J}_\infty(\boldsymbol{y} - \boldsymbol{y}_\infty), \tag{5.8}$$

where the subscript ∞ indicates the equilibrium condition. Considering the physically stable nature of chemical equilibrium, the real parts of the eigenvalues of the chemical Jacobian must be either zero (a conservation mode) or negative (nonexplosive).

A CEM is associated with an eigenvalue with a positive real part. If such a mixture is isolated, an explosive process tends to occur, particularly when the mixture is at the radical explosion stage of ignition, where the radical pool built-up occurs under mostly constant temperature and major species concentrations, such that the system is approximately linear. Note that while theoretically there is a possibility that a CEM can suppress itself due to the basis rotation term in Eq. (5.5), no combustible mixtures have been observed in any previous flame simulations to show such a behavior. However, for an explosive mixture in a diffusive environment or where radical or heat loss is present, a CEM may or may not exponentially grow depending on the relative contribution from the different terms in Eq. (5.5). For instance, in a steady-state stirred reactor (PSR), a CEM may be present near the extinction turning point while the reactor may remain in steady state, as shown in the next section.

It is worth noting that in most flames, when defined based on temperature and species mole concentrations, only one CEM can be observed in a premixture of fuel and oxidizer when preheated, and the CEM vanishes as the mixture ignites and evolves toward chemical equilibrium. However, when defined based on other quantities, such as species mass fractions, multiple explosive modes may be observed in a mixture. This is because changing from mole concentrations to mass fractions may involve nonlinear coordinate transformation, such that a single CEM can affect multiple modes in the new coordinate, and thus result in more than one positive eigenvalues. However, although the number of CEM is small compared with that of nonexplosive modes, the CEM is typically the primary driving force of many combustion processes while the nonexplosive modes simply follow the CEM. As such it is important to understand the roles of CEM in different combustion systems in order to create computational diagnostics based on CEM.

5.2 Role of CEM in Elementary Flames

The presence of CEM is an intrinsic property of a combustible mixture undergoing autoignition processes. To demonstrate this point, Fig. 5.1a shows the temperature profiles of autoigniting H_2/air mixtures at different initial temperatures, with color indicating the sign and magnitude of $\mathrm{Re}(\lambda_e)$, where λ_e is the eigenvalue with the largest real part among the nonconservative chemical modes. Note that when a CEM is not present, λ_e is the eigenvalue with the least negative real part excluding the conservative modes. It is seen in Fig. 5.1a that a CEM is present as indicated by the red color during the entire radical explosion stage at which temperature remains mostly unchanged, and the early portion of the thermal runway stage at which temperature increases rapidly due to the fast heat release. The CEM vanishes in the later stage of thermal runaway, as indicated by the blue color, while the mixture evolves to chemical equilibrium. Figure 5.1b further shows the heat release rate as a function of time centered at the zero-crossing of the CEM eigenvalue, $\mathrm{Re}(\lambda_e) = 0$. It is seen that the peak heat release rates occur almost exactly at the zero-crossing points, which therefore are close to the inflection points on the temperature profiles. Based on this observation, the CEM zero-crossing has been exploited to identify local ignition in previous studies.

In an unpreheated 1-D premixed flame, the fresh mixture entering the domain is approximately chemically frozen due to the low inlet temperature and has to be force-ignited as the mixture enters the flame zone by the radicals and heat back-diffused from the hot product in the reaction zone. Figure 5.2a shows the temperature profiles of 1-D laminar premixed flames for H_2/air under atmospheric pressure and inlet temperature of 300 K at different equivalence ratios. It is seen that the CEM in the fresh reactant is mostly a dormant mode (green color) at inlet and becomes reactive and explosive as the mixture enters the preheat zone, where radical and heat are received through the back-diffusion. Similar to the autoignition cases, the mixture becomes nonexplosive once the mixture ignited as it travels downstream and

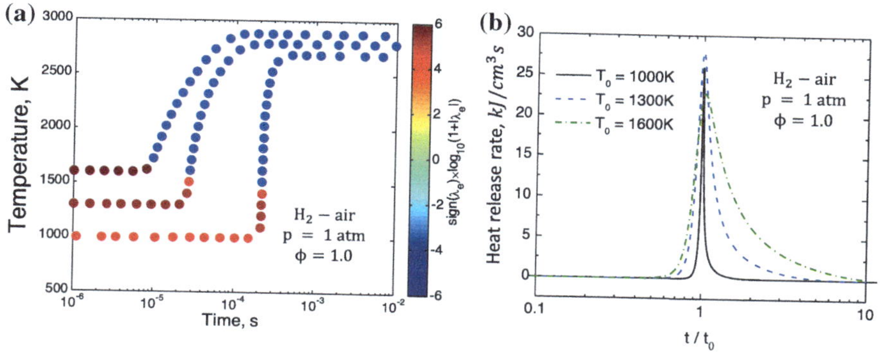

Fig. 5.1 **a** Temperature profile in constant-pressure autoignition of stoichiometric H_2/air under pressure of 1 atm and different initial temperatures. Color indicates the sign and magnitude of the CEM. **b** Heat release rate as a function of time, centered at $\mathrm{Re}(\lambda_e) = 0$

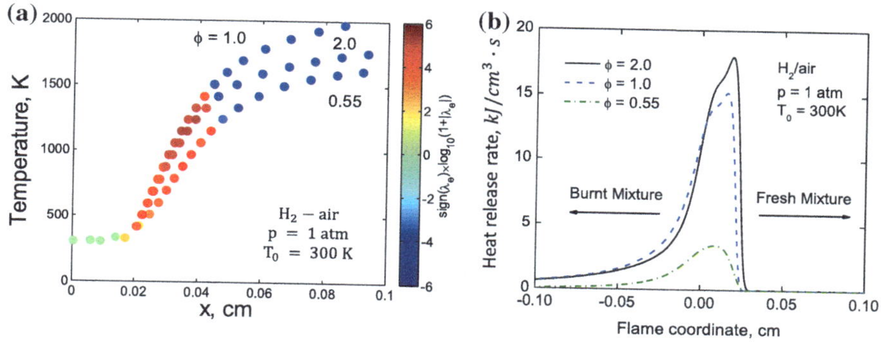

Fig. 5.2 **a** Temperature profile in 1-D freely propagating laminar premixed flames of H_2/air under pressure of 1 atm and inlet temperature of 300 K with different equivalence ratios. Color indicates the sign and magnitude of the CEM. **b** Heat release rate as a function of the spatial coordinate, where $x = 0$ indicates the location of $\mathrm{Re}(\lambda_e) = 0$

evolves to chemical equilibrium. Figure 5.2b further shows that the zero-crossing CEM eigenvalue is located within the thin reaction zones, although not exactly at the peak heat release rate, and thus can indicate the location of the reaction zone in applications where the reaction is treated as a thin surface. Note that the skewed peak heat release rate in the H_2/air flame is primarily induced by the strong differential diffusion effects.

PSR is a representative 0-D system that is relevant to many continuous-flow combustors. The solution of PSR features an S-curve that involves three branches, namely the upper branch for strongly burning states, the lower branch for weakly reacting states, and the physically unstable middle branch, which typically cannot be observed in experiments [10]. The upper turning point on the S-curve is the extinction limit beyond which a steady state strongly burning solution no longer exists. Figure 5.3 shows the upper and middle branches of the S-curves for the stoichiometric H_2/air

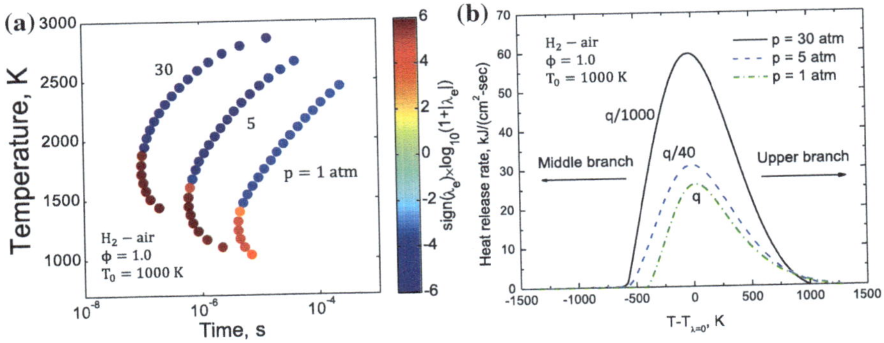

Fig. 5.3 **a** Temperature profile in PSR of stoichiometric H_2/air under inlet temperature of 1000 K and different pressures. Color indicates the sign and magnitude of the CEM. **b** Heat release rate as a function of temperature, centered at $Re(\lambda_e) = 0$

mixture under inlet temperature of 1000 K and different pressures. It is seen that the CEM is absent on most portion of the upper branches and emerges when the system approaches the extinction turning point, and is also present on the middle and lower branches. This is because the mixture on the upper branches with long residence time is near equilibrium, while reactants are less consumed as the system approaches extinction, where the short residence time results in a significant level of incomplete combustion. Therefore, the zero-crossing of CEM eigenvalue is a precursor for extinction and has been used as a rough indicator of PSR extinction in previous studies. Figure 5.3b further shows the heat release rate as a function of temperature centered at the extinction turning points. It is seen that the heat release rate peaks almost exactly at the zero-crossing, indicating that fastest chemical reactions occur near the extinction state, where chemical reactions are pushed to the fast limit by the mixing processes.

These behaviors of CEM in elementary flames indicate the following CEM properties: (1) CEM is present in fresh reactants and absent in hot products and thus can be employed to distinguish reactants and products in complex flames; (2) the CEM zero-crossing is associated with the rapid transition from reactants to products that naturally defines ignition/extinction points and locations of the reaction zones in premixed flames; and (3) CEM is a chemical property and can be calculated with local thermodynamic variables, while the dynamics of CEM depends on both the chemical and the mixing source terms. It is further noted that the above analysis is applicable to both premixed and non-premixed flames, as to be shown in the following for DNS of homogeneous charge compression ignition (HCCI) as an example of premixed flame, and lifted jet flames as examples of non-premixed or partially premixed flames.

5.3 CEMA for DNSs of HCCI Type Combustion

Over the past two decades, numerous experimental and numerical studies have been performed for the development of next-generation IC engines. As one of the promising engines, HCCI engines have drawn great attention from the engine community since they can significantly reduce pollutant emissions including NO_x and particulate matters (PM) by utilizing low-temperature combustion and achieve high-thermal efficiency with high compression ratio like diesel engines [11–13]. HCCI combustion occurs primarily by volumetric autoignition, which leads to excessive pressure rise rate (PRR) and resulting engine knock, ultimately damaging engine integrity. Moreover, it suffers from controlling the ignition timing due to the absence of any explicit ignition sources. To resolve the issues, several variants of HCCI combustion have been proposed thus far. For instance, stratified charge compression ignition (SCCI), reactivity-controlled compression ignition (RCCI), and spark-assisted compression ignition (SACI) combustions have been investigated by introducing composition inhomogeneity, reactivity fluctuation, and spark ignition source in the engine cylinder, respectively. All of these variants are designed to mitigate PRR by spreading out heat release rate (HRR) during the combustion and to control the ignition timing by using direct injection of fuel and spark ignition [11–13].

To understand the details of ignition characteristics in such HCCI type combustion, many direct numerical simulations (DNSs) have been performed and CEMA has been adopted to identify the combustion modes of HCCI type combustion [7, 14–22]. For DNSs of HCCI type combustion, we used Sandia DNS code, S3D [1], which solves the compressible Navier–Stokes, species continuity, and total energy equations using a fourth-order explicit Runge–Kutta method [23, 24] and an eighth-order central differencing scheme [1]. To evaluate thermodynamic and mixture-averaged transport properties, S3D is linked with CHEMKIN and TRANSPORT software libraries [25, 26].

HCCI type combustion has been studied in DNS with fuel/air mixture igniting at a constant volume, which is realized by imposing periodic boundary conditions in all directions of the computational domain [7, 14–19]. Such isochoric process can provide a similar condition of an engine cylinder to that near the top dead center (TDC) when most combustion heat is released, and hence, we can investigate the ignition characteristics of HCCI type combustion without having to describe the engine motion in DNSs. In some recent DNSs [18–20], a compression heating model was adopted to mimic the piston motion of an IC engine, which was incorporated in the code by adding/subtracting mass and momentum in the governing equations. To further realize mixture inhomogeneities in an engine cylinder similar to HCCI, SCCI, RCCI, and SACI combustions, temperature, composition, and reactivity fluctuations are added on top of homogeneous turbulence field [7, 14–22]. For DNSs of HCCI type combustions, we adopted reduced chemical kinetic mechanisms for n-heptane [15, 17], primary reference fuel (PRF) [14, 16, 18–20, 22], biodiesel [7], and ethanol [21] developed by using a strategy combining directed relation graph (DRG)-based methods, isomer lumping, and timescale analysis [10, 27–29].

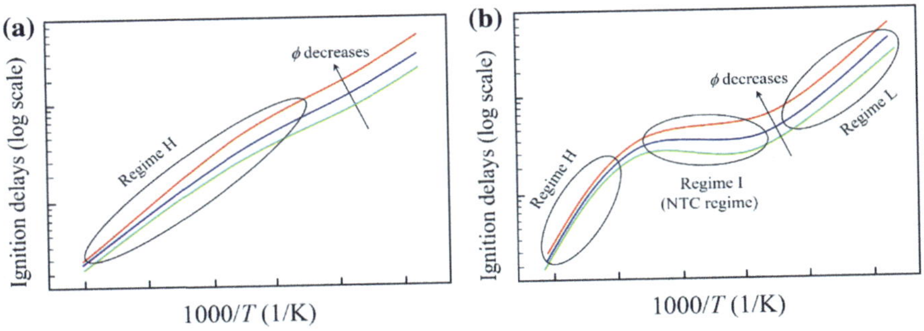

Fig. 5.4 Schematics of 0-D ignition delay variation as a function of the inverse of initial temperature for **a** one-stage ignition fuels and **b** two-stage ignition fuels (from Ref. [7])

Prior to discussing the details of the application of CEMA to DNSs of HCCI type combustion, here we address that we can categorize many fuels into two types based on their ignition characteristics: one-stage ignition and two-stage ignition fuels. For instance, hydrogen, ethanol, and iso-octane feature "one-stage ignition" for which 0-D ignition does not show negative-temperature coefficient (NTC) behavior under high pressures and intermediate temperatures as schematically shown in Fig. 5.4. However, *n*-heptane, biodiesel, and dimethyl-ether (DME) feature "two-stage ignition" for which 0-D ignition exhibits strong NTC behavior associated with the low-temperature chemistry (LTC) at high pressures and intermediate temperatures. As discussed in [7, 14, 16, 17], temperature, composition, and reactivity fluctuations (T', ϕ', and PRF′) have different effects on the HCCI type combustion depending on the fuel types and the mean temperature/equivalence ratio of the initial mixture.

Regardless of fuel types, thermal fluctuations become more effective to spread out HRR and advance the overall combustion than composition stratification when the initial mixture lies within the high-temperature regime (Regime H) where small variation in temperature can significantly change the ignition delay. For two-stage ignition fuels, however, the effect of composition stratifications can be manifested in the NTC regime (Regime I) rather than in Regime H because the ignition delay is more sensitive to equivalence ratio than temperature in Regime I as shown in Fig. 5.4. By applying temperature, composition, and reactivity fluctuations to the fuel/air charge, we can generally induce more deflagration mode of combustion than spontaneous ignition mode of combustion in HCCI type combustion. Such combustion mode analysis was carried out by analyzing DNS data with CEMA. In addition to the combustion mode analysis, CEMA was also adopted to understand the chemical characteristics of ignition of lean fuel/air mixture under HCCI, SCCI, RCCI, and SACI conditions.

Important variables and reactions to the CEM can be further identified by examining the explosion index (EI) and participation index (PI), defined as

$$\mathbf{EI} = \frac{|\boldsymbol{a} \otimes \boldsymbol{b}^T|}{sum(|\boldsymbol{a} \otimes \boldsymbol{b}^T|)} \tag{5.9a}$$

$$\mathbf{PI} = \frac{|(\boldsymbol{b} \cdot \mathbf{S}) \otimes \boldsymbol{R}|}{sum(|(\boldsymbol{b} \cdot \mathbf{S}) \otimes \boldsymbol{R}|)}, \tag{5.9b}$$

where $\boldsymbol{a}$ is the right eigenvector of the CEM, $\mathbf{S}$ is the stoichiometric coefficient matrix, $\boldsymbol{R}$ is the vector of the net rates for the reactions, and $\otimes$ denotes element-wise multiplication of two vectors. It is seen that EI and PI indicate the contribution of a species and reaction to the CEM, respectively [17, 18, 20]. For instance, Fig. 5.5 shows the isocontours of HRR, temperature, n-heptane mass fraction, the log-scale of Re(λ_e), and the EI values of important variables which exhibit relatively large EI values under SCCI condition [17]. From Fig. 5.5a–d, we can readily observe that Re(λ_e) < 0 indicates burnt region while Re(λ_e) > 0 denotes unburnt region where ignition is still underway.

We can also identify thin deflagrations with large HRR and $Da_c \sim O(1)$ between the burnt and unburnt regions based on a Damköhler number, Da_c, defined as:

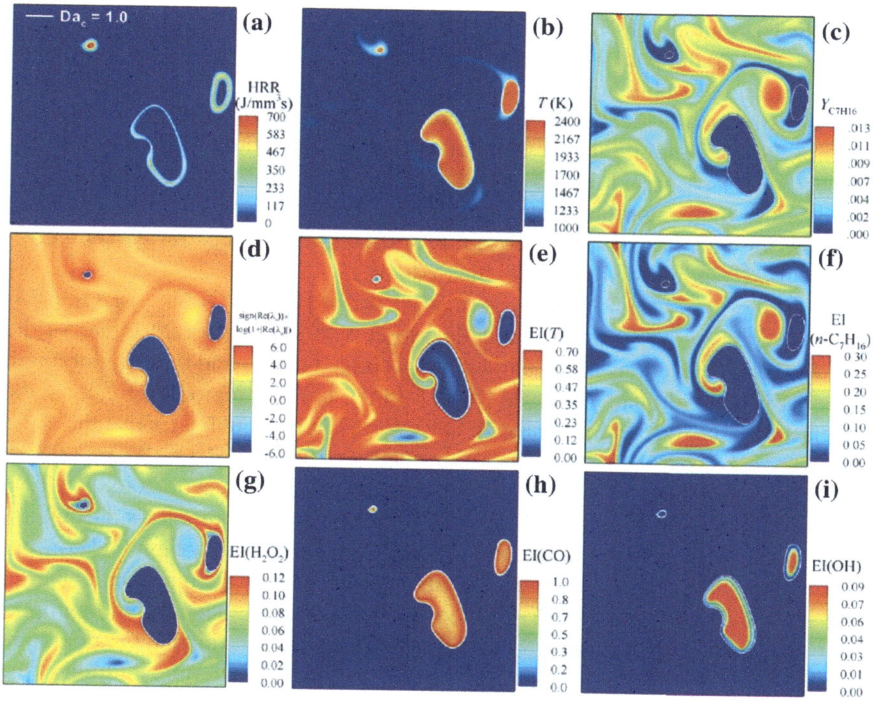

Fig. 5.5 Isocontours of **a** HRR, **b** temperature, **c** $Y_{C_7H_{16}}$, **d** sign(Re(λ_e))×log(1+|Re(λ_e)|), and EI of **e** temperature, **f** n-heptane, **g** H_2O_2, **h** CO, and **i** OH for Case 16 ($T_0 = 933$ K, NC: T′ = 15 K and $\phi' = 0.10$) at $t/\tau_{ig}^0 = 0.67$. The white solid line represents $Da_c = 1.0$ (from Ref. [17])

$$Da_c = Re(\lambda_e)/\chi, \tag{5.10}$$

where χ is the scalar dissipation rate based on progress variable. $Da_c \sim O(1)$ represents deflagration wave in which reaction balances mixing; whereas $Da_c \gg 1$ indicates the dominance of reaction over diffusion, and as such, mixture with $Da_c \gg 1$ will autoignite in the end. Note that Da_c is much larger than unity in the domain except for the burnt regions. CEMA also identifies temperature, *n*-heptane, and H_2O_2 as the main variables controlling the ignition in the unburnt region (see Fig. 5.5e–g. Temperature and n-heptane are identified as the main source of the CEM at the unburnt region with relatively low temperature (T $\sim$ 1000 K) because fuel decomposition still occurs at the relatively low-temperature region with large fuel concentration. Meanwhile H_2O_2 controls the ignition of the unburnt mixture at the unburnt region with relatively high temperature (T $\sim$ 1100 K). This is because the chain branching reaction of H_2O_2 becomes highly reactive, which subsequently results in initiating high-temperature chemistry. In addition, CO and OH are found to be the most important species in the burnt region (see Fig. 5.5h and i) in which the high-temperature chemistry remains controlling the combustion process.

While EI analysis identifies important variables to HCCI type combustion, critical reactions involving the important species can be recognized by PI analysis. Figure 5.6 shows the isocontours of PI values of important reactions with large PI values [17]. The main conversion path of CO to CO_2 (R7: CO + OH $\rightarrow$ CO_2 + H) and the most important chain branching reaction at high temperature (R8: H + O_2 $\rightarrow$ O + OH) are found to be the most critical to the CEM across the deflagrations waves (see Fig. 5.6a, b). Since they are one of the most important high-temperature chemistry (HTC) in hydrocarbon/air premixed flames regardless of specific fuel type [10], PI analysis also identifies their importance within the deflagration waves. Readers are referred to Fig. 5.7 for a better understanding of three different reaction pathways (i.e., LTC, ITC, and HTC) of high-hydrocarbon fuel oxidation.

In addition to R7 and R8, HO_2 formation (R24: H + O_2 + M $\rightarrow$ HO_2 + M) is also found to be important at the deflagration waves because R24 is one of the most exothermic reactions in a hydrogen/air premixed flame, and thus, important in the oxidation of hydrocarbon fuels [10, 17]. In the unburnt region upstream of the deflagration waves (see Fig. 5.6e and f), however, PI analysis indicates that the chain branching of H_2O_2 (R48) and the generation of alkyl radical and H_2O_2 (R292 – R300: RH + (OH, HO_2) $\rightarrow$ R + (H_2O, H_2O_2)) are the most critical reactions to the CEM. Fuel decomposition to alkyl radical and H_2O_2 is found to be important for the unburnt region with relatively low temperature (T $\sim$1000 K) and relatively high fuel concentration, where the low-temperature chemistry (LTC) governs the ignition. However, the chain branching reaction of H_2O_2 produced from the LTC becomes more important for the unburnt region with T $\sim$ 1100 K, where the intermediate temperature chemistry (ITC) governs the ignition, ultimately inducing the thermal ignition or the HTC of the unburnt mixtures.

As discussed above, the spatial distributions of EI and PI values can identify specific chemical information of local mixtures, from which, however, the effects of each variable and each reaction on the overall ignition and their relations cannot be

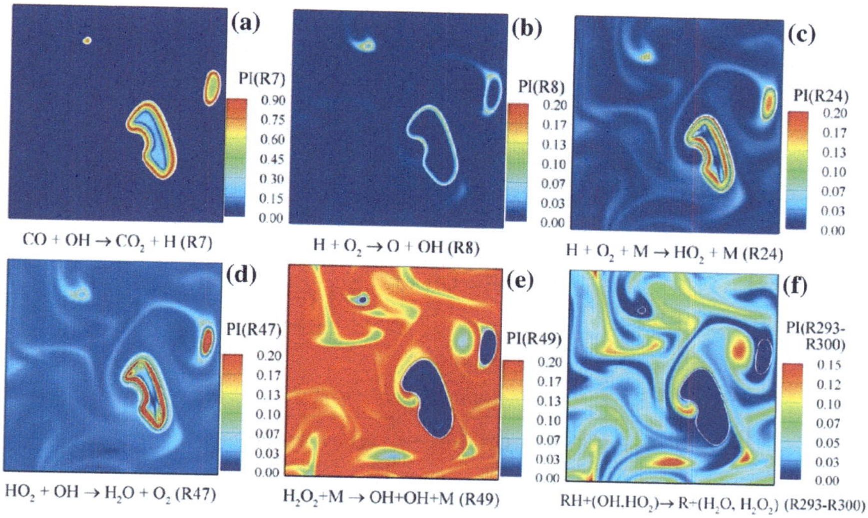

Fig. 5.6 Isocontours of PIs of controlling reactions for Case 16 ($T_0 = 933$ K, NC: $T' = 15$ K and $\phi' = 0.10$) at $t/\tau_{ig}^0 = 0.67$. The white solid line represents $Da_c = 1.0$ (from Ref. [17])

readily elucidated. To identify each contribution of LTC, ITC, and HTC to the overall combustion, therefore, we adopt the conditional means of HRR, λ_e, and EI/PI values of important variables/reactions for RCCI and SCCI combustion at their first-stage ignition delay as shown in Fig. 5.8 [17, 20]. Here, we obtain a conditional mean value by averaging a variable conditioned on temperature. Since it is the early stage of ignition, most heat is released from relatively low-temperature region (830 K $< T <$ 900 K), where the LTC (e.g., the isomerization of RO_2, chain branching reactions of KOOH, and H-atom abstraction from n-heptane) is found to be dominant for both RCCI and SCCI combustion. Meanwhile, the conditional means of EI values identify temperature, n-heptane, and CH_2O as the most important variables to the CEM. It is of interest to note that the intensity of HRR in RCCI case is slightly higher than that in SCCI case, which explains advanced second-stage ignition in RCCI case [12, 13]. In the region with $T > 900$ K, however, the contribution of ITC to the CEM represented by H_2O_2 decomposition (R16) and H-atom abstraction/decomposition of iso-octane becomes predominant and as such, HRR is significantly reduced at this region. Consistent with Fig. 5.7, HO_2 from QOOH decomposition and H_2O_2 via R16 are found to be the critical species to the CEM in the region with ITC. It is also of interest to note that the EI values nearly vanish near 880 K. This result indicates that cool flames develop from initial ignition kernels, banishing the CEM at such a temperature range.

To further identify the ignition characteristics of RCCI and SCCI combustion, we evaluate the conditional means of critical variables and reactions at the time of 10% cumulative heat release rate (CHRR) as shown in Fig. 5.9. Unlike the early stage of

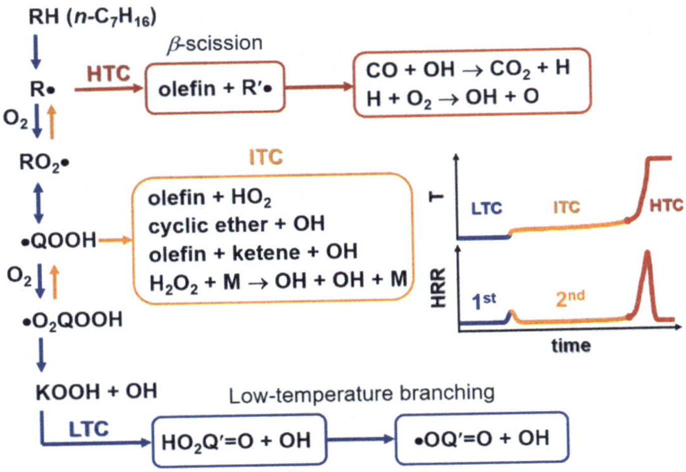

Fig. 5.7 Isocontours of PIs of controlling reactions for Case 16 ($T_0 = 933$ K, NC: $T' = 15$ K and $\phi' = 0.10$) at $t/\tau_{ig}^0 = 0.67$. The white solid line represents $Da_c = 1.0$ (from Ref. [17])

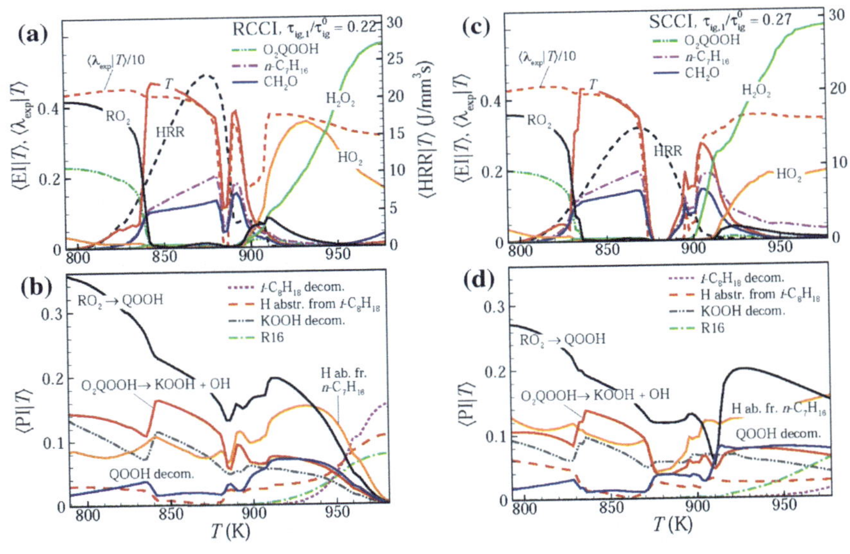

Fig. 5.8 Conditional mean of HRR, λ_e, and EI of critical variables, and PI of critical reactions at the first-stage ignition for RCCI (left column) and SCCI (right column) (from Ref. [18])

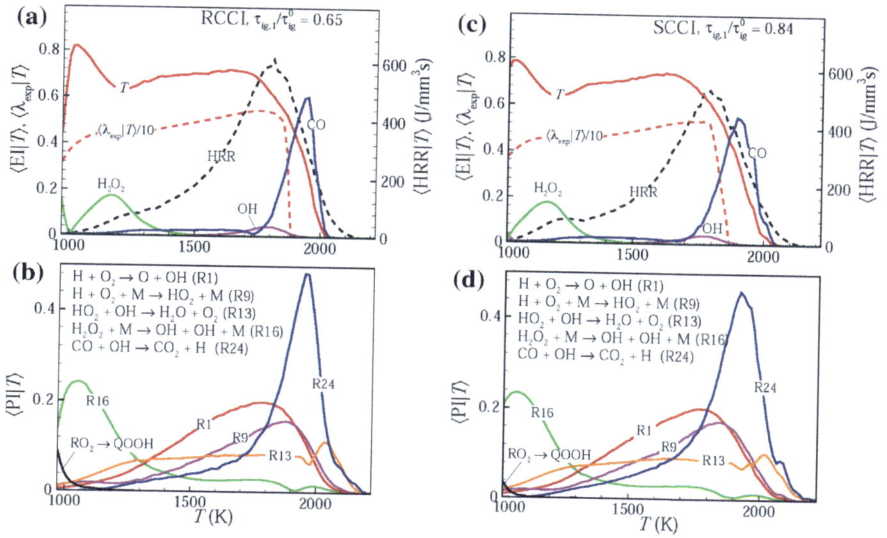

Fig. 5.9 Conditional means of HRR, λ_e, and EI of critical variables, and PI of critical reactions at 10% CHRR for RCCI (left column) and SCCI (right column) (from Ref. [18])

ignition, most HRR occurs at high-temperature region ($T > 1500$ K) where HTC represented by CO oxidation (R24) and hydrogen chemistry (R1, R9, and R13) is found to be critical to the CEM for both RCCI and SCCI combustion. Since HTC governs the ignition at this stage, temperature, CO, and OH are identified as the most important variables to the CEM and the temperature increase at this stage is primarily attributed to high-temperature heat release (HTHR). At the region with $T < 1500$ K, however, the contribution of ITC represented by H_2O_2 decomposition (R16) is found to be still significant as in Fig. 5.8b and d such that HRR from ITC is also marginal at this region. From the conditional means of EI/PI values, we can simply understand the overall RCCI/SCCI combustion by mapping the variation of 2-D RCCI/SCCI combustion in temperature space onto the temporal evolution of 0-D ignition although HTHR in 2-D RCCI/SCCI cases occurs primarily by deflagration mode of combustion rather than by spontaneous ignition mode.

5.4 CEMA for DNSs of Laminar/Turbulent Lifted Jet Flames

Turbulent lifted jet flames can be easily found from many practical combustors including diesel engines, direct-injected spark ignition engines, gas turbine combustors, and commercial boilers. The turbulent lifted jet flame configuration is prevalent in such applications due to their high fuel jet velocities, which can render the flames to be vulnerable to flame blowout. As such, numerous experimental and numerical studies have been performed to elucidate the stabilization mechanisms of turbulent lifted jet flames and ultimately develop novel methods to stabilize them within combustors [11, 30, 31]. The stabilization mechanisms of laminar lifted jet flames have been also investigated to better understand their turbulent counterparts [32]. Since it was first introduced [3], CEMA has been applied to DNSs of various jet flames such as laminar lifted jet flames in heated coflow [33–35], turbulent lifted jet flames in heated coflow [3, 8], turbulent reacting jets in cross-flows [36–39], and turbulent jet flame in a dump combustor [40]. CEMA has been also applied to large-eddy simulations (LESs) [41–46] and one-dimensional turbulence simulations (ODTs) [47, 48] of turbulent flames. In most studies, CEMA was adopted to identify the stabilization mechanisms of turbulent flames and their chemical aspects because it is able to systematically detect important species and reactions in premixed flames and flame ignition/extinction [3–5].

From the studies of laminar lifted DME jet flames at elevated temperatures and pressures [33, 34], the controlling chemistry for different cases was identified with CEMA such that depending on the boundary temperature and the inlet velocity, the stabilization mechanism of laminar lifted DME jet flames in autoignitive conditions is identified as burner-, kinematic-, kinetic-, and multimode-stabilization. Recently, it was found in Ref. [35] that the differential diffusion between methane and hydrogen leads to a decreasing liftoff height behavior of autoignited laminar lifted methane/hydrogen jet flames (see Fig. 5.10) with increasing fuel jet velocity, U_0, while their flame regime changes from a MILD combustion to a transition to a tribrachial edge flame regime. We consider a jet flame to lie in the MILD combustion regime if its maximums allowable temperature increase with respect to the inlet temperature is lower than its autoignition temperature [35]. The controlling chemistry of the lifted flames was also elucidated by identifying the critical variables and reactions upstream of the flame base with CEMA. The flame base is defined as the most upstream point of $\text{Re}(\lambda_e) = 0$ isoline. Figure 5.11 shows the contribution of each chemical reaction to the CEM, or the isocontours of PI of critical elementary reactions to identify the chemical aspects of the autoignited laminar lifted methane/hydrogen jet flames in different regimes [35]. Note that the most upstream point of $\text{Re}(\lambda_e) = 0$ isoline represents the flame base.

It is readily observed that for the MILD combustion regime ($U_0 = 4$ m/s), autoignition of methane/air mixture occurs along the jet centerline upstream of the flame base. After the H abstraction of methane near the fuel nozzle, the conversion of CH_3 to CH_3O/CH_2O to HCO is identified as the following sequence from the PI analysis (not shown here): CH_3 + (O, O_2, HO_2, O_2) → (CH_2O + H, CH_2O + OH, CH_3O + OH, CH_3O + O) (R49, R51 – 53), CH_3O + (O_2, M) → CH_2O + (HO_2, H + M) (R78, R79), and CH_2O + (H, O, OH, HO_2) → HCO + (H_2, OH, H_2O, H_2O_2) (R36 – 38, R40). Figures 5.10a and b show the overall PI values of CH_3 conversion to CH_2O and CH_3O, and CH_2O conversion to HCO by summing up each PI value of the above reactions. Meanwhile, hydrogen peroxide consumption via H_2O_2 + M → OH + OH + M (R16) becomes more active as a part of radical growth for the methane oxidation (see Fig. 5.11c). The conversion of HCO to CO to CO_2 is found

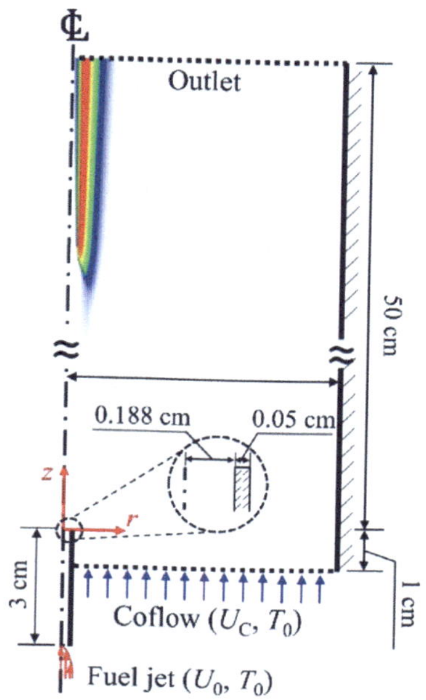

Fig. 5.10 Schematic of the computational domain for simulations of autoignited laminar lifted methane/hydrogen jet flames in heated coflow air (from Ref. [35])

to occur and most heat is released right upstream of the flame base (not shown). It is worth mentioning that compared with many other diagnostics methods, CEMA has advantages in analyzing ignition problems where the CEM, or the interaction of CEM and mixing processes, is the primary driving force.

For the tribrachial edge flame regime ($U_0 = 30$ m/s), however, we can readily identify a significant contribution of hydrogen oxidation reactions to the CEM upstream of the flame base (see Fig. 5.11d–f). The main chain branching reaction of hydrogen chemistry via $H + O_2 \rightarrow O + OH$ (R1) competes with the recombination reaction via $H + O_2 + M \rightarrow HO_2 + M$ (R9) far upstream of the flame base. The chain branching reaction (R1) is found to be more important to the CEM than the recombination reaction (R9) near the flame base. The contribution of the chain branching reaction (R1) and the hydrogen heat release reaction of $H_2 + OH \rightarrow H + H_2O$ (R3) to the CEM become predominant right upstream of the flame base. The PI analysis identifies the occurrence of distinct methane oxidation/hydrogen oxidation processes upstream of the flame bases for the MILD combustion regime/the tribrachial edge flame regime. These distinct chemical features for different regimes verify that the unusual decreasing liftoff height behavior with increasing U_0 is primarily attributed to the hydrogen chemistry intensified by the fast diffusion of H_2 from the fuel jet to the flame base in the tribrachial edge flame regime. It is further substantiated by examining isocontours of heat release rate and EI values critical to the CEM (not shown here) that the oxidation of hydrogen/methane occurs more significantly upstream of the flame base in the tribrachial edge flame regime/the MILD combustion regime.

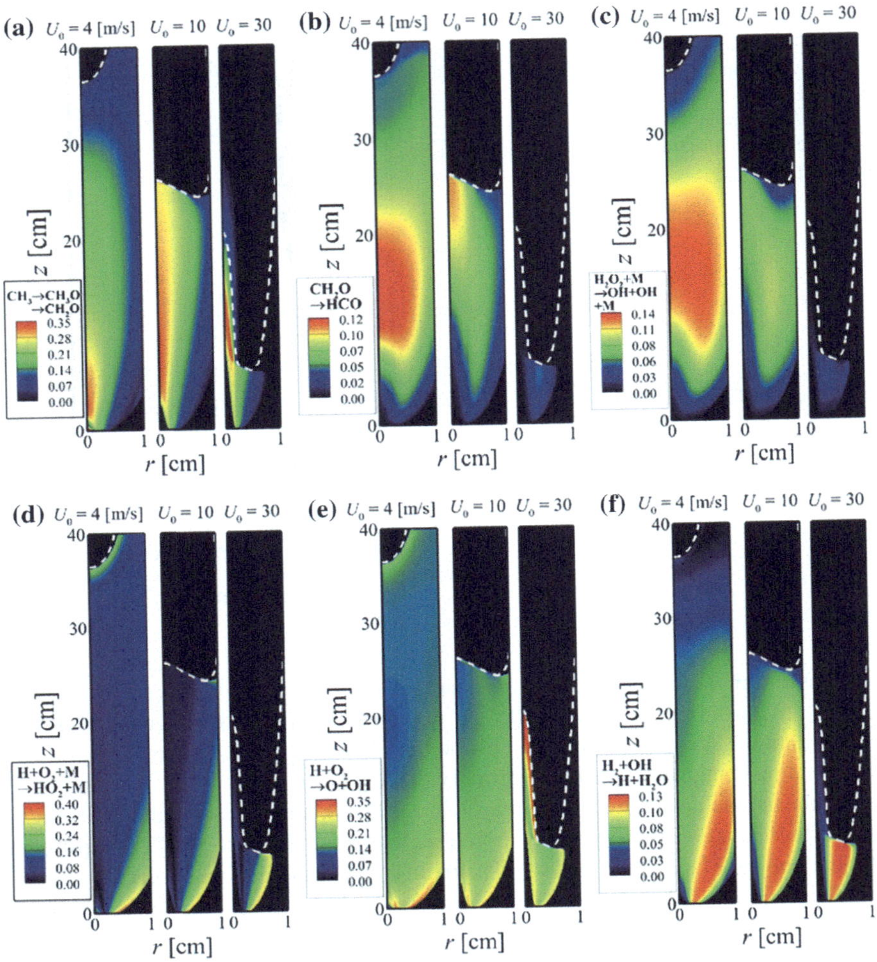

Fig. 5.11 Isocontours of PI of **a** reactions of CH_3 conversion to CH_2O and CH_3O, **b** CH_2O conversion to HCO, **c** $H_2O_2 + M \to OH + OH + M$ (R16), **d** $H + O_2 + M \to HO_2 + M$ (R9), **e** $H + O_2 \to O + OH$ (R1), and **f** $H_2 + OH \to H + H_2O$ (R3) for autoignited laminar lifted methane/hydrogen jet flames for cases with $U_0 = 4$, 10, and 30 m/s. The dashed line represents an isoline of $\mathrm{Re}(\lambda_e) = 0$ (from Ref. [35])

CEMA was also applied to 3-D DNS of a turbulent lifted ethylene jet flame in a highly heated coflow. The DNS was performed in a 3-D slot-burner configuration, where fuel issues from a central jet composed of 18% ethylene and 82% nitrogen by volume at an inlet temperature of 550 K. The central jet is surrounded on either side by coflowing heated air streams at 1550 K, which is high enough for the fuel jet to autoignite rapidly upstream of the flame base. Figure 5.12a shows a typical image of OH and HO_2 mass fraction of the turbulent lifted flame. To figure out the structure

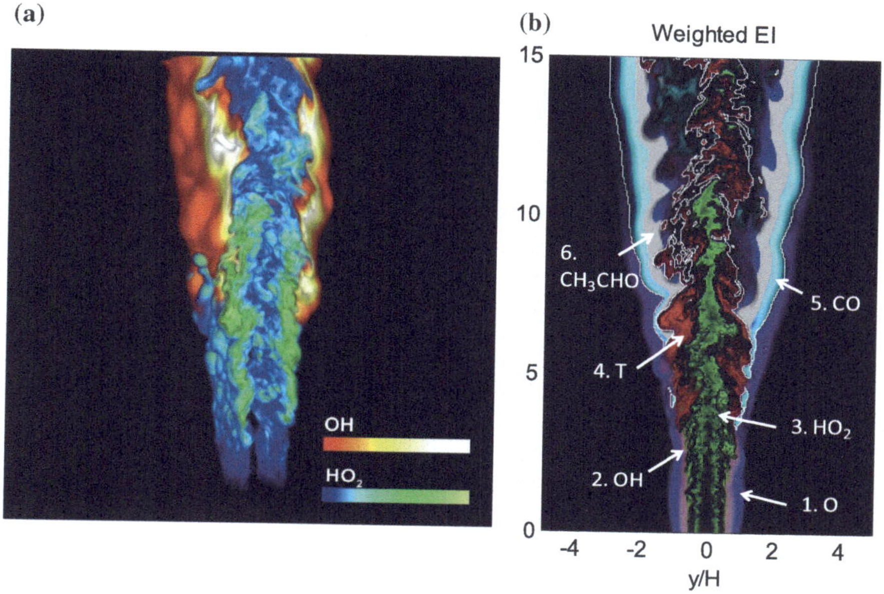

Fig. 5.12 **a** A typical image of mass fractions of OH and HO_2 (from Ref. [8]) and **b** controlling variables in various flame zones of the lifted ethylene jet flame in a heated air coflow, shown in EI-weighted color-mixing of temperature (red), HO_2 (green), O (blue), CH_3 (yellow), OH (magenta), CO (cyan), and CH_3CHO (gray). Zones with dominating chemistry are highlighted by Da on a logarithmic scale. The white isocontour shows the partially premixed flame fronts, where $\mathrm{Re}(\lambda_e) = 0$ (from Ref. [4])

and chemical features of the turbulent lifted jet flame, the CEM eigen zero-crossing was exploited to unambiguously identify the location of the premixed reaction fronts (white isolines in Fig. 5.12b) in this partially premixed flame. The critical variables to the CEM are further identified by the EI analysis in different regions of the lifted jet flame and the result is superposed to the premixed reaction fronts in Fig. 5.12b. The critical reactions to the CEM are also obtained from the PI analysis (not shown here).

It is readily observed that O and OH are critical to the CEM within the autoigniting layers that stabilize the lifted flame (Points 1 and 2), where the conversion of CH_3 to CH_3O/CH_2O (e.g., $CH_3 + O_2 \rightarrow CH_3O + O$ and $CH_3 + O2 \rightarrow CH_2O + OH$) is identified as the most critical reaction, which eventually leads to the formation of CO. Near Point 4, temperature is identified to control the ignition of slightly rich mixtures, for which several reactions such as fuel decomposition, radical branching, and CO formation are found to be critical to the CEM. Near the center of the jet core (Point 3), HO_2 is found to be the most critical species due to low temperature and fuel-rich condition. However, the ignition would not occur in the jet center because chemical reaction in the jet core is significantly slower than that near Points 1 and 2, and hence, is not critical to the flame stabilization. Instead, the rich mixture from the jet center is consumed further downstream when it is mixed with burned mixtures from the flame zone. Downstream of the lean flame fronts, a layer controlled by CO (cyan-colored region) can be readily identified, which corresponds to a post-reaction zone in lean premixed flames for hydrocarbon, where the conversion of CO to CO_2 takes time. The critical reaction to the CEM at Point 5 is the main heat release reaction of hydrocarbon oxidation via $CO + OH \rightarrow CO_2 + H$. Further beyond the CO conversion layer at Point 6, a CH_3OCHO becomes a critical species because of the increasing equivalence ratio. The corresponding controlling reaction then becomes: $CH_3 + HCO + M \rightarrow CH_3CHO + M$. It is thereby demonstrated that rich chemical information can be obtained by CEMA in analyzing flames with complex structures.

5.5 Concluding Remarks

The unique characteristics of CEM as shown above for many combustion problems can probably be best explained by using a simplified one-step reaction model, R→P, where R is the reactant and P is the product. Assuming that density and heat capacity, etc. are constant and the reaction rate is

$$\omega = \frac{dc}{dt} = (1-c)exp\left(-\frac{\alpha}{c}\right), \quad \alpha = \frac{T_a}{T_{ad} - T_0}, \tag{5.11}$$

where c is the progress variable, and T_a, T_{ad} and T_0 are the activation temperature, adiabatic flame temperature, and fresh reactant temperature, respectively. The Jacobian of the single equation thereby degenerates to a scalar, which is also its sole eigenvalue:

$$\lambda = \frac{d\omega}{dc} = exp\left(-\frac{\alpha}{c}\right)\left[\frac{\alpha(1-c)}{c^2} - 1\right]. \tag{5.12}$$

It is seen that for any $\alpha > 1$, which is typical for combustion problems involving large activation energies, $\lambda > 0$ when $c \rightarrow 0$ (fresh reactant), and $\lambda < 0$ when $c \rightarrow 1$ (hot product). For autoignition, the inflection point on the temperature profile can be readily shown to occur at $\lambda = 0$, and for partially premixed cases, the reaction rate peaks at the $\lambda = 0$ in the direction constrained by a constant mixture fraction. In such case, the CEM zero-crossing indicates where reaction becomes fastest, as well as the transition between explosive to nonexplosive mixtures. Depending on the direction of the transition, the zero-crossing defines either an ignition (autoignition or forced ignition) or an extinction state, while the zero-crossing point in 1-D premixed flames as shown in Fig. 5.2 can be explained as a forced-ignition point in the Lagrangian coordinate.

It is seen that CEMA is based on a simple but rather robust property of typical combustible mixtures, that is fast reactions (and many critical flame features) are strongly correlated to the CEM eigen zero-crossing, which is largely independent of flame types and flow conditions. Such a robust property renders CEMA suitable for systematic computational diagnostics of large DNS data involving complex flame configurations. The application of CEMA can be extended by considering the interactions between the CEM and mixing processes, as recently performed in Ref. [6] for identification of different local combustion modes. CEMA-based flame zone segmentation has also been employed to choose appropriate local turbulent combustion models for a lifted diesel jet spray flame [49], and such dynamic modeling efforts for partially premixed flames can be readily extended to other types of flames.

References

1. J.H. Chen, A. Choudhary, B. De Supinski, M. DeVries, E.R. Hawkes, S. Klasky, W.K. Liao, K.L. Ma, J. Mellor-Crummey, N. Podhorszki et al., Comput. Sci. Discov. **2**(1), 015001 (2009)
2. T. Lu, C.K. Law, Prog. Energy Combust. Sci. **35**(2), 192 (2009)
3. T. Lu, C. Yoo, J. Chen, C.K. Law, J. Fluid Mech. **652**, 45 (2010)
4. Z. Luo, C.S. Yoo, E.S. Richardson, J.H. Chen, C.K. Law, T. Lu, Combust. Flame **159**(1), 265 (2012)
5. R. Shan, C.S. Yoo, J.H. Chen, T. Lu, Combust. Flame **159**(10), 3119 (2012)
6. C. Xu, J.W. Park, C.S. Yoo, J.H. Chen, T. Lu, Proc. Combust. Inst. **37**(2), 2407 (2019)
7. M.B. Luong, T. Lu, S.H. Chung, C.S. Yoo, Combust. Flame **161**(11), 2878 (2014)
8. C.S. Yoo, E.S. Richardson, R. Sankaran, J.H. Chen, Proc. Combust. Inst. **33**(1), 1619 (2011)
9. C.S. Yoo, Z. Luo, T. Lu, H. Kim, J.H. Chen, Proc. Combust. Inst. **34**(2), 2985 (2013)
10. C.K. Law, *Combustion Physics* (Cambridge University Press, Cambridge, 2010)
11. J.E. Dec, Proc. Combust. Inst. **32**(2), 2727 (2009)
12. M. Yao, Z. Zheng, H. Liu, Prog. Energy Combust. Sci. **35**(5), 398 (2009)
13. J.E. Dec, *Encyclopedia of Automotive Engineering* (2014), pp. 1–40
14. M.B. Luong, Z. Luo, T. Lu, S.H. Chung, C.S. Yoo, Combust. Flame **160**(10), 2038 (2013)
15. C.S. Yoo, T. Lu, J.H. Chen, C.K. Law, Combust. Flame **158**(9), 1727 (2011)
16. S.O. Kim, M.B. Luong, J.H. Chen, C.S. Yoo, Combust. Flame **162**(3), 717 (2015)
17. M.B. Luong, G.H. Yu, T. Lu, S.H. Chung, C.S. Yoo, Combust. Flame **162**(12), 4566 (2015)
18. M.B. Luong, G.H. Yu, S.H. Chung, C.S. Yoo, Proc. Combust. Inst. **36**(3), 3587 (2017)
19. M.B. Luong, G.H. Yu, S.H. Chung, C.S. Yoo, Proc. Combust. Inst. **36**(3), 3623 (2017)
20. G.H. Yu, M.B. Luong, S.H. Chung, C.S. Yoo, Combust. Flame **208**, 299 (2019)
21. A. Bhagatwala, J.H. Chen, T. Lu, Combust. Flame **161**(7), 1826 (2014)
22. A. Bhagatwala, R. Sankaran, S. Kokjohn, J.H. Chen, Combust. Flame **162**(9), 3412 (2015)
23. C.A. Kennedy, M.H. Carpenter, R.M. Lewis, Appl. Numer. Math. **35**(3), 177 (2000)
24. C.A. Kennedy, M.H. Carpenter, Appl. Numer. Math. **14**(4), 397 (1994)
25. R.J. Kee, F.M. Rupley, E. Meeks, J.A. Miller, Chemkin-iii: a fortran chemical kinetics package for the analysis of gas-phase chemical and plasma kinetics. Technical report (Sandia National Labs., Livermore, 1996)
26. R.J. Kee, G. Dixon-Lewis, J. Warnatz, M.E. Coltrin, J.A. Miller, Sandia National Laboratories Report SAND86-8246 **13**, 80401 (1986)
27. T. Lu, C.K. Law, Combust. Flame **148**(3), 117 (2007)
28. T. Lu, C.K. Law, Combust. Flame **154**(1–2), 153 (2008)
29. T. Lu, C.K. Law, C.S. Yoo, J.H. Chen, Combust. Flame **156**(8), 1542 (2009)
30. K.M. Lyons, Prog. Energy Combust. Sci. **33**(2), 211 (2007)
31. C.S. Yoo, R. Sankaran, J. Chen, J. Fluid Mech. **640**, 453 (2009)
32. S. Chung, Proc. Combust. Inst. **31**(1), 877 (2007)
33. S. Deng, P. Zhao, M.E. Mueller, C.K. Law, Combust. Flame **162**(9), 3437 (2015)
34. S. Deng, P. Zhao, M.E. Mueller, C.K. Law, Combust. Flame **162**(12), 4471 (2015)
35. K.S. Jung, S.O. Kim, T. Lu, S.H. Chung, B.J. Lee, C.S. Yoo, Combust. Flame **198**, 305 (2018)

36. R. Grout, A. Gruber, H. Kolla, P.T. Bremer, J. Bennett, A. Gyulassy, J. Chen, J. Fluid Mech. **706**, 351 (2012)
37. H. Kolla, R.W. Grout, A. Gruber, J.H. Chen, Combust. Flame **159**(8), 2755 (2012)
38. Y. Minamoto, H. Kolla, R.W. Grout, A. Gruber, J.H. Chen, Combust. Flame **162**(10), 3569 (2015)
39. S. Lyra, B. Wilde, H. Kolla, J.M. Seitzman, T.C. Lieuwen, J.H. Chen, Combust. Flame **162**(4), 1234 (2015)
40. K. Aditya, A. Gruber, C. Xu, T. Lu, A. Krisman, M.R. Bothien, J.H. Chen, Proc. Combust. Inst. **37**(2), 2635 (2019)
41. C. Fureby, K. Nordin-Bates, K. Petterson, A. Bresson, V. Sabelnikov, Proc. Combust. Inst. **35**(2), 2127 (2015)
42. I.A. Dodoulas, S. Navarro-Martinez, Combust. Theor. Model. **19**(1), 107 (2015)
43. K. Nordin-Bates, C. Fureby, S. Karl, K. Hannemann, Proc. Combust. Inst. **36**(2), 2893 (2017)
44. L. Cifuentes, E. Fooladgar, C. Duwig, Fuel **232**, 712 (2018)
45. H. Wei, W. Zhao, Z. Lu, L. Zhou, Fuel **241**, 786 (2019)
46. W. Wu, Y. Piao, Q. Xie, Z. Ren, AIAA J. **57**(4), 1355 (2019)
47. J. An, Y. Jiang, M. Ye, R. Qiu, Int. J. Hydrog. Energy **38**(18), 7528 (2013)
48. L. Wang, Y. Jiang, R. Qiu, Energy Fuels **31**(9), 9939 (2017)
49. C. Xu, M.M. Ameen, S. Som, J.H. Chen, Z. Ren, T. Lu, Combust. Flame **195**, 30 (2018)

Chapter 6
Higher Order Tensors for DNS Data Analysis and Compression

Hemanth Kolla, Konduri Aditya, and Jacqueline H. Chen

Abstract We propose the use of higher order tensors, and their decompositions, for efficient analysis of combustion direct numerical simulation (DNS) data. Turbulent combustion DNS data, being inherently multiscale and multivariate, pose many challenges and higher order tensors are a natural abstraction to organise, probe and analyse them. The chapter gives a high-level overview of prominent tensor decomposition methods, their interpretation, algorithmic challenges and desirable properties. Two examples of DNS analysis employing tensor decompositions are then presented. The first analysis, based on truncated higher order singular value decomposition (truncated HOSVD), also known as Tucker decomposition, allows significant, albeit lossy, compression of DNS data, which may be inevitable in the exascale computing era. The factors aiding, and impeding, compression and the implications in terms of element-wise error distributions are presented using three candidate DNS data sets. The second analysis is centred on higher order joint moment tensors, which are richly informative for multivariate non-Gaussian variables. An anomaly detection algorithm based on the decomposition of the fourth moment tensor is presented, and its ability in detecting localised auto-ignition kernels in a homogeneous charge compression ignition (HCCI) data set is examined.

H. Kolla (✉) · J. H. Chen
Sandia National Laboratories, Livermore, CA, USA
e-mail: hnkolla@sandia.gov

J. H. Chen
e-mail: jhchen@sandia.gov

K. Aditya
Department of Computational and Data Sciences, Indian Institute of Science, Bangalore, India
e-mail: konduriadi@iisc.ac.in

H. Pitsch and A. Attili (eds.), *Data Analysis for Direct Numerical Simulations of Turbulent Combustion*, https://doi.org/10.1007/978-3-030-44718-2_6

6.1 Introduction

With ever-increasing computational power and resources the scale of high fidelity combustion direct numerical simulations (DNS) is ever growing. An obvious consequence of this trend is the volumes of data generated by simulations at a rate that threatens to outpace our ability to store, archive, manage and analyse the data. While the scale of combustion DNS has grown dramatically, even state-of-the-art simulations are able to approach Reynolds numbers that are orders of magnitude smaller than encountered in practical devices. Accordingly, combustion DNS, and associated data generation, is certain to keep growing for the foreseeable future. Novel concepts and approaches that allow efficient management and insightful analysis of DNS data are an urgent need of the hour.

Tensors, a generalisation of *multidimensional array*, are a powerful abstraction that provide a natural way of describing multiscale multivariate scientific data. In this chapter we present recent attempts at applying tensors, and tensor decompositions, for efficient analysis of combustion DNS. We present two examples of tensors-based analyses, lossy data compression and in situ anomaly detection, where the power of tensor decompositions has been brought to bear on turbulent combustion data. It is worth noting that these analyses are just as applicable to experimentally generated data as well.

6.2 Tensors and Their Decompositions

A tensor can be defined as a *multidimensional array*; a vector being a first order tensor and a matrix a second order tensor. The 'order' of a tensor is the number of indices required to uniquely reference every element, each index representing a specific 'mode'. Accordingly, a matrix has *row* and *column* modes. It is natural to think of a scalar field in spatially two-dimensional (2D) and three-dimensional (3D) DNS data sets as second and third order tensors, respectively, with the spatial coordinates as the modes. But the concept is more general, and powerful. If the data set comprises temporal snapshots, time can be considered as another independent mode. Combustion DNS are often performed using finite rate chemical kinetics, with tens or even hundreds of chemical species describing the thermo-chemical state, which makes the number of variables describing the state at each grid point rather large. These variables can be considered as an additional mode in a tensor representation of the data set. Potentially, a spatially 3D DNS data set could be viewed as a fifth order tensor, the five modes being the spatial directions, time, and a 'variables' mode. It is to be noted that this view of DNS data applies to structured meshes; for unstructured mesh data treating the spatial directions as independent modes is not straightforward.

Many of the concepts pertaining to matrix linear algebra can be extended to tensors, the so-called *multilinear algebra*. Most importantly, tensors can be factorised and decomposed in a variety of ways which offer a rich set of techniques to handle

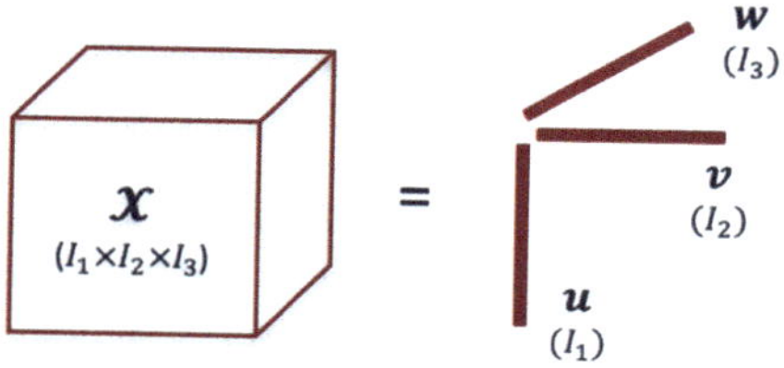

Fig. 6.1 An example of a rank-one tensor of order three

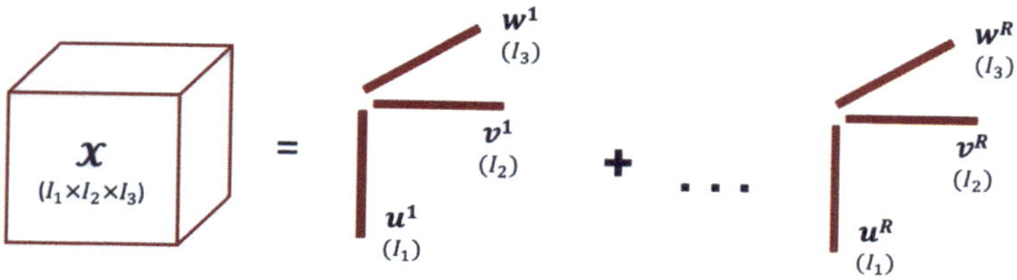

Fig. 6.2 Tensor rank defined in terms of sum of rank-one tensors

large multidimensional data sets. Readers are urged to read the excellent review by Kolda and Bader [1], whose notation and terminology we adopt in this chapter. While tensor factorisations may be thought of as higher order analogues of matrix factorisations, some of the key concepts do not generalise well. For example, the concept of rank of a tensor is not a straightforward extension of the matrix rank, and has many peculiarities. A tensor rank has been defined by various authors in terms of *rank-one tensors* (c.f. Sect. 3 in [1], Sect. 2.2 in [2] and references therein). A rank-one tensor is a tensor that can be simply expressed as an outer product (tensor product) of vectors. For example, a tensor $\boldsymbol{\mathcal{X}} \in \mathbb{R}^{I_1 \times I_2 \times I_3}$, can be expressed using vectors $\mathbf{u} \in \mathbb{R}^{I_1}$, $\mathbf{v} \in \mathbb{R}^{I_2}$, $\mathbf{w} \in \mathbb{R}^{I_3}$ as

$$\boldsymbol{\mathcal{X}} = \mathbf{u} \odot \mathbf{v} \odot \mathbf{w}. \tag{6.1}$$

In index notation,

$$x_{ijk} = u_i v_j w_k, \tag{6.2}$$

for indices $i = 1, \ldots, I_1$, $j = 1, \ldots, I_2$, $k = 1, \ldots, I_3$. This is significant because a rank-one tensor can be represented (stored) using far fewer numbers, $(I_1 + I_2 + I_3)$, than an otherwise full tensor, $(I_1 \times I_2 \times I_3)$. Figure 6.1 illustrates the rank-one tensor, $\boldsymbol{\mathcal{X}}$, of order three.

A tensor rank, accordingly, is defined as the number of rank-one tensors that the original tensor can be expressed in a summation (Fig. 6.2). For the previous example, tensor $\boldsymbol{\mathcal{X}}$ is of rank R if there exist, for $r = 1, \ldots, R$, vectors $\mathbf{u}^r \in \mathbb{R}^{I_1}$, $\mathbf{v}^r \in \mathbb{R}^{I_2}$, $\mathbf{w}^r \in \mathbb{R}^{I_3}$, such that

$$\mathcal{X} = \sum_{r=1}^{R} \mathbf{u}_r \odot \mathbf{v}_r \odot \mathbf{w}_r. \tag{6.3}$$

Kolda and Bader [1] discuss various peculiarities of the tensor rank concept. In general, finding the rank is an NP-hard problem and no straightforward algorithm to determine a tensor rank exists. Other peculiarities include *maximum* and *typical* ranks, which could be different, and the fact that the rank could be different over the real and complex spaces. De Lathauwer et al. [2] introduce the concept of *n-rank*, defined as the rank of the matrix resulting from 'mode-n unfolding' of the tensor. They show that the different *n-ranks* of a tensor may not be equal and further the *n-ranks* may not be equal to the tensor rank. In the rest of this section we will present a brief summary of the two most widely used tensor decompositions. The following sections will present applications specific to combustion DNS.

6.2.1 Canonical Decomposition

Historically, the most widely studied tensor decomposition is the *canonical decomposition* (CANDECOMP), which has also been referred to as 'Polyadic form of a tensor', 'Parallel Factors' (PARAFAC) and 'Topographic components'. Kolda and Bader [1] unify the various names and suggest the use of the term 'CP' decomposition. The canonical decomposition is closely tied to the definition of the tensor rank. Mathematically, the decomposition is nothing but expressing a tensor, approximately, as a sum of rank-one tensors. For an order three tensor, similar to Eq. 6.3,

$$\mathcal{X} \approx \sum_{r=1}^{R} \mathbf{u}_r \odot \mathbf{v}_r \odot \mathbf{w}_r. \tag{6.4}$$

The decomposition would be exact if the exact rank of the tensor were known, which, as stated previously, is not straightforward. Accordingly, the decomposition seeks the best approximation for a specified rank. The vectors could be chosen to have a unit norm, in which case the decomposition, illustrated in Fig. 6.3 could include weights for each rank-one component,

$$\mathcal{X} \approx \sum_{r=1}^{R} \lambda_r \, \mathbf{u}_r \odot \mathbf{v}_r \odot \mathbf{w}_r, \tag{6.5}$$

which, in index notation can be written as

$$x_{ijk} \approx \sum_{r=1}^{R} \lambda_r \, u_{ir} v_{jr} w_{kr}. \tag{6.6}$$

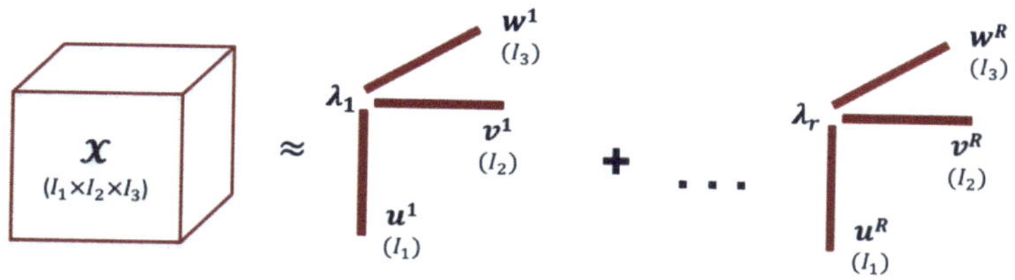

Fig. 6.3 The canonical, or CP, decomposition for an order three tensor

In practice, performing an exact CP decomposition poses many difficulties [1]. As mentioned earlier, the rank of a tensor is not straightforward to determine; hence, knowing the precise number of components required for an exact decomposition is not usually possible. Moreover, tensors can exhibit degeneracy, i.e. for a chosen rank they may have factorisations of lower rank that are equally approximate in terms of the goodness of fit. If the rank is specified somehow, the algorithm most commonly used to perform the decomposition is the *alternating least squares* (ALS) algorithm [3, 4]. Finding the best decomposition is posed as a problem of minimising an objective function which is a norm of the difference between the original tensor and its approximation

$$\text{for } \widetilde{\mathcal{X}} \equiv \sum_{r=1}^{R} \lambda_r \, \mathbf{u}_r \odot \mathbf{v}_r \odot \mathbf{w}_r, \quad f(\widetilde{\mathcal{X}}) = ||\mathcal{X} - \widetilde{\mathcal{X}}||, \quad \underset{\widetilde{\mathcal{X}}}{\text{minimise}} \; f(\widetilde{\mathcal{X}}). \tag{6.7}$$

The tensor (Frobenius) norm, $|| \cdot ||$, is defined as the square root of the scalar (inner) product of a tensor with itself and is equal to the square root of sum of squares of all elements. As per the algorithm, vectors along all modes but one (say $\mathbf{v}_r$ and $\mathbf{w}_r$ in the above example) are fixed, to solve for one set of vectors ($\mathbf{u}_r$), which becomes a linear least squares problem. The modes are then cycled through fixing all but one set of mode vectors and solving for the unfixed mode, and the process repeated iteratively until some termination criterion is satisfied which could be a convergence of the objective function, negligible change in the vectors $\mathbf{u}_r/\mathbf{v}_r/\mathbf{w}_r$, or exceeding a set number of iterations. As noted by Kolda and Bader [1], the ALS algorithm, while robust in general, is susceptible to noisy data and may not converge to a stationary point. It is also not guaranteed to converge to the global minimum, but, depending on the initial guess, only to a solution that satisfies the objective function, which may be a local minimum. Regularisation is sometimes applied to the objective function to aid convergence. Special cases of the CP decomposition have been studied by other authors, such as a partially symmetric decomposition [3] (subset of $\mathbf{u}_r/\mathbf{v}_r/\mathbf{w}_r$ are identical), fully symmetric decomposition of a symmetric tensor [5–7] (i.e. $\mathbf{u}_r = \mathbf{v}_r = \mathbf{w}_r$) and decomposition of sparse tensors [8].

6.2.2 Higher Order Singular Value Decomposition (HOSVD)

As the name suggests, Higher Order Singular Value Decomposition (HOSVD) is the higher order extension to matrix singular value decomposition (SVD). De Lathauwer et al. [2] establish HOSVD as a generalisation of the SVD to arbitrary order tensors and examine various similarities and points of departure between properties of matrix SVD and HOSVD. This decomposition was originally proposed by Tucker [9] for order three tensors and has been known as Tucker decomposition in the psychometrics community. To present the decomposition, it is necessary to introduce the concept of an *n-mode* product of a tensor by a matrix and the notation for a mode-n tensor unfolding (matricising). For an order N tensor, $\mathcal{A} \in \mathbb{R}^{I_1 \times I_2 \times \cdots \times I_N}$, a mode-$n$ unfolding yields a matrix, denoted by $\mathbf{A}_{(n)} \in \mathbb{R}^{I_n \times (I_{n+1} I_{n+2} \cdots I_N I_1 I_2 \cdots I_{n-1})}$. The *n-mode* product of tensor, $\mathcal{A}$, by a matrix, $\mathbf{U} \in \mathbb{R}^{J \times I_n}$, denoted by $\mathcal{A} \times_n \mathbf{U}$, is tantamount to premultiplying the mode-n unfolded matrix $\mathbf{A}_{(n)}$ by $\mathbf{U}$, and refolding the resulting matrix back into an order N tensor. The result is a tensor of dimensions $I_1 \times I_2 \times \cdots \times I_{n-1} \times J \times I_{n+1} \times \cdots \times I_N$. For the definition in index notation, as well as visual illustration of this product, see Sect. 2.4 of De Lathauwer et al. [2] and Sect. 2.5 of Kolda and Bader [1].

For simplicity, we present the concept for an order three tensor, although the definition generalises for any arbitrary order tensor. Formally, the HOSVD of the order three tensor $\mathcal{X} \in \mathbb{R}^{I_1 \times I_2 \times I_3}$, is a decomposition of form (Fig. 6.4)

$$\mathcal{X} = \mathcal{G} \times_1 \mathbf{U} \times_2 \mathbf{V} \times_3 \mathbf{W}, \tag{6.8}$$

where $\mathcal{G} \in \mathbb{R}^{I_1 \times I_2 \times I_3}$ is the *core tensor* and $\mathbf{U} \in \mathbb{R}^{I_1 \times I_1}$, $\mathbf{V} \in \mathbb{R}^{I_2 \times I_2}$ and $\mathbf{W} \in \mathbb{R}^{I_3 \times I_3}$ are orthonormal *factor matrices*. Although presented for the real space, the analogous properties hold if $\mathcal{X}, \mathcal{G}, \mathbf{U}, \mathbf{V}, \mathbf{W}$ are complex tensors and matrices. The key properties of the HOSVD are:

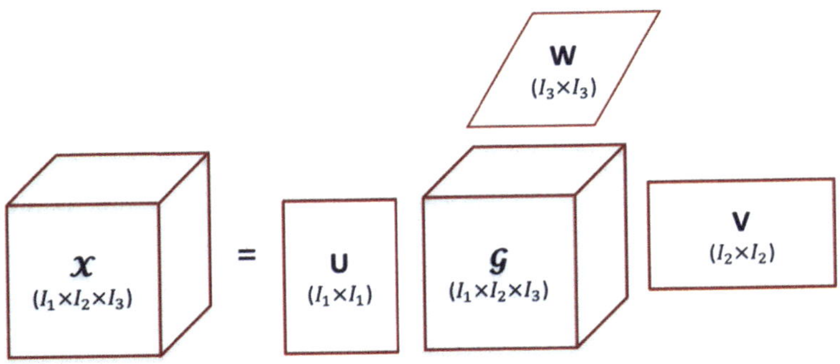

Fig. 6.4 The full-rank Higher Order Singular Value Decomposition for an order three tensor

- The column vectors of $\mathbf{U}$ are the 1-mode singular vectors of $\boldsymbol{\mathcal{X}}$, i.e. the left singular vectors of $\mathbf{X}_{(1)}$, the matrix resulting from mode-1 unfolding of $\boldsymbol{\mathcal{X}}$. Similarly, $\mathbf{V}$ and $\mathbf{W}$ are 2-mode and 3-mode singular vectors of $\boldsymbol{\mathcal{X}}$, left singular vectors of $\mathbf{X}_{(2)}$, $\mathbf{X}_{(3)}$ respectively.
- In contrast with matrix SVD, the core tensor is dense in general and not pseudo-diagonal.
- The core tensor $\boldsymbol{\mathcal{G}}$ has a property of *all-orthogonality*, i.e. any two subtensors, obtained by fixing a specific mode to fixed values, say a, b, have an inner product equal to 0 if $a \neq b$, see Sect. 3 of [2].
- The core tensor $\boldsymbol{\mathcal{G}}$ is related to the original tensor by the relation

$$\boldsymbol{\mathcal{G}} = \boldsymbol{\mathcal{X}} \times_1 \mathbf{U}^T \times_2 \mathbf{V}^T \times_3 \mathbf{W}^T. \tag{6.9}$$

De Lathauwer et al. [2] examine other properties of HOSVD in detail, in particular the special cases of symmetry and pairwise symmetry, connections between eigenvalue decomposition (EVD) and higher order EVD, obtaining 'maximal diagonality' of the core tensor, etc.

Note that the definition in Eq. 6.8 corresponds to a full-rank decomposition. Due to the property of the factor matrices being made up of mode-wise singular vectors, computing the HOSVD boils down to computing the SVD repeatedly by unfolding the tensor along each mode, which yields the factor matrices $\mathbf{U}$, $\mathbf{V}$ and $\mathbf{W}$, and computing the core tensor via Eq. 6.9. Recall, also, that the concept of *n-rank* suggests that the matrix resulting from mode-n unfolding of $\boldsymbol{\mathcal{X}}$ may not be full rank. Let $J_1 \leq I_1$, $J_2 \leq I_2$, $J_3 \leq I_3$ be the *1-rank*, *2-rank* and *3-rank* of $\boldsymbol{\mathcal{X}}$, respectively. The HOSVD then becomes the same as in Eq. 6.8, but with $\boldsymbol{\mathcal{G}} \in \mathbb{R}^{J_1 \times J_2 \times J_3}$, $\mathbf{U} \in \mathbb{R}^{I_1 \times J_1}$, $\mathbf{V} \in \mathbb{R}^{I_2 \times J_2}$ and $\mathbf{W} \in \mathbb{R}^{I_3 \times J_3}$. Alternatively, extending the concept of truncated SVD, a truncated HOSVD can be performed by setting J_1, J_2 and J_3 according to a suitable criterion, resulting in an approximate decomposition

$$\widetilde{\boldsymbol{\mathcal{X}}} \approx \boldsymbol{\mathcal{G}} \times_1 \mathbf{U} \times_2 \mathbf{V} \times_3 \mathbf{W}. \tag{6.10}$$

The criterion is, typically, ensuring that the norm of the difference between $\boldsymbol{\mathcal{X}}$ and $\widetilde{\boldsymbol{\mathcal{X}}}$ is below a specified threshold. If the various mode-n unfoldings of $\boldsymbol{\mathcal{X}}$ have significant low-rank structure, then J_1, J_2 and J_3 of the approximate decomposition can be much smaller than I_1, I_2 and I_3 respectively. This presents an opportunity for significant compression of the original tensor, since the storage cost of the approximate HOSVD, to leading order, scales with the storage of the core tensor, $(J_1 \times J_2 \times J_3)$, which can be significantly smaller than the storage cost of the original tensor, $(I_1 \times I_2 \times I_3)$.

6.3 Applications of Tensor Decompositions to DNS Data Analysis

Linear algebra and matrix decompositions have long been applied to analyse DNS data from various perspectives, e.g. proper orthogonal decomposition [10], principal component analysis (PCA) for dimensionality reduction [11–13], dynamic mode decomposition (DMD) [14, 15]. We propose that, owing to the inherently multi-scale multidimensional nature of combustion DNS, tensor decomposition offers a variety of analyses. We focus on two applications here: DNS data compression and anomalous event detection.

6.3.1 Compression by Truncated HOSVD

Kolla and Chen [16] highlighted the trend that, roughly over the past twenty years, the computational problem sizes tackled by representative combustion DNS have grown exponentially with time, keeping pace with the exponential growth in computing resources. Yet, for leading DNS simulations today, the turbulent Reynolds numbers affordable remain at least an order of magnitude lower than encountered in practical devices, indicating that the simulations are likely to get larger in the exascale computing era. An unfortunate impediment to this growth is the cost of long-term data storage, which is likely to severely constrain either the problem sizes or the amount of information from simulations that may be saved for a posteriori analysis. It appears that efficient ways of compressing DNS data is the need of the hour. Lossless compression, while ideal may not offer significant reduction, and lossy compression might have to be employed.

If the data has low-rank structure along each mode, a truncated HOSVD (Tucker decomposition) yields a core tensor that is orders of magnitude smaller than the original tensor, making this an attractive approach for data compression. From Eq. 6.10, the storage cost of the decomposed model is largely that of the core tensor, $\boldsymbol{\mathcal{G}}$, (the storage for the factor matrices will be comparatively very small) and the compression ratio is $\approx (I_1 \times I_2 \times I_3)/(J_1 \times J_2 \times J_3)$. Conceptually, the decomposition exploits low-rank structure along each mode, and the more the number of modes (higher the order of tensor) greater the potential for compression, as the overall reduction is multiplicative. Austin et al. [17] have demonstrated compression by multiple orders of magnitude for combustion DNS. They also address parallelisation of the Tucker decomposition in a distributed setting, and identify key kernels and optimisations for an efficient parallel implementation. The compression is posed as a decomposition to a user-specified error threshold, ϵ, which guarantees that the compressed model satisfies

$$||\boldsymbol{\mathcal{X}} - \widetilde{\boldsymbol{\mathcal{X}}}|| \equiv ||\boldsymbol{\mathcal{X}} - (\boldsymbol{\mathcal{G}} \times_1 \mathbf{U} \times_2 \mathbf{V} \times_3 \mathbf{W})|| \leq \epsilon ||\boldsymbol{\mathcal{X}}||. \tag{6.11}$$

Two algorithms were considered for parallel implementation (c.f. Sect. of [17]):

1. *Sequentially Truncated HOSVD (ST-HOSVD)*: The algorithm consists of a sequence, looping over all modes and determining, for each mode, the *n-rank*, $J_n \leq I_n$, by performing an SVD of $\mathbf{X}_{(n)}$. The rank, J_n, is chosen such that the singular values $\sigma^i_{(n)}$ of $\mathbf{X}_{(n)}$ satisfy the condition

$$\sum_{i=J_n+1}^{I_n} (\sigma^i_{(n)})^2 \leq \epsilon^2 \; ||\boldsymbol{\mathcal{X}}||^2 \; / \; N. \tag{6.12}$$

This criterion ensures that, summed over all modes, the resulting approximation satisfies [18] the inequality $||\boldsymbol{\mathcal{X}} - \widetilde{\boldsymbol{\mathcal{X}}}|| \leq \epsilon||\boldsymbol{\mathcal{X}}||$, which is the desired accuracy as per Eq. 6.11. The factor matrix, $\mathbf{U}^{(n)}$, is set to be the leading J_n singular vectors. *Sequentially truncated* refers to the specification [18] that, when going from one mode (n) to the next in the sequence, the input tensor be truncated to dimension J_n from I_n by performing an *n-mode* product of the current input tensor with $\mathbf{U}^{(n)^T}$. In this manner, the size of the tensor being operated upon is successively reduced making the computation progressively less expensive.
2. *Higher order Orthogonal Iteration (HOOI)*: ST-HOSVD produces a decomposition that, while satisfying the accuracy constraint Eq. 6.11, is not guaranteed to be the most optimal. Furthermore, the result depends on the order in which the modes are processed. The HOOI, starting from an initial decomposition provided by ST-HOSVD, improves upon the model accuracy iteratively by looping over all modes and recomputing the factor matrices. For mode n in the loop, a tensor $\boldsymbol{\mathcal{Y}} = \boldsymbol{\mathcal{X}} \times_m \{\mathbf{U}^{(m)^T}\}_{m \neq n}$ is formed by multiplying the input tensor with factor matrices along all modes except n. The leading J_n singular vectors of $\mathbf{Y}_{(n)}$ are recomputed and set as $\mathbf{U}^{(n)}$. This procedure monotonically improves the accuracy and is continued until no further accuracy improvement is observed.

For tensors that are distributed across multiple nodes parallelisation of the three main kernels: mode-wise unfolding (matricisation), tensor-matrix product and the SVD of the unfolded matrix, is discussed in [17]. In lieu of SVD, which can be for a potentially large matrix $\mathbf{X}_{(n)}$, Austin et al. [17] suggest forming the Gram matrix $\mathbf{X}_{(n)}\mathbf{X}^T_{(n)}$ and computing its eigen decomposition instead, since the size of the Gram matrix, $I_n \times I_n$, is almost always much smaller than the size of $\mathbf{X}_{(n)}$ for DNS data sets. The resulting implementation is available as an open software library TuckerMPI [19], which was employed for compressing combustion DNS data sets presented here.

We consider data sets generated by S3D [20], a massively parallel DNS code that solves gas-phase reacting flow Navier–Stokes equations in the compressible form using explicit high order accurate numerical schemes; eighth order central difference scheme for spatial derivatives and fourth order six-stage low-storage Runge–Kutta scheme for temporal derivatives. S3D employs finite rate chemical kinetics, mixture-averaged transport properties and Navier–Stokes characteristic boundary conditions (NSCBCs) to simulate laboratory-scale turbulent flames in rectangular Cartesian domains. From a compression standpoint DNS data sets can have attributes that are dependent on the problem configuration; statistically stationary (statistical properties are independent of origin in time) or homogenous (statistical properties are independent of origin in space) or both. A stationary configuration is amenable for compression along the time mode. Similarly, spatial modes (directions) that do not contain significant multiscale structure allow significant compression. The field variables comprise the three components of fluid velocity, temperature, pressure and a list of chemical species concentrations that represent the thermo-chemical state. In the limit, the number of chemical species could be as small as two, which is an extremely reduced representation of the chemical kinetics, or as large as a hundred which warrants a reduction in dimensionality.

For the present study we consider three data sets that represent the extremes of combustion DNS:

- *HCCI* The homogeneous charge compression ignition (HCCI) case [21] represents ignition processes in an ethanol–air mixture under conditions similar to piston compression in an internal combustion (IC) engine with exhaust gas recirculation. It is a spatially two-dimensional (2D) temporally evolving (non-stationary) configuration and the ethanol combustion is represented by 28 chemical species; temperature contour plots at four instants are shown in Fig. 6.5. The size of the full tensor is $672 \times 672 \times 32 \times 626$.

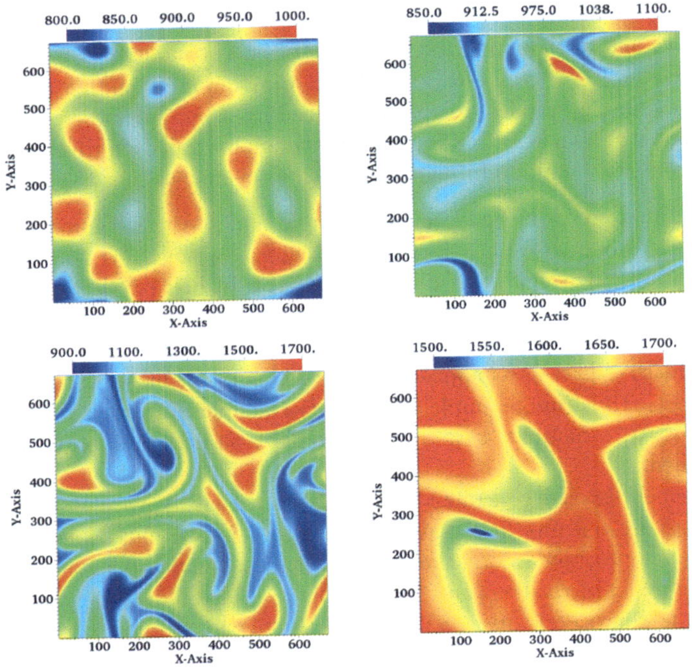

Fig. 6.5 Contour plots of temperature (Kelvin) in the HCCI case at four time instants: initial (top left), 1×10^{-3} s (top right), 2×10^{-3} s (bottom left) and 3×10^{-3} s (bottom right). The colour ranges for each plot are shown above

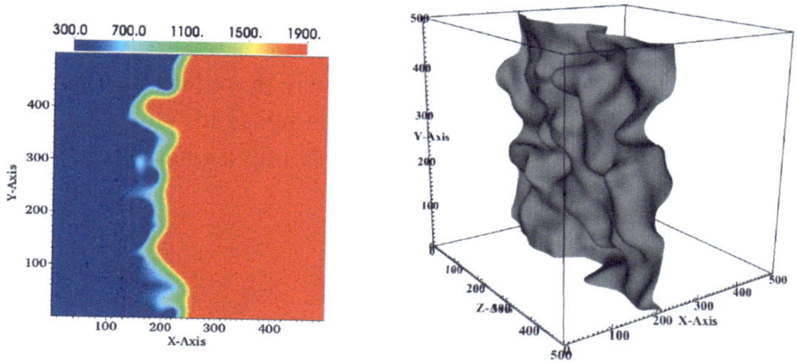

Fig. 6.6 (left) Contour plots of temperature (Kelvin) in the SP case on midplane slice along the Z direction at a selected time instant. The plot on right shows a 3D rendering of the isocontour corresponding to 1200 K

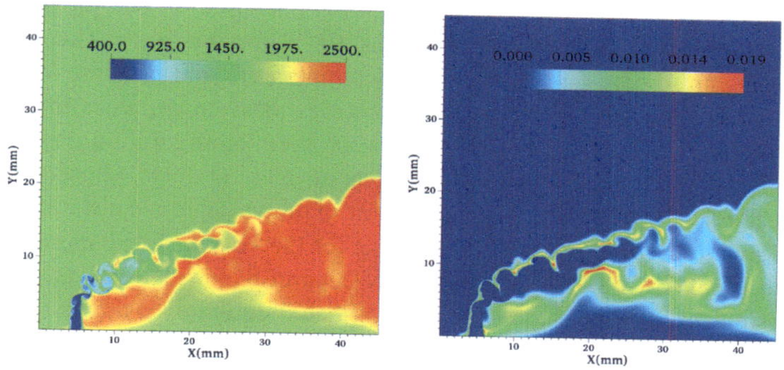

Fig. 6.7 Contour plots of temperature (left) and OH mass fraction (right) in the JICF case on midplane slice along the Z direction at a selected time instant

- *SP* The statistically planar (SP) case [22] pertains to a three-dimensional (3D) statistically stationary planar turbulent premixed methane–air flame represented by a reduced combustion mechanism involving only 6 chemical species. The size of the tensor is $500 \times 500 \times 500 \times 11 \times 150$ and contour plots at a selected instant are shown in Fig. 6.6.
- *JICF* The jet-in-cross-flow (JICF) corresponds [23] to auto-ignition of a hydrogen fuel jet in vitiated cross-flow of methane–air combustion products, representing conditions in lean premixed stationary gas-turbine combustors. While over 400 time snapshots of this case were available, only 10 were chosen for the compression study. The size of this data is $1500 \times 2080 \times 1500 \times 18 \times 10$ (Fig. 6.7).

In all cases the last mode is the number of discrete time snapshots, the penultimate is the number of variables and the preceding modes are the spatial dimensions (number of grid points). Intuitively, we anticipate larger compression for the SP and JICF cases, along time due to stationarity, and along the spatial modes due to portions of the domain being devoid of any structure. On the other hand, the HCCI case is temporally evolving, has only two spatial dimensions and contains scalar field structure spanning the entire spatio-temporal domain. Conversely, the field variable dimension in the SP case is smaller (11) compared to that of the HCCI case (33). The JICF case represents a state-of-the-art simulation possible with current computing resources and considers an experimental burner configuration that is representative of gas-turbine combustor.

6.3.1.1 Data Preprocessing

S3D solves the governing PDEs in a non-dimensionalised form by normalising all variables by suitable reference quantities. Even so, in multivariate data sets, the raw solution data (normalised values of field variables) span orders of magnitude. The chemical species can be broadly classified as major (e.g. CH_4, O_2, CO_2, H_2O) and minor (e.g. OH, CO, HO_2 etc.). The maximum concentration is typically of order unity for the former but can be as small as 10^{-10} for the latter. The other solution variables are the velocity components, temperature and pressure which, after being normalised by suitable reference quantities, have magnitudes that typically vary between 10^3 and 10^{-3} depending on the problem configuration. When considered together each data set invariably consists of field variables whose magnitudes span over ten decades, if not more, as evident in the variable-wise statistics (mean, standard deviation, minimum and maximum) for the HCCI and SP cases presented in Tables 6.1 and 6.2. When performing a SVD, careful consideration is required for preprocessing the data so that all fields are treated equitably. Parente and Sutherland [24] consider different ways of preprocessing combustion data in the context of

Table 6.1 Variable-wise statistics of the SP case. Variables #1 - #6 are the species mass fractions, #7 temperature, #8 pressure, #9 - #11 are the three velocity components. Temperature, pressure and velocities are normalised internally in S3D by suitable reference quantities

Var. id	Mean	Standard deviation	Min	Max
1	1.541e-02	1.830e-02	6.365e-15	3.920e-02
2	1.290e-01	7.300e-02	6.753e-02	2.239e-01
3	6.380e-05	7.402e-05	1.036e-15	4.606e-04
4	6.518e-02	5.015e-02	1.036e-15	1.074e-01
5	5.344e-02	4.112e-02	1.036e-15	8.806e-02
6	7.369e-01	3.523e-06	7.369e-01	7.369e-01
7	1.033e+01	5.970e+00	2.499e+00	1.537e+01
8	7.144e-01	5.077e-05	7.135e-01	7.152e-01
9	3.118e-03	2.737e-03	−1.818e-02	1.997e-02
10	−9.323e-05	1.845e-03	−1.859e-02	2.025e-02
11	4.871e-06	1.774e-03	−1.650e-02	1.569e-02

Table 6.2 Variable-wise statistics of the HCCI case

Var. id	Mean	Standard deviation	Min	Max
1	2.754e-09	6.198e-09	2.925e-20	4.591e-08
2	4.165e-06	9.194e-06	1.422e-18	5.253e-05
3	3.748e-07	6.366e-07	1.590e-20	4.171e-06
4	1.183e-01	2.624e-02	7.549e-02	1.403e-01
5	1.797e-05	2.787e-05	1.609e-16	1.033e-04
6	3.987e-02	1.553e-02	2.633e-02	6.493e-02
7	3.537e-05	5.881e-05	7.250e-12	3.235e-04
8	1.758e-04	4.129e-04	7.099e-14	2.619e-03
9	3.355e-03	7.309e-03	4.312e-19	2.785e-02
10	2.525e-01	2.368e-02	2.346e-01	2.931e-01
11	3.225e-04	6.662e-04	6.585e-17	3.223e-03
12	3.413e-08	1.156e-07	4.785e-17	1.505e-06
13	7.269e-06	2.125e-05	4.669e-22	1.730e-04
14	5.399e-07	1.189e-06	4.427e-19	9.109e-06
15	5.326e-06	1.039e-05	6.501e-17	5.498e-05
16	1.086e-04	2.662e-04	6.832e-17	1.302e-03
17	1.669e-06	3.926e-06	6.707e-17	1.989e-05
18	2.146e-08	6.933e-08	6.181e-17	5.451e-07
19	2.691e-04	6.414e-04	6.629e-17	3.272e-03
20	1.187e-08	3.283e-08	6.324e-17	2.071e-07
21	4.280e-04	9.111e-04	7.608e-17	4.257e-03
22	1.615e-02	1.354e-02	7.644e-17	2.882e-02
23	3.478e-06	7.707e-06	6.090e-17	5.354e-05
24	1.963e-14	3.750e-14	8.122e-22	2.624e-13
25	1.767e-08	3.276e-08	4.042e-17	1.764e-07
26	7.548e-10	1.275e-09	3.928e-17	1.534e-08
27	4.814e-07	7.853e-07	3.617e-17	3.251e-06
28	5.685e-01	2.291e-04	5.660e-01	5.698e-01
29	1.020e+01	2.692e+00	6.689e+00	1.476e+01
30	4.830e+01	1.144e+01	3.241e+01	6.669e+01
31	3.971e-06	1.632e-03	−4.634e-03	4.835e-03
32	1.650e-07	1.704e-03	−5.400e-03	4.831e-03

Principal Component Analysis (PCA) and emphasise the importance of 'centering and scaling'. The focus of that work was to identify a low-dimensional manifold underlying the state space comprised of species concentrations in a manner that preserves the data spread as much as possible. Centering usually involves subtracting a mean and scaling requires dividing the samples by a metric that quantifies the data spread. Accordingly, preprocessing techniques were assessed in terms of the implications for the covariance matrix and outliers in the data. If each variable is normally distributed then centering by the mean and scaling by the standard deviation is the most natural choice. Combustion variables, however, are far from being normally distributed and literature suggests that bimodal [25] or beta [26] PDFs better represent the distributions of these variables.

The preprocessing choices for combustion data were guided chiefly by the following, not unrelated, factors:

- The ranges of variables differ by orders of magnitude.
- Some variables are positive (e.g. species mass fractions, temperature, pressure) while others are not (e.g. velocity).
- Along each variable slice SVD provides a tighter fit for the 'significant' elements relative to the 'insignificant' ones.

The first factor implies that for all variables to be treated equitably a scaling is essential. Scaling, when performed in conjunction with centering, can have undesirable consequences. For variables that are positive, centering maps elements in the vicinity of the centering quantity (e.g. mean) to zero, thereby diminishing their magnitude, while smaller elements get mapped to negative values. In the extreme the smallest raw elements get mapped to potentially large negative values amplifying their magnitude. For instance, in the case of variable 5 of the HCCI case (see Table 6.2), centering by the mean and scaling by the standard deviation maps the smallest elements in the raw data to around 0.64 and the largest elements to around 3.1, both of order unity. It thus seemed appropriate to only scale the variables but not centre them.

For a scaling-only preprocessing the only choice is that of the normalising quantity. The most natural choice is to scale by the maximum of the absolute values of all elements. This maps all variables to an equitable range of [0, 1] or [−1, 1] and retains the significance/insignificance of the original elements relative to each other (both within one variable and across variables). In data sets that contain a few extreme outliers normalising by the maximum would not be appropriate and a better choice would be to normalise by a suitably large percentile value. However, this was not a concern since the combustion data, which are from extremely well-resolved simulations, do not contain isolated outliers.

6.3.1.2 Compression Results and Error Analysis

The three data sets were compressed using the parallel library TuckerMPI [19]. The library provides an efficient MPI-based parallel implementation of the ST-HOSVD and the HOOI algorithms for performing the Tucker decomposition of a distributed

Table 6.3 Summary of compressions of the three datasets

Dataset	Input tensor size	Compression error (ϵ)	Compressed tensor size	Compression ratio
HCCI	$672 \times 672 \times 32 \times 626$	10^{-4}	$330 \times 310 \times 31 \times 199$	1e+01
		10^{-2}	$111 \times 105 \times 22 \times 46$	7e+02
SP	$500 \times 500 \times 500 \times 11 \times 400$	10^{-4}	$95 \times 129 \times 125 \times 7 \times 125$	4e+02
		10^{-2}	$30 \times 38 \times 35 \times 6 \times 11$	2e+05
JICF	$1500 \times 2080 \times 1500 \times 18 \times 10$	10^{-4}	$424 \times 387 \times 261 \times 18 \times 10$	1e+02
		10^{-2}	$90 \times 61 \times 48 \times 13 \times 6$	4e+04

data set. It requires the user to specify the dimensions of the input tensor, the relative error threshold ϵ, and a 'processor grid' that prescribes the number of processes (MPI ranks) along each mode that establish the domain decomposition of the overall tensor. The library provides an interface to the key kernels that could be invoked *in situ* by a running solver, or read and compress data from files written to persistent storage in a post-processing mode using efficient MPI-I/O. A more detailed discussion of the communication, computation and storage costs of TuckerMPI is presented in [27].

The compressions were performed on Rhea at the Oakridge Leadership Computing Facility (OLCF), a cluster comprising Intel Xeon compute nodes with 16 cores and 128 GB of memory per node. The HCCI case was compressed using 144 cores (processor grid $12 \times 12 \times 1 \times 1$), the SP case using 4000 cores (processor grid $10 \times 10 \times 10 \times 1 \times 4$) and the JICF case using 5600 cores (processor grid $10 \times 8 \times 7 \times 1 \times 10$). Two error thresholds, 10^{-4} and 10^{-2}, were considered to explore the accuracy-compression tradeoff for each of the cases. The compression results are summarised in Table 6.3. It is evident that, for the same specified compression error, the three cases compress by vastly different amounts. As anticipated, the HCCI case compresses the least, at most by two orders of magnitude for the higher error threshold. On the other hand, the SP case and the JICF cases compress to much smaller sizes, by five orders of magnitude for the SP case with an error threshold of 10^{-2}. This can be attributed to two factors. Both cases are statistically stationary and contain large portions of the spatial domain where no structure in any of the scalar fields is present in the data, and it is possible that the HOSVD is able to represent these portions with very few modes. This is apparent from the sizes of the compressed tensor listed in Table 6.3, which show that the first three modes, which are the spatial modes for the SP and JICF cases, account for most of the compression.

It is worth noting that Tucker compression is inherently lossy and it is important to examine the implications of the loss of information. Recall that the error threshold, the only parameter that specifies the accuracy of the compressed format, is the global norm of the tensor error relative to the norm of the original tensor (Eq. 6.11). By construction, this parameter provides no guarantees on errors of individual tensor elements. Figures 6.8 and 6.9 compare contour plots of temperature and OH mass

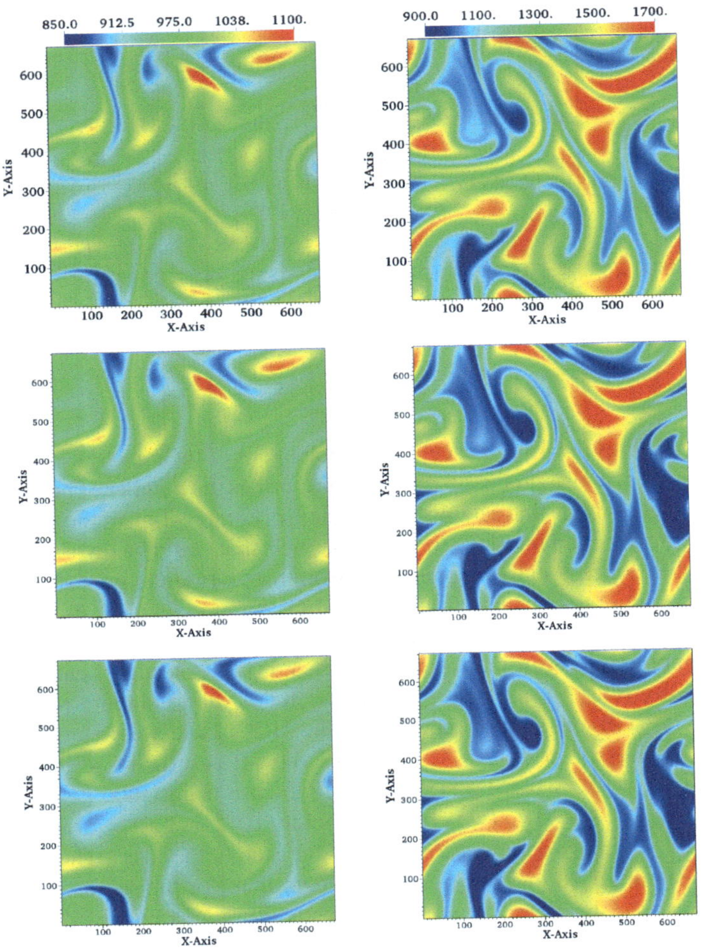

Fig. 6.8 Contour plots of temperature (Kelvin) in the HCCI case from the original data (top), from data compressed to $\epsilon = 10^{-4}$ (middle), and data compressed to $\epsilon = 10^{-2}$ (bottom). The plots are shown at two instants: 1×10^{-3}s (left) and 2×10^{-3}s (right)

fraction, respectively, from the original as well as the compressed (and reconstructed) HCCI data sets for two selected time instants. The statistics in Table 6.2 indicate that all temperature values (variable #29), upon normalising by the global maximum are in the range $O(10^{-1}) - O(10^{0})$, while normalised OH mass fractions (variable #5) are in the range $O(10^{-12}) - O(1)$. Visually, the temperature contour plots from the compressed data look nearly identical to the original data for both error thresholds. On the other hand, the OH contour plots show visible differences between the original and compressed data, particularly for the earlier snapshot at the higher error threshold. The same trend is apparent from contour plots of temperature and OH mass fraction

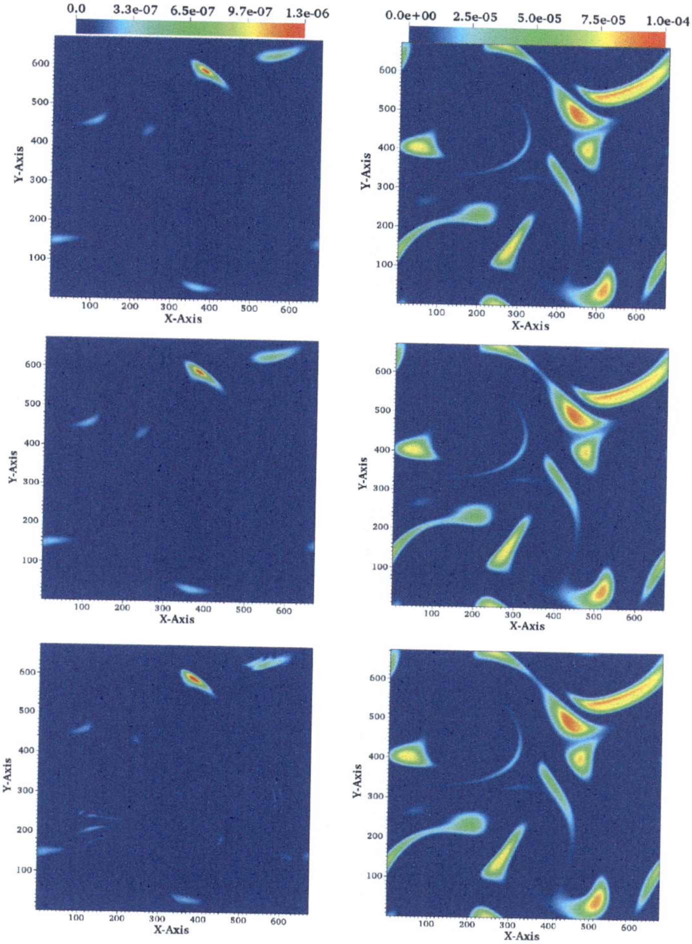

Fig. 6.9 Contour plots of OH mass fraction in the HCCI case from the original data (top), from data compressed to $\epsilon = 10^{-4}$ (middle), and data compressed to $\epsilon = 10^{-2}$ (bottom). The plots are shown at two instants: 1×10^{-3} s (left) and 2×10^{-3} s (right)

on the midplane slice along the Z direction for the JICF case, shown in Fig. 6.10. Noticeable differences are evident for the OH contour plots, especially for the higher error threshold. These comparisons indicate that tensor elements that are larger in magnitude are, potentially, better preserved (have lower relative error) than elements with lower magnitude. Even with judicious normalisation of the data this seems unavoidable and is an inherent limitation of the HOSVD based compression.

An analysis of the distribution of absolute error across the elements of the compressed tensor clarifies this effect. Figure 6.11 shows, for OH mass fraction and temperature fields of the compressed HCCI data, stacked histograms of logarithm

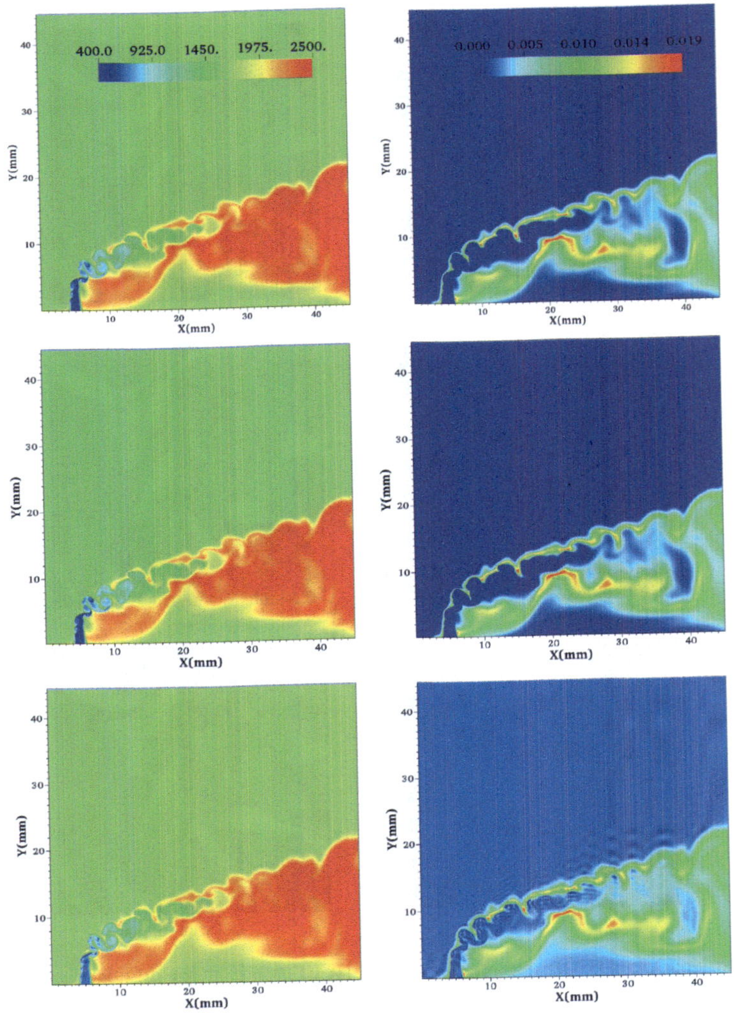

Fig. 6.10 Contour plots of temperature (left) and OH mass fraction (right) in the JICF case on mid-plane slice along the Z direction at a selected time instant. The original data (top) are compared with compressed data corresponding to the error thresholds $\epsilon = 10^{-4}$ (middle) and $\epsilon = 10^{-2}$ (bottom)

of absolute error of elements, further separated by bins representing the original element magnitudes. For the higher error threshold, $\epsilon = 10^{-2}$, the histograms are shifted to the right compared to those for $\epsilon = 10^{-4}$, indicating a larger absolute error, as expected. The histograms also show that for temperature, all elements are within the largest magnitude bin, $10^{-2} - 10^{0}$, while for OH mass fraction a considerable number of elements are in the $10^{-10} - 10^{-8}$ bin. For elements with such small magnitudes, even an absolute error as small as 10^{-5} translates to a large relative error

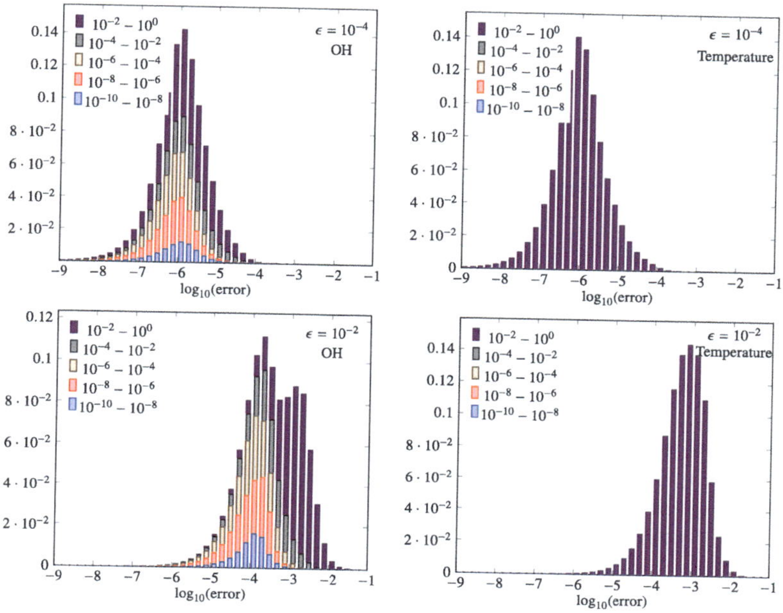

Fig. 6.11 Stacked histograms of logarithm of absolute error of individual tensor elements from the compressed HCCI data are shown for two variables: OH mass fraction (left), and temperature (right). The top row plots are for error threshold $\epsilon = 10^{-4}$, and the bottom row for $\epsilon = 10^{-2}$. The bars are further coloured by bins representing various ranges of the original element magnitudes (after normalising)

$\sim 10^5$. Since truncated SVD cannot provide control of reconstruction accuracy for individual elements, this is a peril of Tucker compression, particularly for data sets with a large spread of element magnitudes. Another consideration, particularly for combustion DNS data, is that truncated HOSVD is not guaranteed to preserve fundamental realisability constraints such as ensuring the mass fractions stay positive and are bounded by unity, or even that the mass fractions at a fixed grid point sum to unity.

6.3.2 Anomaly Detection Based on Higher Moment Tensors

Another application of tensors uniquely suited for combustion data involves higher order joint moments. It is well known that quantities central to the modelling and description of turbulent combustion have probability density functions (PDFs) that are not Gaussian, e.g. mixture fraction for non-premixed flames [28] and progress variable [29] for premixed flames are characterised by beta PDFs while scalar dissipation rate by a log-normal PDF. While many analyses involving joint moments focus on covariance, for multivariate non-Gaussian random variables there is important statistical information in the higher order joint moments, which are symmetric higher order tensors (e.g. co-skewness is an order three tensor, co-kurtosis is an order four tensor). Recently, we proposed a method for detecting anomalous events in distributed scientific data by decomposing and examining the fourth order co-kurtosis tensor [30]. The principles underpinning the anomaly detection method are as follows:

- Scientific simulation data may contain anomalous, but physically valid, events which manifest in the state of a collection of points being different from a 'normal'. Examples could be ignition kernels in combustion data, hurricanes in weather simulations, cracks in fracture mechanics.
- The signature of such anomalous events is discernible from higher joint moments. The fourth moment, kurtosis, is a reliable measure of existing outliers or propensity to produce outliers, according to Westfall [31].
- For multivariate data the joint moments, i.e. the co-kurtosis tensor, rather than the marginal moments, contain the signature of anomalous events.
- One way to extract this signature is to construct, by analogy to PCA and covariance, 'principal vectors of kurtosis' from the co-kurtosis tensor. A larger data set, with an anomalous subset of data points, can be hypothesised to have principal kurtosis vectors that are distinct.
- For data generated from parallel computing, such as combustion DNS, anomalous events can be assumed to be spatially and temporally local. Hence, an algorithm that constructs, and compares, principal kurtosis vectors among subdomains of the simulation can identify the anomalous event at its inception.

In the design of an anomaly detection algorithm based on these principles, a key step is the construction of the principal kurtosis vectors (PKVs), which we have defined as higher order analogues to PCA vectors. The PCA vectors are, by definition, eigenvectors of the covariance matrix and constitute an orthogonal basis. However, there is no straightforward higher order analogue of matrix eigendecomposition applicable to tensors. A few works have studied tensor eigenpairs [6, 32] but the algorithms pose problems with convergence, robustness to perturbations, and missed eigenpairs. Alternatively, application of a symmetric CP decomposition is not straightforward since the rank of the decomposition has to be determined independently, which is difficult to determine. Comon et al. [5] show that theoretically, at best, the rank of a symmetric CP decomposition lies within certain bounds, but is not unique. Moreover, while enforcing symmetry of the CP decomposition is possible, in general the decomposition does not enforce orthogonality and it is unclear how the algorithms would behave under imposed orthogonality constraints [33].

Considering the various challenges, we adopted a simpler approach, which has roots in Independent Component Analysis (ICA), also known as 'blind sources separation' problem. For a set of observed random variables, ICA identifies a linear transformation to a set of orthogonal basis vectors that are statistically independent, as opposed to linearly uncorrelated basis vectors identified by PCA. Conceptually, the observed variables can be thought of as resulting from a linear superposition (mixing) of independent non-Gaussian random variables, and the goal of ICA is the identification of the transformation (mixing) matrix and ultimately the underlying independent sources. Fourth order moment tensors, specifically fourth order cumulant tensors, are at the heart of the mathematical framework of ICA and identifying the transformation matrix is posed as a problem of diagonalising the fourth order cumulant tensor [34–36]. For a vector of N random variables, x_i, the fourth order cumulant tensor, in index notation, is

$$\begin{aligned}[\boldsymbol{C}]_{ijkl} &= \mathbb{E}[x_i x_j x_k x_l] - \mathbb{E}[x_i x_j]\,\mathbb{E}[x_k x_l] \\ &-\mathbb{E}[x_i x_k]\,\mathbb{E}[x_j x_l] - \mathbb{E}[x_i x_l]\,\mathbb{E}[x_j x_k], \quad 1 \le i, j, k, l \le N, \end{aligned} \tag{6.13}$$

where $\mathbb{E}$ is the expectation operator. Note that the last three terms on the r.h.s involve covariances. Assuming a statistical model, where the x_i are a result of linear mixing of statistically independent sources, s_j, with superimposed white noise, i.e.

$$x_i = \sum_{j=1}^{N} a_{ij} s_j + n_i, \tag{6.14}$$

ICA seeks a diagonalisation of the cumulant tensor in the form

$$[\boldsymbol{C}] = \sum_{j=1}^{N} \kappa_j \; a_j \odot a_j \odot a_j \odot a_j \tag{6.15}$$

where a_j are the row vectors of the mixing matrix and κ_j is the excess kurtosis of the jth source, s_j. While our interest is not in ICA, *per se*, our objective of performing an eigendecomposition of the symmetric fourth moment tensor is identical to the cumulant tensor diagonalisation problem of ICA, and the mixing matrix can be interpreted as comprising the PKVs (c.f. Sect. 2.4 of [30] for more details). While many iterative algorithms for ICA have been proposed in literature, a much simpler approach, suggested by De Lathauwer et al. [2] and Anandkumar et al. [37] identifies the property that since the tensor is symmetric it unfolds identically along all modes, and its HOSVD yields identical factor matrices (in the case of real tensors). Alternatively, the SVD of the unfolded tensor yields the desired principal vectors. This approach may be unstable if any of the singular values are close or are identical [37], but it has much smaller computational complexity compared to iterative algorithms which makes it attractive. Accordingly, our proposed approach for determining the PKVs involves the steps:

1. Construct the fourth order cumulant tensor, defined as per Eq. 6.13, from the data.
2. Unfold the tensor (along any one mode) and perform SVD. The resulting left singular vectors are the PKVs.

6.3.2.1 Anomaly Detection Algorithm for Distributed Computing

Since DNS are performed with parallel computing, employing domain decomposition, the objective is to devise an algorithm for distributed data to detect anomalous events in relevant subdomains of the simulation. The guiding principle, as stated earlier, is that a subdomain that contains the anomalous events has a PKVs that are 'distinct' (in space and/or time). The natural approach is to construct the set of PKVs on each subdomain, at each time step, and compare them with a nominal. To facilitate such a comparison quantitatively, we proposed 'moment metrics', defined for each original variable (feature), that encode the orientation of the PKVs (with respect to the original variable space). For a data set with N variables and as many PKVs, $\hat{v}_j$, $j = 1, \cdot\cdot, N$, the moment metric for the ith variable is defined as

$$F_i = \frac{\sum_{j=1}^{N} \kappa_j \, \hat{v}_j^i}{\sum_{j=1}^{N} \kappa_j}. \tag{6.16}$$

The ith element of the jth PKV, $\hat{v}_j^i$, is the dot product of $\hat{v}_j$ with the unit vector representing the direction of the ith variable in the N-dimensional multivariate space, and κ_j is the singular value corresponding to $\hat{v}_j$ from the SVD of the unfolded tensor $\boldsymbol{C}$, see Eq. 6.15. The moment metrics satisfy two properties: (1) they signify the contribution of each variable to the overall outlierness in the data and (2) when summed over all variables, the result is unity, i.e. $\sum_{i=1}^{N} F_i = 1$, which arises from the properties of SVD. Note that if the PKVs in Eq. 6.16 were replaced by the PCA vectors, the metric quantifies the contribution of each variable to the spread in the data, as opposed to an outlierness. Essentially, we have defined a rather small vector, of size N, to be a signature of the outlierness in the data corresponding to a subdomain. Comparing such small vectors across all subdomains is relatively inexpensive in message-passing (MPI) based parallel computing simulations. Furthermore, the property ensuring F_i sum to unity allows us to interpret them as a discrete probability distribution, and we can employ f-divergence metrics for comparison against a nominal. The outline of the proposed algorithm is [30]:

1. On each subdomain (a single MPI rank or a subgroup of spatially contiguous MPI ranks) construct the fourth order cumulant tensor, $\boldsymbol{C}$ (Eq. 6.13).
2. Compute the PKVs, $\hat{v}_j$, by unfolding $\boldsymbol{C}$ and performing an SVD.
3. Compute the vector of moment metrics, F_i (Eq. 6.16).
4. Compute the nominal moment metric vector, $\langle F_i \rangle$, which we define as the average of F_i over all subdomains (spatial average).
5. On every subdomain compute a divergence metric, the Hellinger distance, between its F_i and the nominal $\langle F_i \rangle$.
6. If a subdomain has a distance metric above a threshold, flag it as anomalous. The property that the Hellinger distance is bounded from above by unity guides the choice of a suitable threshold.

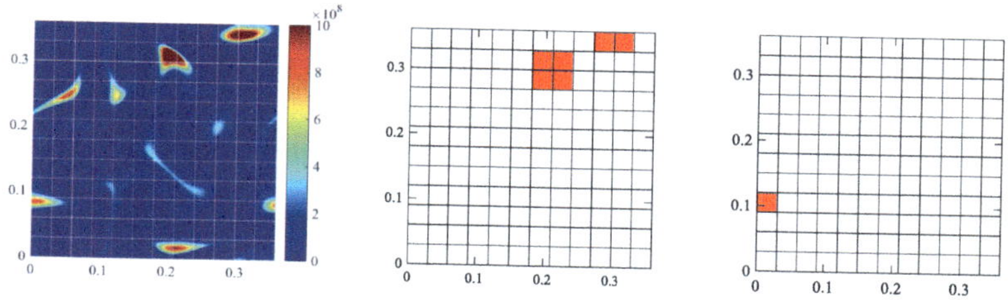

Fig. 6.12 Ignition event maps from the HCCI simulation at a time instant of 0.845 ms. The left panel shows heat release rate ($J/m^3/s$) across various subdomains, the middle panel shows the subdomains where the heat release rate crosses the threshold corresponding to auto-ignition, and the right panel shows subdomains flagged as anomalous by the detection algorithm

7. Computing the distance metric between F_i from one time step and a previous time step detects temporal anomalies.

Note that the most computationally intensive steps in the algorithm are steps (1) and (2). Indeed the size of the cumulant tensor is $O(N^4)$, which can grow quickly with the number of variables (N), even though, owing to symmetry, the number of unique elements of the tensor would be smaller. Efficient algorithms for computing the cumulant tensor, and accelerating them by exploiting symmetry and porting to heterogeneous architectures (accelerators, GPUs) is a topic of future study.

The anomaly detection algorithm was employed on the HCCI data set. The data set was partitioned into a set of 12×12 spatial subdomains, and the moment metrics were computed on each subdomain to flag the anomalous ones that contain an auto-ignition event. For the conditions of the simulation, the heat release rate crossing a value of 1×10^9 $J/m^3/s$ was an appropriate threshold signifying auto-ignition [21]. Figure 6.12 shows an early instance with ignition pockets appearing in the simulation. The subdomains where the auto-ignition is occurring (above the heat release rate threshold) are shown in the middle panel of Fig. 6.12 and those flagged by the anomaly detection algorithm are shown in the right panel. The detection is mostly accurate with only one subdomain flagged as a false positive, and two flagged as false negatives. While there are two parameters implicit in the comparison in Fig. 6.12 (the heat release rate threshold and the distance metric threshold), it was demonstrated in [30] that the algorithm is robust to the choice of these parameters and has a desirable behaviour in terms of flagging false/true positives and negatives.

6.4 Concluding Remarks

Considering the inherently multidimensional nature of combustion DNS data, multilinear algebra and tensor decompositions offer a natural framework for analysing and organising combustion data. Many aspects of matrix linear algebra do not generalise to higher order tensors; nonetheless, the two major classes of tensor decompositions, the CP decomposition and HOSVD, offer a rich set of applications for combustion data. We show that the Tucker decomposition (a form of truncated HOSVD) offers orders of magnitude, albeit lossy, compression. This can be extremely useful in sharing reasonably accurate representations of the original DNS data, since they fit on desktop size resources. Such reduced representations can offer a glimpse of the underlying physics, which can be the basis for selective *a posteriori* analyses of the original data, alleviating the need to share or transfer the complete data which can be impractical. Another application we demonstrate is the analysis of higher order moment tensors which can contain important statistical information beyond just the covariance, especially if the variables have distributions that are far from Gaussian. We suggest an approach of constructing the principal kurtosis vectors to identify anomalous events (e.g. ignition kernels) at their inception. Another interesting application of the higher order tensors is to consider a dimensionality reduction based on such principal vectors, as an alternative to PCA based dimensionality reduction, which may capture dynamics that manifest principally in the higher statistical moments. This is akin to considering "independent components" of the data as opposed to principal components, and this is a topic for future study.

Acknowledgements We thank Dr. Tamara Kolda from Sandia National Laboratories for extensive help and guidance in applying tensor decompositions to combustion data. We also thank Dr. Grey Ballard at Wake Forest University and Dr. Alicia Klinvex at Naval Nuclear Laboratory for allowing us the use of the TuckerMPI library. Technical inputs from Dr. Prashant Rai, Dr. Warren Davis, Mr. Timothy Shead, and Dr. Philip Kegelmeyer of Sandia National Laboratories, and Dr. Julia Ling of Citrine Informatics, are gratefully acknowledged. This work was funded through U.S. Department of Energy Advanced Scientific Computing Research (ASCR) grant FWP16-019471. Sandia National Laboratories is a multi-mission laboratory managed and operated by National Technology and Engineering Solutions of Sandia, LLC., a wholly owned subsidiary of Honeywell International, Inc., for the U.S. Department of Energy's National Nuclear Security Administration under contract DE-NA-0003525. The views expressed in the article do not necessarily represent the views of the U.S. Department of Energy or the United States Government.

References

1. T.G. Kolda, B.W. Bader, Tensor decompositions and applications. SIAM Rev. **51**(3), 455–500 (2000)
2. L.D. Lathauwer, B.D. Moor, J. Vandewalle, A multilinear singular value decomposition. SIAM J. Matrix Anal. Appl. **21**(4), 1253–1278 (2000)
3. J.D. Carroll, J.J. Chang, Analysis of individual differences in multidimensional scaling via an N-way generalization of "Eckart-Young" decomposition. Psychometrika **35**, 283–319 (1970)
4. R.A. Harshman, Foundations of the PARAFAC procedure: models and conditions for an "explanatory" multi-modal factor analysis. in *UCLA Working Papers in Phonetics*, vol. 16, pp. 1–84 (1970)
5. P. Comon, G. Golub, L.H. Lim, B. Mourrain, Symmetric tensors and symmetric tensor rank. SIAM J. Matrix Anal. Appl. **30**(3), 1254–1279 (2008)
6. E. Kofidis, P.A. Regalia, On the best rank-1 approximation of higher order supersymmetric tensors. SIAM J. Matrix Anal. Appl. **23**, 863–884 (2002)

7. T.G. Kolda, Numerical optimization for symmetric tensor decomposition. Math. Program. **151**(1), 225–248 (2015). https://doi.org/10.1007/s10107-015-0895-0
8. T.G. Kolda, B.W. Bader, J.P. Kenny, Higher-order web link analysis using multi-linear algebra, in *ICDM 2005: Proceedings of the 5th IEEE International Conference on Data Mining*, pp. 242–249 (2005)
9. L.R. Tucker, Some mathematical notes on three-mode factor analysis. Psychometrika **31**, 279–311 (1966)
10. P. Moin, R.D. Moser, Characteristic-eddy decomposition of turbulence in a channel. J. Fluid Mech. **200**, 471–509 (1989). https://doi.org/10.1017/S0022112089000741
11. C. Frouzakis, Y. Kevrekidis, J. Lee, K. Boulouchos, A. Alonso, Proper orthogonal decomposition of direct numerical simulation data: data reduction and observer construction. Proc. Combust. Inst. **28**(1), 75 – 81 (2000). https://doi.org/10.1016/S0082-0784(00)80197-6. http://www.sciencedirect.com/science/article/pii/S0082078400801976
12. O. Owoyele, T. Echekki, Toward computationally efficient combustion dns with complex fuels via principal component transport. Combust. Theory Model. **21**, 770–798 (2017)
13. J.C. Sutherland, A. Parente, Combustion modeling using principal component analysis. Proc. Combust. Inst. **32**(1), 1563 – 1570 (2009). https://doi.org/10.1016/j.proci.2008.06.147. http://www.sciencedirect.com/science/article/pii/S1540748908002630
14. T. Grenga, J.F. MacArt, M.E. Mueller, Dynamic mode decomposition of a direct numerical simulation of a turbulent premixed planar jet flame: convergence of the modes. Combust. Theory Model. **22**, 795–811 (2018)
15. P.J. Schmid, Dynamic mode decomposition of numerical and experimental data. J. Fluid Mech. **656**, 5–28 (2010). https://doi.org/10.1017/S0022112010001217
16. H. Kolla, J.H. Chen, Turbulent Combustion Simulations with High-Performance Computing (Springer, Singapore, 2018), pp. 73–97
17. W. Austin, G. Ballard, T.G. Kolda, Parallel tensor compression for large-scale scientific data, in *IPDPS'16: Proceedings of the 30th IEEE International Parallel and Distributed Processing Symposium* (2016), pp. 912–922 . https://doi.org/10.1109/IPDPS.2016.67
18. N. Vannieuwenhoven, R. Vandebril, K. Meerbergen, A new truncation strategy for the higher-order singular value decomposition. SIAM J. Sci. Comput. **34**(2), A1027–A1052 (2012). https://doi.org/10.1137/110836067
19. TuckerMPI. https://gitlab.com/tensors/TuckerMPI
20. J.H. Chen et al., Terascale direct numerical simulations of turbulent combustion using S3D. Comput. Sci. Discov. **2**(1), 015001 (2009). https://doi.org/10.1088/1749-4699/2/1/015001
21. A. Bhagatwala, J.H. Chen, T. Lu, Direct numerical simulations of HCCI/SACI with ethanol. Combust. Flame **161**(7), 1826–1841 (2014). https://doi.org/10.1016/j.combustflame.2013.12.027
22. H. Kolla, X.Y. Zhao, J.H. Chen, N. Swaminathan, Velocity and reactive scalar dissipation spectra in turbulent premixed flames. Combust. Sci. Technol. (2016). https://doi.org/10.1080/00102202.2016.1197211
23. S. Lyra, B. Wilde, H. Kolla, J.M. Seitzman, T. Lieuwen, J. Chen, Structure of hydrogen-rich transverse jets in a vitiated turbulent flow. Combust. Flame **162**, 1234–1248 (2015)
24. A. Parente, J.C. Sutherland, Principal component analysis of turbulent combustion data: data pre-processing and manifold sensitivity. Combust. Flame **160**, 340–350 (2013)
25. K.N.C. Bray, P.A. Libby, J.B. Moss, Unified modeling approach for premixed turbulent combustion - part I: general formulation. Combust. Flame **61**, 87–102 (1985)
26. N. Peters, Laminar diffusion flamelet models in non-premixed turbulent combustion. Prog. Energy Combust. Sci. **10**, 319–339 (1984)
27. G. Ballard, A. Klinvex, T.G. Kolda, TuckerMPI: a parallel C++/MPI software package for large-scale data compression via the Tucker tensor decomposition (2019)
28. N. Peters, Laminar diffusion flamelet models in non-premixed turbulent combustion. Prog. Energy Combust. Sci. **10**(3), 319 – 339 (1984). https://doi.org/10.1016/0360-1285(84)90114-X. http://www.sciencedirect.com/science/article/pii/036012858490114X

29. Y.C. Chen, R.W. Bilger, Experimental investigation of three-dimensional flame-front structure in premixed turbulent combustion–i: hydrocarbon/air bunsen flames. Combust. Flame **131**(4), 400–435 (2002). https://doi.org/10.1016/S0010-2180(02)00418-2. http://www.sciencedirect.com/science/article/pii/S0010218002004182
30. K. Aditya, H. Kolla, W.P. Kegelmeyer, T.M. Shead, J. Ling, W.L. Davis, Anomaly detection in scientific data using joint statistical moments. J. Comput. Phys. **387**, 522–538 (2019)
31. P.H. Westfall, Kurtosis as Peakedness, 1905–2014. R.I.P. Am. Stat. **68**(3), 191–195 (2014)
32. T. Kolda, J.R. Mayo, Shifted power method for computing tensor eigenpais. SIAM J. Matrix Anal. Appl. **32**(4), 1095–1124 (2011)
33. T. Kolda, Orthogonal tensor decompositions. SIAM J. Matrix Anal. Appl. **23**, 243–255 (2001)
34. J.F. Cardoso, Source separation using higher order moments, in *International Conference on Acoustics, Speech, and Signal Processing*, vol. 4 (1989)
35. P. Comon, Independent component analysis, a new concept? Signal Process. **36**(3), 287–314 (1994)
36. P. Comon, J.F. Cardoso, Eigenvalue decomposition of cumulant tensor with applications, in *SPIE Conference on Advanced Signal Processing Algorithms*, Architectures and Implementations (San Diego, United States, 1990), pp. 361–372
37. A. Anandkumar, R. Ge, D. Hsu, S.M. Kakade, M. Telgarsky, Tensor decompositions for learning latent variable models. J. Mach. Learn. Res. **15**(1), 2773–2832 (2014)

Chapter 7
Data-Driven Modal Decomposition Techniques for High-Dimensional Flow Fields

Nicholas Arnold-Medabalimi, Cheng Huang, and Karthik Duraisamy

Abstract Data-driven decomposition techniques are presented for the analysis and development of reduced-order models of complex flow dynamics. The Proper Orthogonal Decomposition (POD) produces optimal representations of the dynamics in the sense of the energy norm. Alternatively, Dynamic Mode Decomposition (DMD) efficiently extracts coherent dynamics based on eigendecompositions of linearized dynamics. An extension to the latter, the Higher Order Dynamic Mode Decomposition (HODMD) method uses time delays to develop efficient reduced models to represent complex dynamics in a nonintrusive manner. High-fidelity simulation results of a laboratory-scale single-element gas turbine combustor are used to demonstrate and evaluate the capabilities of the aforementioned decomposition techniques.

7.1 Introduction

In complex fluid systems such as aircraft engines, a wide variety of physical phenomena are present, including acoustics, turbulence, combustion, and interactions therein. The coupling of these physics is of primary concern in the study of thermoacoustic instabilities (or combustion instabilities) [1]. While these instabilities have been well-investigated, [2–5], it is generally challenging to quantitatively characterize the underlying triggering mechanisms. Conventional diagnostics of combustion instability has been mainly focused on signal processing of local pressure variations in a chosen region of interest, or statistical analysis of combustion signals. To improve

N. Arnold-Medabalimi · C. Huang · K. Duraisamy (✉)
Department of Aerospace Engineering, University of Michigan, 1320 Beal Ave, Ann Arbor, MI 48109, USA
e-mail: kdur@umich.edu

N. Arnold-Medabalimi
e-mail: narnoldm@umich.edu

C. Huang
e-mail: huangche@umich.edu

H. Pitsch and A. Attili (eds.), *Data Analysis for Direct Numerical Simulations of Turbulent Combustion*, https://doi.org/10.1007/978-3-030-44718-2_7

the level of insight required to diagnose instability mechanisms, there is a need to move past traditional analysis methods.

Data-driven decomposition methods are being used to obtain reduced dimensional descriptions and approximations of high-dimensional data [6–9], and are capable of providing direct physical insight into the system dynamics.

The most common decomposition technique is the Proper Orthogonal Decomposition (POD) [8–11], which organizes spatial modes in the order of energy content. This method has previously been applied to the study of turbulence [8, 9] and combustion problems [12].

The dynamic mode decomposition (DMD) [13] aims to extract the spectral properties of the linear approximation of the Koopman operator [14–16]. DMD is especially useful for the analysis of combustion dynamics, as the response corresponding to specific frequencies can be examined for the entire flow field. DMD has been applied to non-reacting [13, 17] as well as reacting flow cases [18].

Unlike DMD, POD analysis is not able to provide a systematically consistent interpretation of the underlying dynamics. To resolve this shortcoming, Spectral Proper Orthogonal Decomposition(SPOD) [19], has been formulated as a space-time version of the traditional space-only POD. SPOD has shown to combine the advantages of POD and DMD by generating DMD-like modes with the added benefit of having an intrinsic ranking.

Though DMD has been used in the analysis of reacting flow problems, it is fundamentally a linear approximation to nonlinear dynamics and can yield poor results when applied to ill-conditioned or incomplete data matrices. More recently, techniques such as the High-Order Dynamic Mode Decomposition (HODMD) [20–22], have been developed which have shown promise in overcoming this limitation as well as showing potential in developing efficient nonintrusive reduced models from high-dimensional datasets.

This chapter will discuss POD, DMD, and HODMD, including the algorithms and practical applications to a representative complex high-dimensional dataset.

7.2 Modal Decompositions

Decomposition techniques are designed to seek low-rank representations of complex dynamics, and attempt to provide valuable insight into the physical phenomena. Moreover, decomposition techniques typically represent the starting point toward developing efficient reduced-order models [23, 24] for complex systems.

In the context of this chapter, we focus on decomposition methods that can approximate a spatio-temporal function as a summation of k components.

$$A(x, t) \approx \sum_{k}^{N_p} y_k(t)\psi_k(x). \tag{7.1}$$

The kth spatial mode is $\psi_k(x)$ and the kth temporal coefficient is $y_k(t)$. The function of interest is expressed as

$$A(x,t) = A_0(x) + A'(x,t) \approx A_0(x) + \sum_k y_k(t)\psi_k(x), \tag{7.2}$$

with $A_0(x)$ representing the time average. The entire dataset is organized into a matrix $\boldsymbol{A}$

$$\boldsymbol{A} = \left[\mathbf{a_1}\ \mathbf{a_2}\ \ldots\ \mathbf{a_M}\right] \in \mathbb{R}^{N\times M}, \tag{7.3}$$

where a single temporal realization of the flow field is organized into a vector $\mathbf{a}_i$, where N is the number of observables, and M is the number of available snapshots. Since the ultimate goal is to aid in the extraction of meaningful physics, a discrete notation is useful

$$\boldsymbol{A} = \sum_i^r \mathbf{u_i}\mathbf{v_i}^T, \quad \mathbf{u_i} \in \mathbb{R}^N, \quad \mathbf{v_i} \in \mathbb{R}^M. \tag{7.4}$$

Here, the spatio-temporal data is decomposed into many spatial and temporal vectors whose dimension is limited by the rank r of the given dataset. Once decomposition is complete, it is possible to reconstruct the response of an isolated kth temporal and spatial mode set:

$$\boldsymbol{A}_k = \mathbf{u_k}\mathbf{v_k}^T. \tag{7.5}$$

7.2.1 Proper Orthogonal Decomposition

The proper orthogonal decomposition [9] (POD) is formulated such that the spatial functions $\psi_k(x)$ are orthonormal

$$\int_\Omega \psi_n(x)\psi_m(x)dx = \delta_{mn}. \tag{7.6}$$

This can be calculated in matrix form as

$$\boldsymbol{A}\boldsymbol{A}^T\boldsymbol{U} = \boldsymbol{\Lambda}\boldsymbol{U}. \tag{7.7}$$

The corresponding temporal coefficients can then be constructed as

$$\mathbf{v_i} = \mathbf{u_i}^T A. \tag{7.8}$$

The most common way to solve Eq. 7.7 is the singular value decomposition (SVD). Considering a data matrix, $\boldsymbol{A} \in \mathbb{R}^{N\times M}$, organized as in Eq. 7.3, the economy SVD is given by $\boldsymbol{A} = \boldsymbol{U}\boldsymbol{\Sigma}\boldsymbol{V}^T$, with

$$\boldsymbol{U} \in \mathbb{R}^{N \times min(M,N)}, \boldsymbol{\Sigma} \in \mathbb{R}^{min(M,N) \times min(M,N)}, \boldsymbol{V} \in \mathbb{R}^{M \times min(M,N)},$$

where $\boldsymbol{U}$ and $\boldsymbol{V}$ are unitary matrices, and $\boldsymbol{\Sigma}$ is a diagonal matrix, which contains the singular values $\sigma_i >= 0$ ordered in descending fashion. One can construct a rank r reduced-order approximation of the data by taking the first r outer-products

$$\boldsymbol{A}_r = \sum_r \boldsymbol{u}_i \sigma_i \boldsymbol{v}_i^T. \tag{7.9}$$

This approximation represents a reduction in dimension by $max(M, N) - r$ and has a reconstruction error

$$||A - A_r||_F = \sqrt{\sum_r^{min(m,n)} \sigma_i}. \tag{7.10}$$

An alternative method to the SVD is the Method of Snapshots (MOS) [25], which determines the temporal modes using a reduced eigensolve of dimension set by the number of snapshots M. It has to be noted that for large datasets, matrix operations can become extremely memory intensive to implement. This requirement can be somewhat mitigated by using the method of snapshots to reduce the overall size of the underlying eigensolve. However, this approach comes with the caveat of requiring fast access to each of the snapshots in memory.

7.2.2 Dynamic Mode Decomposition

Dynamic mode decomposition (DMD) [13] identifies coherent dynamics in the data based on temporal frequencies. DMD assumes that an approximate linear map $\boldsymbol{C} \in \mathbb{R}^{N \times N}$ exists between consecutive snapshots, $\mathbf{a}_i$ and $\mathbf{a}_{i+1}$,

$$\mathbf{a}_{i+1} \approx \boldsymbol{C}\mathbf{a}_i, \tag{7.11}$$

The standard DMD method follows an Arnoldi procedure with the assumption that there exists a specific number, M, beyond which the snapshot, $\mathbf{a_M}$, can be expressed as a linear combination of the previous snapshots,

$$\mathbf{a_M} = b_1\mathbf{a_1} + b_2\mathbf{a_2} + \cdots + b_{M-1}\mathbf{a_{M-1}} + \mathbf{r}, \tag{7.12}$$

or

$$\mathbf{a_M} = \boldsymbol{A}_1\mathbf{b} + \mathbf{r}, \mathbf{b} \in \mathbb{R}^{M-1}, \tag{7.13}$$

where $\mathbf{b} = \begin{bmatrix} b_1 \ b_2 \ \dots \ b_{M-1} \end{bmatrix}^T$ are the coefficients, and $\mathbf{r}$ is the residual error if the target dataset does not satisfy the two assumptions in Eqs. 7.11 and 7.12. Equation 7.12

is an over-determined system for which the residual vector $\mathbf{r}$ may be expected to decrease as the number of snapshots is increased.

$$\boldsymbol{C A}_1 = \boldsymbol{A}_2 = \boldsymbol{A}_1\boldsymbol{S} + \mathbf{r}, \quad \boldsymbol{S} \in \mathbb{R}^{(M-1)\times(M-1)}, \tag{7.14}$$

where $\boldsymbol{A}_1 = \left[\mathbf{a}_1\ \mathbf{a}_2\ \ldots\ \mathbf{a}_{M-1}\right]$ and $\boldsymbol{A}_2 = \left[\mathbf{a}_2\ \mathbf{a}_3\ \ldots\ \mathbf{a}_M\right]$. Here $\boldsymbol{S}$ is a companion matrix that shifts each snapshot

$$\boldsymbol{S} = \begin{bmatrix} 0 & & & & c_1 \\ 1 & 0 & & & c_2 \\ & \ddots & \ddots & & \vdots \\ & & 1 & 0 & c_{M-2} \\ & & & 1 & c_{M-1} \end{bmatrix} \in \mathbb{R}^{(M-1)\times(M-1)} \tag{7.15}$$

This matrix can be extracted from the dataset and provides a low dimensional representation of the dynamical system. The characteristic solution of this system approximates the eigenvalues of the full system $\boldsymbol{C}$ [26].

In contrast to POD, DMD creates modes that are associated with single frequencies. However, in practice, this companion matrix $\boldsymbol{S}$ can be of poor quality and yield an ill-conditioned eigenvalue problem. As a result, we introduce a preprocessing method used by Schmidt [13] to improve the condition of the eigensolve of $\boldsymbol{S}$ when implementing this for a practical dataset.

Starting from Eq. 7.14, we apply SVD to $\boldsymbol{A}_1$

$$\boldsymbol{A}_1 = \boldsymbol{U\Sigma V}^T, \tag{7.16}$$

and substitute Eq. 7.16 to Eq. 7.14,

$$\boldsymbol{CU\Sigma V}^T = \boldsymbol{A}_2, \tag{7.17}$$

$$\boldsymbol{CU} = \boldsymbol{A}_2\boldsymbol{V\Sigma}^{-1}, \tag{7.18}$$

We introduce a projection onto $\boldsymbol{U}$, the orthogonal basis of $\boldsymbol{A}_1$ to form a similarity matrix $\tilde{\boldsymbol{S}}$ which is similar to $\boldsymbol{S}$. Additionally, this method allows us to improve the conditioning should we choose by removing singular values below a certain threshold.

$$\boldsymbol{U}^T\boldsymbol{CU} = \boldsymbol{U}^T\boldsymbol{A}_2\boldsymbol{V\Sigma}^{-1} = \tilde{\boldsymbol{S}} \tag{7.19}$$

$$\tilde{\boldsymbol{S}} \in \mathbb{R}^{(M-1)\times(M-1)}$$

To recognize that this is in similarity matrix form recall Eq. 7.14.

$$\boldsymbol{U}^T\boldsymbol{A}_2\boldsymbol{V\Sigma}^{-1} = \tilde{\boldsymbol{S}} \tag{7.20}$$

$$\boldsymbol{U}^T(\boldsymbol{U\Sigma V}^T)\boldsymbol{SV\Sigma}^{-1} = \tilde{\boldsymbol{S}} \tag{7.21}$$

$$(\boldsymbol{\Sigma V}^T)\boldsymbol{S}(\boldsymbol{\Sigma V})^{-1} = \tilde{\boldsymbol{S}} \tag{7.22}$$

With the formulation, we can approximate the eigenvectors of $\boldsymbol{S}$ by taking the eigen-decompositon

$$\tilde{\boldsymbol{S}} = \tilde{\boldsymbol{T}}\tilde{\boldsymbol{\Delta}}\tilde{\boldsymbol{T}}^{-1}. \tag{7.23}$$

$$\tilde{\boldsymbol{T}} \in \mathbb{C}^{(M-1)\times(M-1)}, \tilde{\boldsymbol{\Delta}} \in \mathbb{C}^{(M-1)\times(M-1)} \tag{7.24}$$

Recall that, to create a similarity matrix form of $\boldsymbol{S}$, we projected onto its POD basis. Thus, we calculate the spatial modes as

$$\boldsymbol{u}_i = \boldsymbol{U}\tilde{\boldsymbol{y}}_i, \quad \mathbf{u_i} \in \mathbb{C}^N. \tag{7.25}$$

Here $\tilde{\mathbf{y}}_i$ is the ith eigenvector, and the temporal modes as

$$\boldsymbol{v}_i = (u_i)^T \boldsymbol{A} = \boldsymbol{V}\boldsymbol{\Sigma}(\tilde{\mathbf{y}}_i^{-1})^T, \quad \mathbf{v_i} \in \mathbb{C}^{M-1}. \tag{7.26}$$

The reconstruction of the data matrix corresponding to each of these modes can be determined (Eq. 7.5) as

$$\boldsymbol{A}_i = real(\mathbf{u_i}\mathbf{v_i}^T), \quad \boldsymbol{A}_i \in \mathbb{R}^{N\times(M-1)}. \tag{7.27}$$

7.2.3 Higher Order Dynamic Mode Decomposition

The ability of DMD to extract frequency-specific responses and quantify their relative weights is inherently helpful in identifying essential physics. However, this methodology still assumes a linear Markovian evolution and is limited by the availability of snapshots.

DNS or LES datasets can have a far higher spatial resolution with limited snapshot counts, leading to tall and skinny data matrices. On the other hand, experimental probe data usually have rich temporal content but may be limited to the number of installed sensors. In both of these cases, data matrices can become irregular, reducing the effectiveness of standard DMD. This concern is more apparent if the physics of interest is not completely captured in the partially observed snapshot set.

The general concept of using time delays to construct a rich feature space for the reconstruction of nonlinear dynamics is justified by the Takens embedding theorem [27]. Time-delay analysis has been of interest in the dynamical systems community [21], and is leveraged in the so-called higher order DMD. For ergodic dynamical systems, high-order DMD has been shown to provide a convergent approximation to the Koopman spectrum [14, 28].

Taking the format of the standard snapshots matrix, shown in Eq. 7.3, we have the equivalent time-delay form for d time delays

$$A_d = \begin{bmatrix} \mathbf{a_1} & \mathbf{a_2} & \dots & \mathbf{a_{M-d+1}} \\ \mathbf{a_2} & \mathbf{a_3} & \dots & \mathbf{a_{M-d+2}} \\ \vdots & \vdots & & \vdots \\ \mathbf{a_d} & \mathbf{a_{d+1}} & \dots & \mathbf{a_M} \end{bmatrix} \in \mathbb{R}^{dN\times(M-d+1)} \quad (7.28)$$

Extending this concept to the DMD algorithm, we aim to describe the transformation $S_d \in \mathbb{R}^{(d\times N)\times(d\times N)}$.

$$\begin{bmatrix} \mathbf{a_2} & \dots & \mathbf{a_{M-d+1}} \\ \mathbf{a_2} & \dots & \mathbf{a_{M-d+2}} \\ \vdots & \vdots & \vdots \\ \mathbf{a_{d+1}} & \dots & \mathbf{a_M} \end{bmatrix} = S_d \begin{bmatrix} \mathbf{a_1} & \dots & \mathbf{a_{M-d}} \\ \mathbf{a_2} & \dots & \mathbf{a_{M-d+1}} \\ \vdots & \vdots & \vdots \\ \mathbf{a_d} & \dots & \mathbf{a_{M-1}} \end{bmatrix} \quad (7.29)$$

$$A_{d1} = S_d A_{d2} \quad (7.30)$$

Here we can observe that for $d = 1$, this reduces to the same formulation as the standard DMD algorithm. Similar to the standard DMD, conditioning is required, and thus we rewrite A_d as a projection onto the POD basis of A,

$$\hat{A}_d = \begin{bmatrix} \mathbf{w_1} & \dots & \mathbf{w_{M-d+1}} \\ \mathbf{w_2} & \dots & \mathbf{w_{M-d+2}} \\ \vdots & \vdots & \vdots \\ \mathbf{w_d} & \dots & \mathbf{w_M} \end{bmatrix} \in \mathbb{R}^{d(M-d+1)\times(M-d+1)}, \quad (7.31)$$

where $\mathbf{w_i} = U^T \mathbf{a_i}$.

Linking this back to the original form of the linear solution, we obtain a similar form as a classical DMD approximation of $\tilde{S}$

$$\hat{A}_{d2} = \tilde{S}_d \hat{A}_{d1}, \quad (7.32)$$

where

$$\tilde{S}_d = \begin{bmatrix} U^T & & \\ & \ddots & \\ & & U^T \end{bmatrix} S_d \begin{bmatrix} U & & \\ & \ddots & \\ & & U \end{bmatrix}. \quad (7.33)$$

Recalling the time-delay form

$$\hat{A}_d = \hat{U}_d \hat{\Sigma}_d \hat{V}_d^T, \quad (7.34)$$

we apply the same strategy,

$$\hat{A}_{d2} = \tilde{S}_d \hat{A}_{d1}, \quad (7.35)$$

$$\hat{U}_d^T \hat{A}_{d2} = \hat{U}_d^T \tilde{S}_d \hat{U}_d \hat{U}_d^T \hat{A}_{d1}. \quad (7.36)$$

which can be written as a compact expression,

$$\bar{A}_{d2} = \bar{S}_d \bar{A}_{d1} \tag{7.37}$$

with

$$\bar{A}_{d2} = \hat{U}_d^T \hat{A}_{d2}, \quad \bar{S}_d = \hat{U}_d^T \hat{S}_d \hat{U}_d, \quad \bar{A}_{d1} = \hat{U}_d^T \hat{A}_{d1},$$

where $\bar{A}_{d2}$ and $\bar{A}_{d1}$ are the projected time-delayed matrices. Thus, we can solve this linear system for $\bar{S}_d$. It should be noted that practical implementations may require the truncation of small singular values prior to the linear solution.

Finally, we can perform the eigendecomposition of $\bar{S}_d$ and generate the corresponding temporal and spatial modes in a similar format as before

$$\hat{S}_d = \hat{U}_d \bar{T}_d \bar{\Lambda}_d \bar{T}_d^{-1} \hat{U}_d^T \tag{7.38}$$

with

$$\Psi = U \hat{U}_d \bar{T}_d \tag{7.39}$$

$$\mathbf{y} = \bar{T}_d^{-1} \hat{U}_d^{\,T} \hat{X}_d \tag{7.40}$$

Pan et al. [20] provide an analytical expression for the optimal number of time delays to reconstruct dynamics on an attractor. However, for large practical problems, the number of delays is a problem-dependent tuning parameter. Despite the additional memory requirements, the ability of linear time-delay embedding to approximate the Koopman spectrum [28], and the associated improvement conditioning makes it appealing. We will comment on these effects and performance in the following section.

7.3 Application: Lean Direct Injection (LDI) Combustor

In this section, high-fidelity simulation data of a real laboratory-scale gas turbine combustor is used to examine and evaluate the different techniques introduced in the previous section.

In recent decades, emissions reduction has been an important topic in aerospace propulsion. One of the significant challenges in achieving these are lean operating conditions that tend to induce combustion instability. Lean direct injection (LDI) is an existing concept for lean burning, low NOx, liquid-fueled gas turbine combustors. In this configuration, liquid fuel is injected into a highly inertial and swirling airflow and accelerated through a convergent–divergent subsonic venturi [29]. The primary flow feature is the swirling airflow that promotes fuel vaporization and provides the mechanism for flame stabilization. The introduction of fuel as a liquid spray leads to a change in flame structure compared to gaseous injection due to inhomogeneity in the mixture fraction.

Computational [30] and experimental [31] studies have been performed at Purdue University on a single-element model LDI combustor to diagnose and examine the underlying combustion dynamics. The studies adopted the practical conditions and aimed to replicate essential physics in liquid-fueled swirl-stabilized combustor, for example, the well-known recirculation zone dynamics from the swirling flow and the Precessing Vortex Core (PVC), a fundamentally three-dimensional helical feature [32]. The couplings between these dynamics and the natural acoustics of the combustor make the system susceptible to thermoacoustic instability, especially under fuel-lean conditions.

Data used in this work was generated by simulations from the GEMS (General Equation and Mesh Solver) [33] an unstructured, finite volume solver with second-order numerical scheme and an implicit, dual time-stepping procedure. The solver is used in Detached Eddy Simulation (DES) mode utilizing a $k - \omega$ turbulence model in the RANS regime. Combustion modeling is achieved using 6 equation finite-rate chemistry in a simplified 2-step reaction mechanism for Jet A Fuel. Further details of the simulation methodology are described in Huang et al. [30].

A schematic of the combustor is shown in Fig. 7.1, which consists of an air plenum feeding into a swirler. The swirled air is then focused through a converging–diverging section where liquid fuel (Jet A) is introduced. The swirling flow rapidly mixes the fuel and oxidizer and reactions occur. The inflow and outflow are choked to enforce consistent acoustic boundaries for both the experiment and the simulation. The combustor is designed to withstand acoustic frequencies with a nominal chamber pressure of 1 MPa. The single-element LDI combustor is designed for a fundamental longitudinal frequency, denoted as 1L, between 300–400 Hz, dependent on the operating conditions. The 1 L acoustic frequency corresponds to a 1/2-wave mode shape in the combustor. Higher harmonics are also present. Thus the 2 L mode (700 Hz) corresponds to a full-wave shape, the 3 L mode corresponds to a 3/2-wave, and so on.

While a variety of operating conditions were examined [31], the low equivalence ratio case ($\phi = 0.36$) is selected for demonstrative purposes. In general, thermoacoustic instability is more prevalent in lean operating conditions, which makes the current case more practically relevant. Moreover, the low equivalence ratio case exhibits rich physical complexity, as shown in the power density spectrum (PSD) of

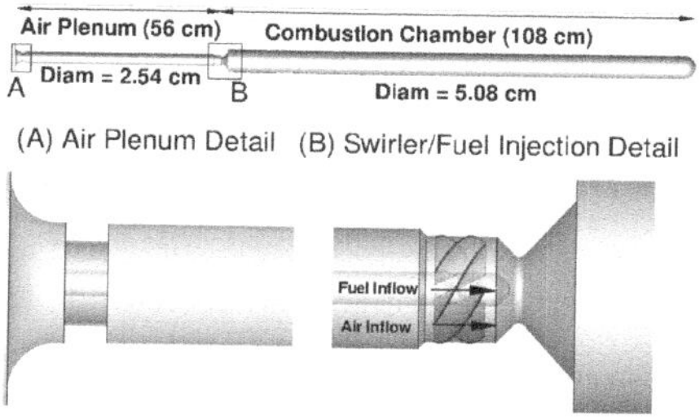

Fig. 7.1 Schematic of LDI combustor

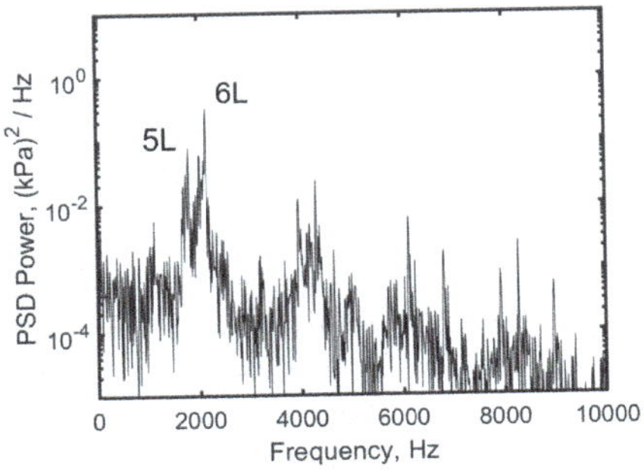

Fig. 7.2 Power spectral density(PSD) for LDI combustor located along wall 1 mm from venturi feature

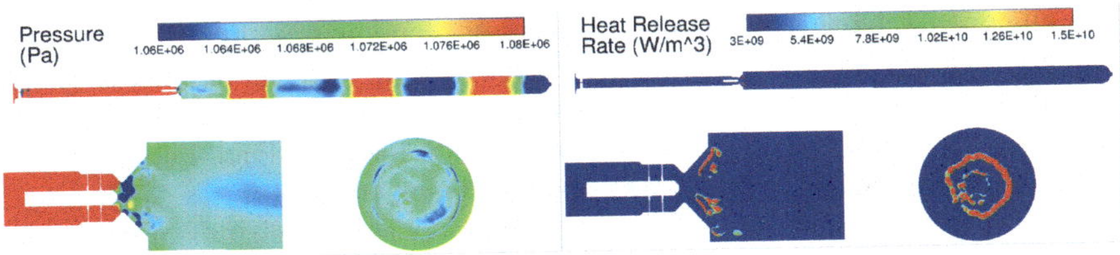

Fig. 7.3 Representative instantaneous snapshots of pressure and heat release rate

pressure in Fig. 7.2. It can be readily seen that the dominant observed acoustic modes for the current case are 5 and 6 L modes with many other secondary peaks standing out. Additionally in Fig. 7.3, we show a representative snapshot of the pressure and heat release rate to present an idea of the physical complexity. In the pressure field, a six half-wave acoustic mode shape can be seen spanning the entire combustion chamber, indicating the dominance of 6 L acoustic mode. The heat release field presents a very compact flame restricted to the diverging section. The couplings between the global acoustics and the local compact flame are key to investigate thermoacoustic instability in gas turbine engines.

For the following results, we take a contiguous set of snapshots separated by a fixed time interval with the mean subtracted. For this specific case, 1200 snapshots are used. No additional post-processing is performed. The temporal ordering of the data is not strictly required for the POD analysis but is used for consistent preprocessing with the DMD and HODMD analysis.

7.3.1 Proper Orthogonal Decomposition

The relative importance of each mode and its cumulative contribution is given by

$$\text{Mode Power} = \frac{\sigma_i^2}{\sum \sigma_i^2} \times 100\% \tag{7.41}$$

Fig. 7.4 Singular value decay

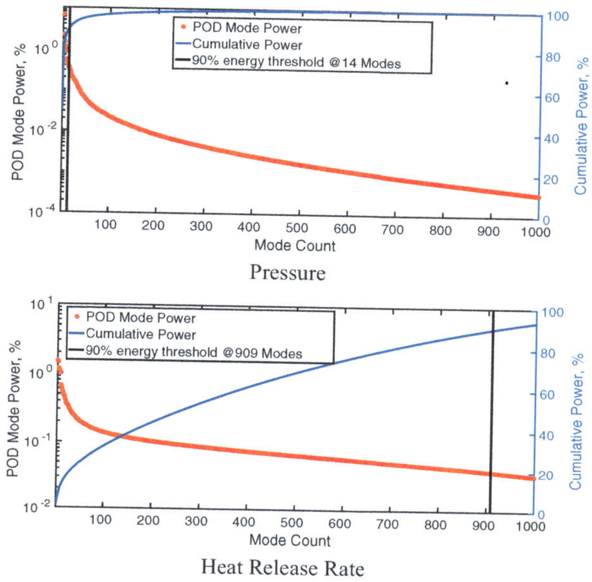

$$\text{Cumulative power} = \frac{\sum_1^k \sigma_i^2}{\sum \sigma_i^2} \times 100\% \qquad (7.42)$$

and shown in Fig. 7.4. For the Pressure data, 90% of the system energy is contained in approximately the first 20 modes. For organized and coherent signals such as pressure, POD can capture a majority of the space-time information in a relatively small set of modes. Due to the nonlinearity of the kinetics and the distributed nature of the features, the heat release rate requires a large number of modes (15% of energy contained in 20 modes) for accurate representation. In this case, the poor energy decay is primarily due to the nonlinearity of the chemical kinetics. However, there are turbulent flow cases with sufficient richness that lead to similar poor singular decay. It is therefore difficult to prescribe a general statement other than the understanding that for this case we observe the kinetics as being the more difficult physics to capture using modal decomposition. Similar observations have been reported by Huang et al. [34]

In Figs. 7.5 and 7.6, we show the spatial modes alongside a Power Spectral Density (PSD) of the temporal modes of pressure and heat release, respectively. The PSD of the temporal modes provides a direct measurement of the dominant frequency content in each individual mode. The spatial modes are given as a slice along the axial centerline. In addition, a spanwise slice located 1 mm downstream of the venturi is given to elucidate any potential three-dimensional characteristics. From the POD modes, it can be observed that each mode contains multiple frequencies. The global acoustic response of the system is captured in the lower mode counts. For pressure, we

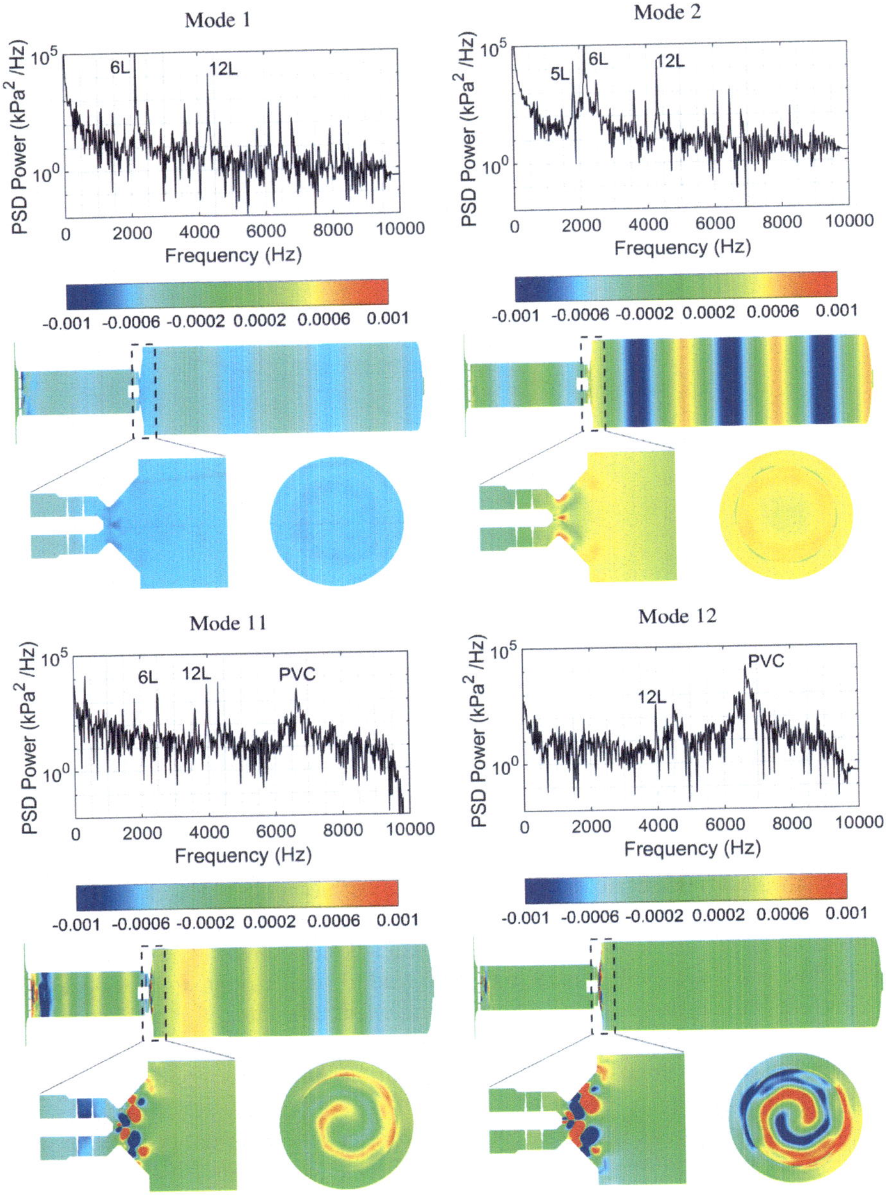

Fig. 7.5 POD modes of pressure

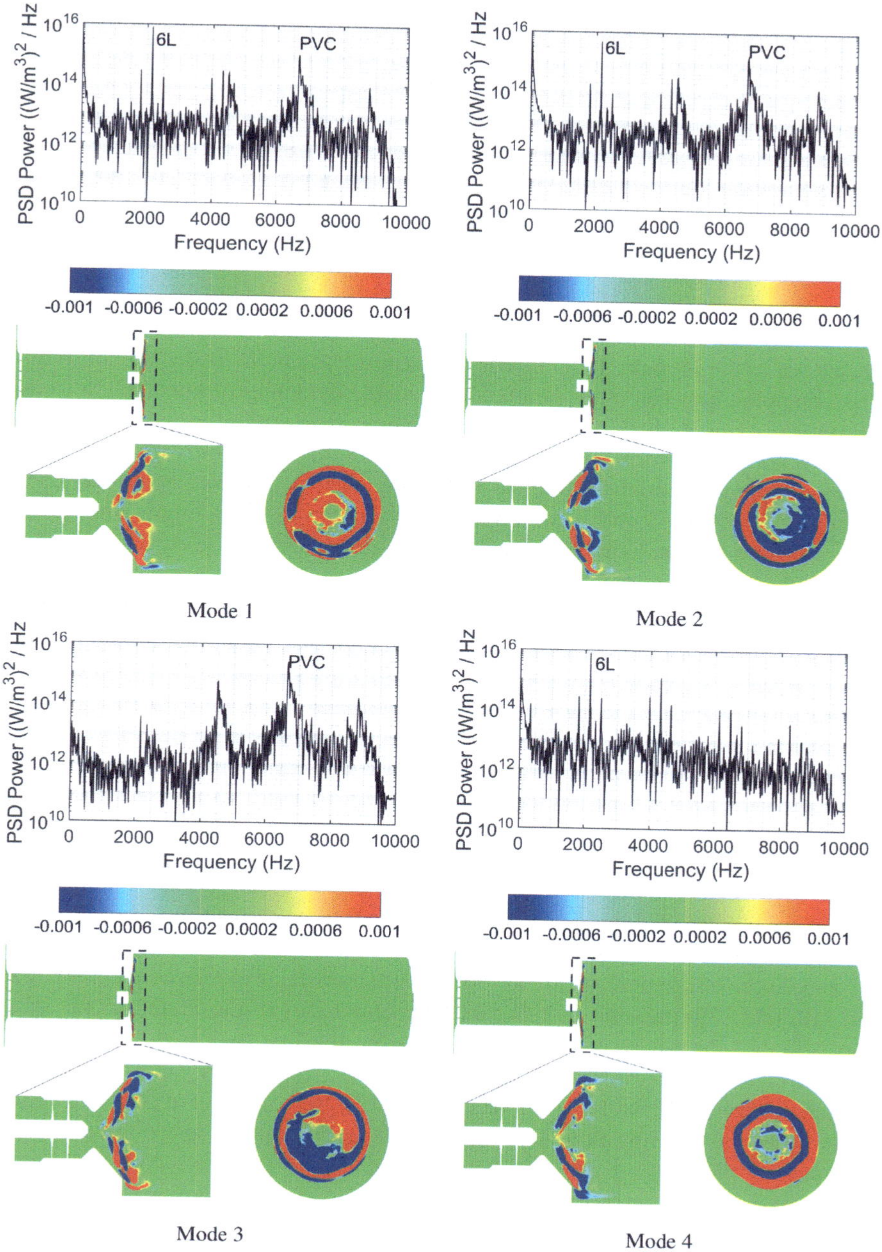

Fig. 7.6 POD modes of heat release rate

can see that in modes 1 and 2, the primary peak is associated with the 6 L frequency. In the corresponding side view, a six 1/2-wave feature is observed. Local acoustic features are likewise observed at the higher mode counts. The PSD of modes 11 and 12 show a dominant frequency around 7000 Hz, which corresponds to the so-called Precessing Vortex Core (PVC) hydrodynamics. In contrast to modes 1 and 2, the dominant responses of modes 11 and 12 are restricted within the venturi section featured with a precessing helical structure seen from streamwise and spanwise views, which is the major characteristic of PVC [32]. In addition to PVC, we can also observe a weaker contribution of the acoustic features, 6 and 12 L, both in temporal and spatial modes.

Similarly, we can examine the POD modes of heat release rate as shown in Fig. 7.6. In contrast to the pressure POD modes, the heat release POD modes do not show a direct mapping between mode number and specific dynamics of interest. For example, Modes 1 and 4 show dominant frequencies corresponding to the 6 L response with a relatively uniform distribution in the azimuthal direction with variation in the longitudinal direction. While the PVC frequency is observed as the dominant frequency in Modes 2 and 3, which show a variation in the azimuth while being more consistent in the longitudinal direction. We can observe a PVC response peak in mode 1 and likewise, a 6 L peak in mode 2.

Referring back to the energy decay of the POD (Fig. 7.4) we again note that the mode power decay for heat release is quite weak when compared to the pressure. While this can be partially accounted for by the fundamentally nonlinear reaction kinetics, it is mostly a consequence of the nature of turbulent reacting flow.

From this POD analysis, we have been able to observe two main coherent structures present in the combustion system. A qualitative understanding of the various spatial mode contributions can be obtained. However, it is also found that specific dynamics of interest can be spread in several POD modes and more importantly, there is no one-to-one correlation between the pressure and heat release rate based on the POD mode numbers, which makes it difficult to establish physical insight.

Having identified significant modes, we can use the space-time information to reconstruct the flow field response (Eq. 7.5). In Fig. 7.7 we show the response of the 1st and 2nd pressure modes, and in Fig. 7.7 we show the response corresponding to the 11th and 12th pressure modes. Here the limitations of the POD become more apparent. For the 6 L response constructed from the first two modes, we observe the expected 6 L 1/2-wave spatial variations. In contrast, when we reconstruct the pressure response of the PVC (modes 11 and 12), we can observe that in addition to the expected swirling flow features, acoustic waves appear less coherent in the upstream air plenum.

Likewise, for the identified heat release modes, we reconstruct the flow field response. In Fig. 7.8 we show the response of the 1st and 4th heat release rate modes, and in Fig. 7.8 we show the response corresponding to the 3rd and 4th heat release rate modes. Most apparent in the spanwise slices, we can observe the swirling flow feature. Similarly, we can observe the acoustic 6 L response in the heat release as the heat release fluctuation at the characteristic frequency.

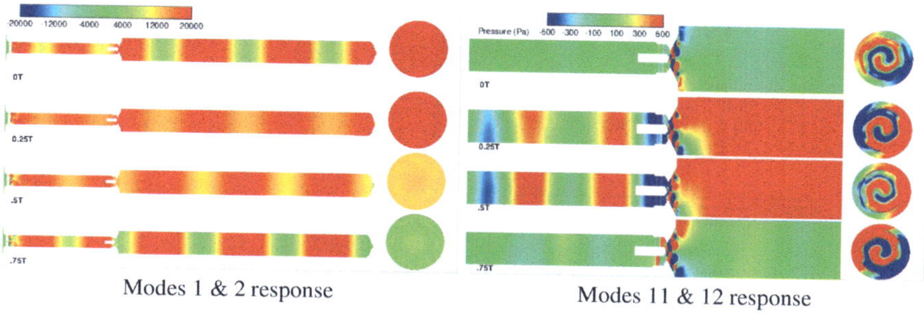

Fig. 7.7 POD Pressure Response

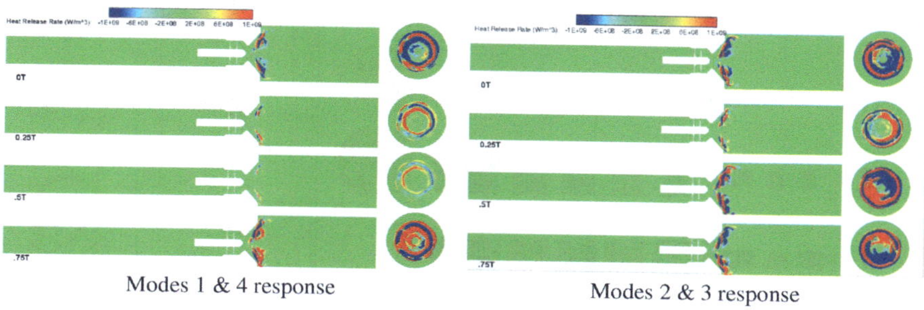

Fig. 7.8 POD heat release rate response

While it can be argued that this technique has been successful in extracting complex features from the data, there is no guarantee as to which group of modes will capture the dynamic features of interest as it is not possible to extract specific frequency information.

7.3.2 *Dynamic Mode Decomposition*

Dynamic Mode Decomposition (DMD) provides a direct approach to extract the dynamics of interest at each frequency. The DMD spectrum for both pressure and heat release rate is shown in Fig. 7.9.

The dominant 6 L acoustic mode response is shown in Fig. 7.10. Similar to the POD response, a standing wave and a swirling PVC feature are observed. In contrast to the POD, however, the DMD mode is less contaminated and does not have additional acoustic features and noise noted in the POD response of the PVC feature.

Similarly, we attempt DMD on the heat release data. Examining the frequency spectrum in Fig. 7.9, we can observe the usual 5L, 6L, and PVC peaks. Generating the responses for the 6 L frequency further confirms our observations from the POD analysis. We can observe in Fig. 7.11 that the response is less noisy compared to the

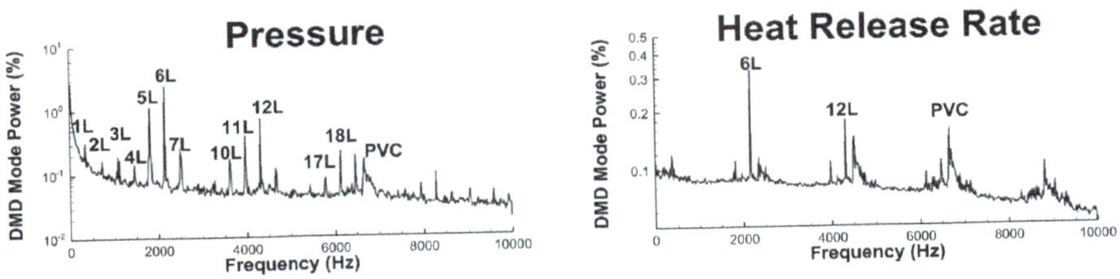

Fig. 7.9 DMD frequency spectrum of pressure and heat release rate

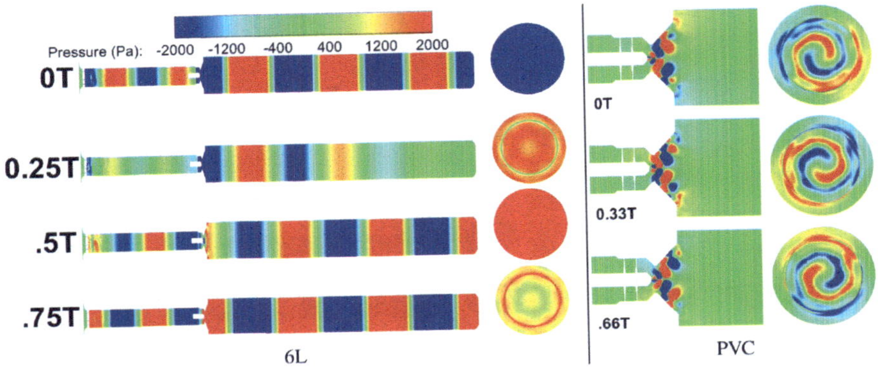

Fig. 7.10 DMD responses

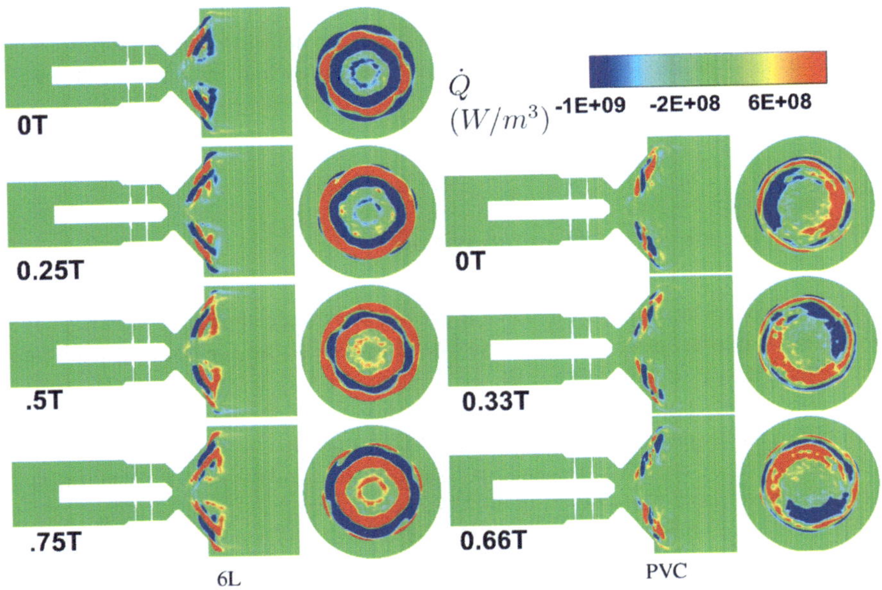

Fig. 7.11 Heat release rate DMD modes

corresponding POD mode. As a result, it is much easier to observe the longitudinal variation and axial consistency. Likewise for the PVC frequency response in Fig. 7.11, we can observe the swirling mode shape which has longitudinal consistency with an axial variation.

7.3.3 Higher Order Dynamic Mode Decomposition & Predictive Modeling

In the previous sections, modal decompositions were used to provide insight into the underlying dynamics. In this section, we will apply the HODMD methodology (Sect. 7.2.3) to the temporal mode coefficients generated in Sect. 7.3.1 to estimate an operator that will advance the coefficients in time. Thus, one can reconstruct and predict the associated flow field.

The eigenvalues of the approximated linear operator (also called Ritz values [26]) contain information about the growth rate and frequency of the corresponding mode. As seen in Fig. 7.12, increasing the number of time delays increases the number of neutrally stable Ritz values. This is because HODMD has the capability to describe dynamics on an attractor with limited data. Our ultimate goal is to analyze the temporal mode set, i.e., the training data, and predict the future state past that training time. A subset of the temporal modes is shown in Fig. 7.15.

In Fig. 7.13, we show the result of the action of the derived operator on the initial condition. The training region is shown in red and is 80% of the total time domain. The region in blue is the true unobserved future state. We observe that applying the standard DMD analysis to these temporal modes and attempting to predict the

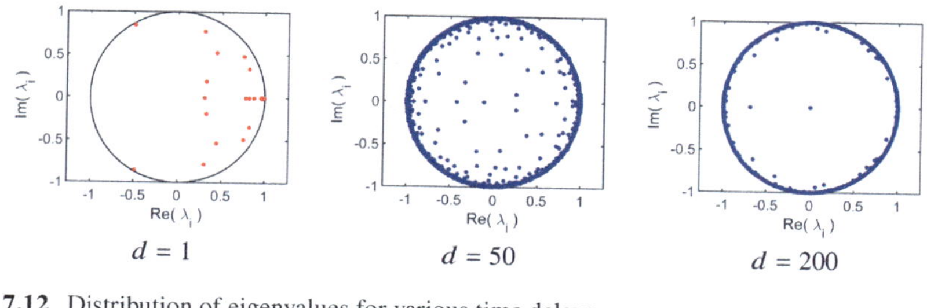

Fig. 7.12 Distribution of eigenvalues for various time delays

Fig. 7.13 Standard DMD prediction of 3rd temporal mode

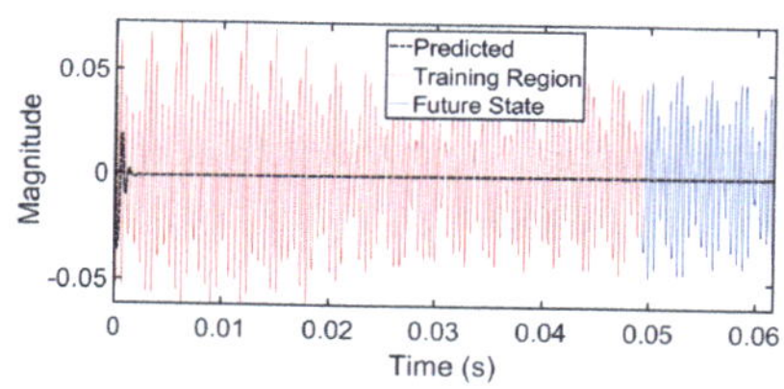

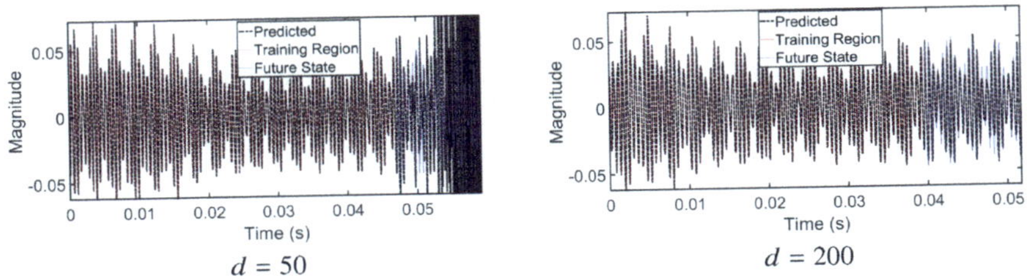

Fig. 7.14 HODMD prediction of 3rd temporal mode with 50 time delays and 200 time delays

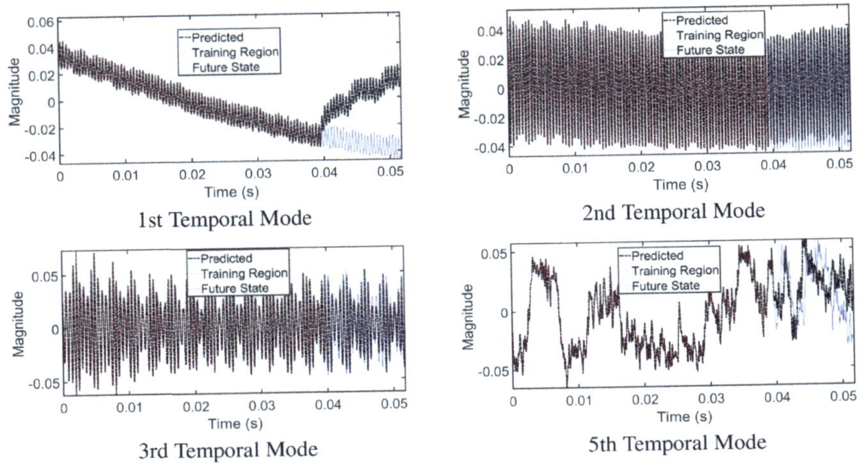

Fig. 7.15 HODMD prediction of various temporal modes

dynamics does not show a correct reconstruction of even the training regime of the temporal mode.

In Fig. 7.14, we show the prediction of the 3rd temporal mode using 50 and 200 time delays. While the reconstruction of the dynamics in the training region is quite acceptable, instabilities are noticeable in the prediction region.

With this observation, the number of time delays is increased. Shown in Fig. 7.14 is the 3rd temporal mode with 200 time delays. A significant improvement is noticeable with good prediction through the training region and even into the future state.

With the success in applying the time-delay method on the 3rd mode, we examine the prediction of other temporal modes in Fig. 7.15. It is seen that the training data is well-reconstructed. However, future state prediction is of variable quality. For the 1st temporal mode, the high-frequency signal is well-predicted, but the low-frequency transient trend near $t = 0.04$ s is not captured. The second and third temporal modes are dominated by a high-frequency response and for these cases, the HODMD is seen to perform well.

It can be additionally seen that the 5th temporal mode begins to deviate from the true future state. HODMD acts in a way to enhance features in the data that might

Fig. 7.16 Predictive properties as a function of time-delay count

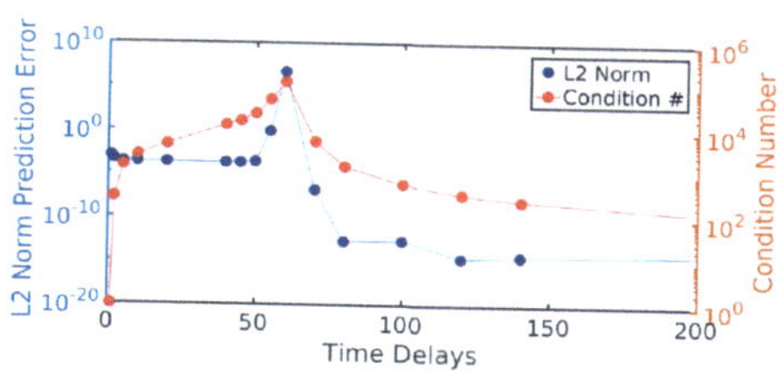

otherwise be indistinguishable from the noise. However, predictions may be poor if the features in the future state are not described by the training data.

Finally, we show the quality of the future state prediction as a function of the number of time delays in Fig. 7.16. We can see that the error quickly spikes at 60 time delays which is observed in Fig. 7.14 to correspond to a peak in condition number. We observe that for 1 time delay, the condition number approaches unity, corroborating the HODMD theory outlined by Pan et al. [20]. Further, the condition number is maximized at TD $= max(M, N)/min(M, N)$ which corresponds to a square time-delay matrix. This relationship should be viewed in concert with the performance of the algorithm at various time delays. At low TD counts, we observe similar performance to classic DMD. The predicted temporal mode is overdamped and has little resemblance to the true data. As we increase the number of time delays, the prediction can become unstable and as can be observed by the spike in the L2 error. Finally after passing the crossover time-delay count, the error again stabilizes at a smaller value. This remaining error cannot be reduced by increasing the number of time delays and is most likely due to features in the future state not present in the training region.

The predicted temporal mode coefficients are used with Eq. 7.5 to recover the flow field and a representative snapshot is shown in Fig. 7.17. As expected from the temporal mode prediction (Fig. 7.15), high-frequency acoustic features are predicted well.

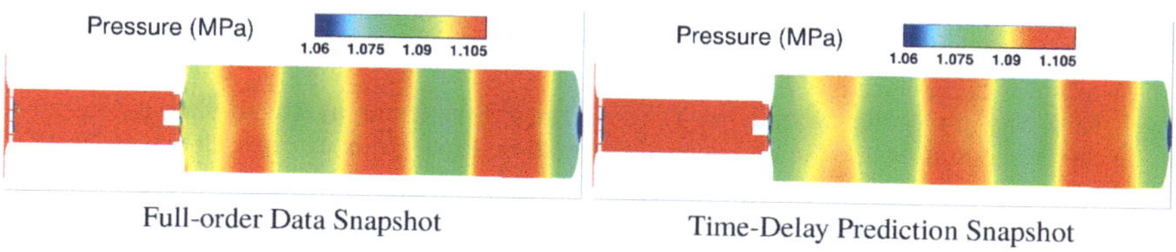

Fig. 7.17 Snapshot comparison of true future state (left) and predicted future state (right)

7.4 Conclusions

This chapter is focused on the algorithmic aspects and application of data-driven modal decomposition techniques for complex flows. A sample dataset from the simulation of a single-element lean direct injection (LDI) gas turbine combustor is chosen for the investigations. Proper orthogonal decomposition (POD) and dynamic mode decomposition (DMD) are applied to both pressure and heat release variables of the evolving flow field. POD generates a series of modes that provide the optimal representation of the dataset in the sense of the energy norm. In contrast, the DMD methodology examines the eigenstructure of approximated linear dynamics. The modes generated by this method correspond to discrete frequencies. This makes DMD a good alternative to POD in extracting coherent dynamics from complex fluid systems, especially around an attractor. We then introduce the Higher Order Dynamic mode Decomposition (HODMD). Utilizing time-delay embedding, this method is shown to accurately reconstruct the evolution of temporal modes and produce reasonable future state prediction.

Acknowledgements This work is supported by funding from the National Science Foundation(NSF) Grant #1634709. Computational resources were provided by the Air Force Research Laboratories (AFRL). The authors would also like to acknowledge Shawou Pan for help with HODMD methodology.

References

1. J.W. Strutt, B. Rayleigh, *The Theory of Sound*, vol. 2 (Macmillian, New York, 1896)
2. T. Lieuwen, H. Torres, C. Johnson, B.T. Zinn, A mechanism of combustion instability in lean premixed gas turbine combustors. *Volume 2: Coal, Biomass and Alternative Fuels; Combustion and Fuels; Oil and Gas Applications; Cycle Innovations* (1999), p. V002T02A001
3. T.C. Lieuwen, V. Yang, *Combustion Instabilities in Gas Turbine Engines: Operational Experience, Fundamental Mechanisms, and Modeling* (American Institute of Aeronautics and Astronautics, Inc., 2010)
4. T. Feldman, M. Harvazinski, C. Merkle, W. Anderson, Comparison between simulation and measurement of self-excited combustion instability, in *48th AIAA/ASME/SAE/ASEE Joint Propulsion Conference and Exhibit* (2012)
5. T. Kim, M. Ahn, J. Hwang, S. Kim, Y. Yoon, The experimental investigation on the response of the Burke-Schumann flame to acoustic excitation. Proc. Combust. Inst. **36**(1), 1629–1636 (2017)
6. K. Taira, S.L. Brunton, S.T.M. Dawson, C.W. Rowley, T. Colonius, B.J. McKeon, O.T. Schmidt, S. Gordeyev, V. Theofilis, L.S. Ukeiley, Modal analysis of fluid flows: an overview. AIAA J. **55**(12), 4013–4041 (2017)
7. K. Taira, M.S. Hemati, S.L. Brunton, Y. Sun, K. Duraisamy, S. Bagheri, S.T.M. Dawson, C.-A. Yeh, Modal analysis of fluid flows: applications and outlook, 1–36 (2019), arXiv:1903.05750
8. C.W. Rowley, Model reduction for fluids, using balanced proper orthogonal decomposition. Int. J. Bifurc. Chaos **15**(03), 997–1013 (2005)
9. G. Berkooz, P. Holmes, L. John, The proper orthogonal decomposition in the analysis of turbulent flows. Ann. Rev. Fluid Mech. **25**(1), 539–575 (1993)
10. K. Willcox, J. Peraire, Balanced model reduction via the proper orthogonal decomposition. AIAA J. **40**(11), 2323–2330 (2012)

11. A. Chatterjee, An introduction to the proper orthogonal decomposition: Rensselaer libraries quick search. Current Sci. **78**(7), 808–817 (2000)
12. P. Iudiciani, C. Duwig, S.M. Husseini, R.Z. Szasz, L. Fuchs, E.J. Gutmark, Proper orthogonal decomposition for experimental investigation of flame instabilities. AIAA J. **50**(9), 1843–1854 (2012)
13. P.J. Schmid, Dynamic mode decomposition of numerical and experimental data. J. Fluid Mech. **656**, 5–28 (2010)
14. H. Arbabi, I. Mezić, Ergodic theory, dynamic mode decomposition and computation of spectral properties of the Koopman operator. SIAM J. Appl. Dyn. Syst. **16**(4), 2096–2126 (2017)
15. J. Nathan Kutz, S.L. Brunton, B.W. Brunton, J.L. Proctor, *Dynamic Mode Decomposition: Data-Driven Modeling of Complex Systems* (SIAM, 2016)
16. J.H. Tu, C.W. Rowley, D.M. Luchtenburg, S.L. Brunton, J. Nathan Kutz, On dynamic mode decomposition: theory and applications, 1–30 (2013), arXiv:1312.0041
17. A. Seena, H.J. Sung, Dynamic mode decomposition of turbulent cavity flows for self-sustained oscillations. Int. J. Heat Fluid Flow **32**(6), 1098–1110 (2011)
18. C. Huang, W.E. Anderson, M.E. Harvazinski, V. Sankaran, Analysis of self-excited combustion instabilities using decomposition techniques. AIAA J. **54**(9), 2791–2807 (2016)
19. A. Towne, O.T. Schmidt, T. Colonius, Spectral proper orthogonal decomposition and its relationship to dynamic mode decomposition and resolvent analysis. J. Fluid Mech. **847**, 821–867 (2018)
20. S. Pan, K. Duraisamy, On the structure of time-delay embedding in linear models of non-linear dynamical systems (2019), arXiv:1902.05198
21. S. Le Clainche, J.M. Vega, Higher order dynamic mode decomposition. SIAM J. Appl. Dyn. Syst. **16**(2), 882–925 (2017)
22. S.L. Brunton, B.W. Brunton, J.L. Proctor, E. Kaiser, J. Nathan Kutz, Chaos as an intermittently forced linear system. Nat. Commun. **8**(1), 1–8 (2017)
23. M. Mohebujjaman, L.G. Rebholz, T. Iliescu, Physically-constrained data-driven, filtered reduced order modeling of fluid flows, 1–21 (2018), arXiv:1806.00350
24. B.R. Noack, M. Morzynski, G. Tadmor, *Reduced-Order Modelling for Flow Control*, vol. 528 (Springer Science & Business Media, Berlin, 2011)
25. L. Sirovich, Turbulence and the dynamics of coherent structures part i: coherent structures. Q. Appl. Math. **45**(3), 561–571 (1987)
26. C.W. Rowley, I. Mezi, S. Bagheri, P. Schlatter, D.S. Henningson, Spectral analysis of nonlinear flows. J. Fluid Mech. **641**, 115–127 (2009)
27. F. Takens, Detecting strange attractors in turbulence dynamical systems and turbulence, Warwick 1980. Dyn. Syst. Turbul. **898**, 366–381 (1981)
28. H. Arbabi, I. Mezić, Study of dynamics in post-transient flows using Koopman mode decomposition. Phys. Rev. Fluids **2**(12) (2017)
29. G.E. Andrews, H.S. Alkabie, M.M.A. Aziz, U.S. Abdul Hussain, N.A. Al Dabbagh, N.A. Ahmad, A.A. Shaikly, M. Kowkabi, A.R. Shahabadi, High-intensity burners with low NOx emissions. Proc. Inst. Mech. Eng., Part A: J. Power Energy **206**(1), 3–17 (1992)
30. C. Huang, C. Yoon, R. Gejji, W. Anderson, V. Sankaran, Computational study of combustion dynamics in a single- element lean direct injection gas turbine combustor, in *52nd Aerospace Sciences Meeting*, vol. 298 (2014), pp. 1–14
31. R.M. Gejji, C. Huang, C. Fugger, C. Yoon, W. Anderson, Parametric investigation of combustion instabilities in a single-element lean direct injection combustor. Int. J. Spray Combust. Dyn. **0**(0)175682771878585 (2018)
32. S. Candel, D. Durox, T. Schuller, J.-F. Bourgouin, J.P. Moeck, Dynamics of swirling flames. Ann. Rev. Fluid Mech. **46**(1), 147–173 (2013)
33. D. Li, G. Xia, V. Sankaran, C.L Merkle, Computational framework for complex fluids applications, in *3rd International Conference on Computational Fluid Dynamics* (2004)
34. C. Huang, R. Gejji, W. Anderson, C. Yoon, V. Sankaran, Combustion dynamics in a single-element lean direct injection gas turbine combustor. Combust. Sci. Technol. **0**(0), 1–28 (2019)

Chapter 8
Dynamic Mode Decomposition: A Tool to Extract Structures Hidden in Massive Datasets

T. Grenga and M. E. Mueller

Abstract Dynamic Mode Decomposition (DMD) is able to decompose flow field data into coherent modes and determine their oscillatory frequencies and growth/decay rates, allowing for the investigation of unsteady and dynamic phenomena unlike conventional statistical analyses. The decomposition can be applied for the analysis of data having a broad range of temporal and spatial scales since it identifies structures that characterize the physical phenomena independently from their energy content. In this work, a DMD algorithm specifically created for the analysis of massive databases is used to analyze three-dimensional Direct Numerical Simulation of spatially evolving turbulent planar premixed hydrogen/air jet flames at varying Karlovitz number. The focus of this investigation is the identification of the most important modes and frequencies for the physical phenomena, specifically heat release and turbulence, governing the flow field evolution.

8.1 Introduction

The Navier–Stokes equations are a set of nonlinear partial differential equations describing the evolution of fluid flow. However, when used to evaluate flow evolution in complex geometries and challenging flow regimes (e.g., high speed, turbulent, and/or featuring chemical reactions), analytical solutions, even using strong simplifications, are not possible. The study of such flows must rely on experiments and, moreover, on detailed computational simulations. As experiments and computations become more advanced, they generate ever-increasing amounts of data. The availability of these huge datasets creates new challenges: how to manipulate and analyze such massive data to find meaningful results and physical insights. This is leading

T. Grenga (✉)
RWTH Aachen University, Templegraben 64, 52062 Aachen, Germany
e-mail: t.grenga@itv.rwth-aachen.de

M. E. Mueller
Princeton University, Olden Street, Princeton, NJ 08544, USA
e-mail: muellerm@princeton.edu

H. Pitsch and A. Attili (eds.), *Data Analysis for Direct Numerical Simulations of Turbulent Combustion*, https://doi.org/10.1007/978-3-030-44718-2_8

to data-driven methods based on mathematical techniques and/or on artificial intelligence that can take a dataset and characterize it in meaningful ways with minimal guidance.

Modal decompositions have recently become popular as data-driven analysis tools in several fields including fluid dynamics. Generally, modal decompositions take a set of data and from it computes a set of modes, or characteristic features. The meaning of the modes depends on the particular type of decomposition used. However, in all cases, the basic premise is that the modes identify features of the data that elucidate the underlying flow physics.

Dynamic mode decomposition (DMD) is a method developed with the aim to analyze the temporal dynamics of fluid flows by decomposing the flow field into modes sharing the same characteristic frequency. DMD was first introduced in 2008 by Schmid [1, 2]. The method was presented as a generalization of linear stability analysis, relying on the linear tangent approximation of the nonlinear equations between consecutive instances of time. The algorithm presented by Schmid based on Singular Value Decomposition (SVD) has become the standard for DMD practitioners. In parallel, Rowley et al. [3] showed that DMD can also be interpreted as an approximation to Koopman decomposition.

Recently, the use of DMD has been adopted also in other fields in order to characterize the evolution of quantities of interest at different timescales. Some examples of DMD applications include: in medicine for the study of hemodynamics [4], aerosols [5], and disease diffusion [6]; in economics to predict the variation of stock market [7]; and in automation to detect target movements [8].

In the following section, the relationship between DMD and other modal decompositions is briefly discussed. In Sect. 8.3, DMD is described in detail: the theoretical framework (Sect. 8.3.1), a short excursus of published applications (Sect. 8.3.2), and the algorithm created for the analysis of massive databases (Sect. 8.3.3). This material is followed by the application of DMD to two large datasets from Direct Numerical Simulation (DNS) of turbulent premixed planar jet flames.

8.2 Modal Decompositions: POD and DMD

Proper orthogonal decomposition (POD), perhaps the most common modal decomposition in the fluids community, is a data analysis method that identifies the dominant structures in a dataset in terms of energy content [9–12]. POD identifies the set of orthogonal modes that best reconstructs a dataset. Stated differently, the first n orthogonal POD modes optimally reconstruct a dataset from an energy content perspective. In other research fields POD is also known as Principal Component Analysis [13], the Hotelling transform [14], the Karhunen–Loève transform [15], and Empirical Orthogonal Function analysis [16]. However, the POD modes are not necessarily optimal for modeling flow dynamics. This limitation has been partially overcome through the development of Balanced POD (BPOD) [17, 18] but is only applicable to fluid flows in the context of flow control due to a requirement of linear input–output dynamics.

Koopman spectral analysis [19] is even older than POD and is able to provide a spatio-temporal characterization of the dynamics capturing the regular components of a dataset containing both chaotic and regular behavior. Therefore, the Koopman eigenfuctions completely characterize the dynamics of a nonlinear system, and each of them is associated with an eigenvalue. When the dataset is sufficiently large and rich, DMD modes and eigenvalues are an approximation of the Koopman modes and eigenvalues. The capability to represent the dynamics with modes having a specific frequency is the main similarity between Koopman decomposition and DMD.

8.3 Dynamic Mode Decomposition

8.3.1 Theory

Schmid [2] presented the theoretical basis of DMD as the idea of representing nonlinear dynamics through analysis of some approximating linear system. The idea builds on the theory of Krylov subspaces [20], due to the original derivation of DMD as a variant of the Arnoldi algorithm [21]. This approach apparently limits the capability of DMD to analyze nonlinear system. However, Rowley et al. [3] addressed this limitation presenting the DMD as a particular numerical algorithm that computes an approximate Koopman decomposition. In detail, the Koopman operator is an infinite-dimensional linear operator associated with full nonlinear system of finite dimension and does not rely on linearization of the dynamics. The capability of this operator to capture the full information of the nonlinear system has been shown by Mezic and Banaszuk [22] and Mezic [23]. The spectral analysis of the Koopman operator determines the Koopman modes, which are a decomposition describing a complex nonlinear system. They are computed directly from the data using a variant of the standard Arnoldi method which is identical to the procedure proposed by Schmid for DMD.

Considering two data matrices X and Y with arbitrary but matching sizes (typically whose columns are taken from a sequential time series), the matrix A in the system $AX = Y$ defined using a right pseudoinverse yields the best-fit linear operator relating X and Y, in a least-squares/minimum-norm sense. A DMD algorithm relying on this concept, which is simply the Singular Value Decomposition (SVD), is able to produce unique modes and eigenvalues even if the elements of X are not linearly independent.

When X has linearly independent columns, DMD can be interpreted as an approximation to Koopman spectral analysis, ensuring that the modes characterize the nonlinear dynamics. This equivalence allows for alternative DMD algorithms, for instance, with concatenated datasets and nonuniformly sampled data. DMD is also related to Linear Inverse Modeling (LIM), a method developed in the climate science community [24].

Goulart et al. [25] introduced Optimal Mode Decomposition, which involves a simultaneous optimization of a modal subspace and a dynamic matrix. DMD can be interpreted as a similar optimization where the modal subspace is constrained to be spanned by POD modes. Kutz et al. [26] integrated DMD with a multi-resolution analysis with the aim to separate complex systems into a hierarchy of multi-resolution timescale components. More recently, Recursive Dynamic Mode Decomposition [27] has been proposed with the aim to comprise key features of POD and DMD. The review article by Mezic [28] offers a nice overview of DMD variants.

8.3.2 Applications in Fluid Dynamics

Many publications have used DMD to study the nonlinear phenomena in fluid dynamics and also the instabilities in reacting flows. Schmid [2] illustrated applications of his methodology in several configurations, such as plane channel flow, flow over a two-dimensional cavity, wake flow behind a flexible membrane, and a jet passing between two cylinders. Schmid and collaborators also published several papers where experimental data for various jet flows has been analyzed [29–31]. Several flow configurations have been investigated by other research groups, such as instabilities in annular liquid sheets [32], flows behind one or more cylinder [33–35], cavity flows [36–38], detonation waves [39, 40], and flow around a high-speed train [41] and behind a space launcher [42], with a limited amount of data due to both the capability to perform the analysis and the availability and/or resolution of the data itself. A discussion of the relevance of these limitations in experimental data from turbulent combustion has been discussed by Richecoeur et al. [43]. Other studies on reacting flows focused on some specific phenomena localized in a narrowband of scales: flickering in a flame [44], flames in a swirl-stabilized dump combustor [45], inducted transverse combustion instability [46], rocket combustors with self-excited longitudinal combustion instabilities [47, 48], a combustor with a nozzle [49], industrial gas turbine combustion chamber [50], and in a V-flame in an afterburner-type configuration with self-excited longitudinal and transverse modes [51].

Conversely, DNS used in this investigation provides data for the complete range of broadband phenomena at any temporal and spatial scale, and DMD is used to find the relevant coherent structures and their frequencies [52]. In addition, computational data overcomes the limitation with experimental data, such as quantity of interest, sampling frequency, dimensionality, etc., providing any variable value in any location in the domain. However, the amount of data generated makes the analysis impractical to realize without using an algorithm designed to minimize the memory footprint and the computational cost.

8.3.3 Algorithm

In order to overcome computational limitations in the analysis of very large datasets, Tu et al. [53] created a memory-efficient algorithm that only requires loading into memory two snapshots at a time. It is based on the method of snapshots introduced by Sirovich [54] which allows replacing the evaluation of the SVD of X with the inexpensive operations described below.

For data vectors arranged into two matrices of dimensions $n \times m$, $X = [\mathbf{x}_1, \ldots, \mathbf{x}_{m-1}]$ and $Y = [\mathbf{x}_2, \ldots, \mathbf{x}_m]$, where each column $\mathbf{x}_k \in \mathbb{R}^n$ contains the variables (e.g., velocity components, temperature, etc.) in the multidimensional domain in a single snapshot in time, m is the number of snapshots forming the sequence of data, and n is the dimension of the vector $\mathbf{x}_k$, the linear tangent approximation would give the system $Y = AX$. However, the linear mapping A, connecting a data vector from of one snapshot to the next one, to this point, remains unknown.

In the original DMD algorithm [2, 3], the DMD modes and Ritz values are evaluated using an approximation of A, $\tilde{A}$. This is defined, using the SVD of X, as

$$\tilde{A} = U^* Y V \Sigma^{-1}, \tag{8.1}$$

where U^* is the conjugate transpose of the orthonormal basis of X, V is the orthonormal basis of X^H, and Σ is a diagonal matrix containing the singular values of X. For very large datasets this algorithm is inefficient or even impractically expensive even using massively parallel computation. The data snapshot matrices contain tens of millions of elements and hundreds of snapshots in order to include all the scales typical of broadband phenomena, such as turbulent combustion.

Tu et al. [55] proposed an exact DMD, based on the definition $A \equiv YX^+$, where X^+ is the pseudoinverse of X. This formulation does not require that the snapshots be stored in a sequential set of data vectors but only that their elements form pairs that satisfy the linear tangent approximation. This formulation, included in the memory-efficient algorithm [53] and further improved by Belson et al. [56], makes possible overcoming the memory limitations of typical HPC machines using the method of snapshots.

The eigenvectors and eigenvalues of the correlation matrix [54], defined as the inner product of the data vectors

$$[H]_{i,j} = \langle \mathbf{x}_i, \mathbf{x}_j \rangle \tag{8.2}$$

where the indices i and j have the range $1, \ldots, m-1$, are computed by solving the eigensystem $HW = W\tilde{\Sigma}$. They allow rewriting the product U^*Y in the definition of $\tilde{A}$ (Eq. 8.1) in terms of X^*Y using the POD modes of X given by $U = XW\tilde{\Sigma}^{1/2}$,

$$\tilde{A} = \tilde{\Sigma}^{-1/2} W^* [H' H''] W \tilde{\Sigma}^{-1/2}. \tag{8.3}$$

where $H' = [H]_{1:m-1,2:m-1}$ and $H'' = <\mathbf{x}_m, \mathbf{x}_j>$ with $j = 1, \ldots, m-1$.

After solving the eigensystem for $\tilde{A}$, $\tilde{A}\tilde{V} = \Lambda\tilde{V}$, the scaled DMD modes are given by

$$\phi_{k,j} = \sum_{i=1}^{m} \lambda_i^{m-1} \mathbf{x}_{k,i} [T]_{i,j} \tag{8.4}$$

where λ_i are eigenvalues of $\tilde{A}$ stored as diagonal elements of Λ, $T = W^* \tilde{\Sigma}^{-1/2} \tilde{V} D$, and D is a diagonal matrix having diagonal elements

$$d = (\tilde{V}^* \tilde{V}) \tilde{V} \tilde{\Sigma}^{-1/2} \tilde{W}^* [H]_{1:m-1,1}, \tag{8.5}$$

with $j = 1, \ldots, m-1$, and $k = 1, \ldots, N_{var}$, which identify the mode and the variable, respectively.

Note that the correlation matrix H, Eq. 8.2, and the matrix $\tilde{A}$, as defined in Eq. 8.3, are relatively small having the dimension of the number of snapshots, typically hundreds or a few thousand. Therefore, computation of their eigenvectors and eigenvalues is computationally inexpensive.

The implementation of this algorithm into a parallel software able to efficiently optimize the I/O workload is described in detail in [52].

8.4 Dynamic Mode Decomposition Analysis

8.4.1 Test Cases

Two spatially developing turbulent premixed planar jet flames at $Re = 5,000$ and having different Karlovitz numbers [57] are considered in this work. They are composed of a central jet of bulk velocity U_0 and height H_0, which is separated by thin walls from primary coflow jets of bulk velocity U_c and height H_c, the values of which are reported in Table 8.1. Flow in the central jet is initialized from auxiliary DNS calculations of fully developed turbulent channel flow. Secondary low-velocity bulk coflows isolate the central and coflow jets from the domain boundaries. For these cases, fully developed laminar velocity profiles are specified at the inlet for the primary coflow jets, and the use of laminar coflows or turbulent coflows was found not to significantly affect the turbulence budgets within the flame.

The central jet consists of a gaseous mixture of hydrogen and oxygen at stoichiometric equivalence ratio, diluted 80.9% by mass with nitrogen, at $T_0 = 300$ K and $p_0 = 1$ atm. Equilibrium products of combustion of the same mixture issue from the coflow jets at $T_c = 2047.5$ K and $p_c = 1$ atm. A nine species hydrogen chemical kinetic model [58] is used.

The critical Karlovitz number separates two different regimes [59, 60]. For $Ka \ll Ka_{cr}$, pressure-dilatation induced by the flame is the primary source of turbulent kinetic energy because it represents the small scale source; conversely, for $Ka \gg$

Table 8.1 Simulation parameters for the low- (K1) and high-Karlovitz number (K2) cases. Karlovitz number is reported at the centerline and at $\widetilde{C} = 0.5$ for axial location $x/H_0 = 3$

case	K1	K2
H_0 (mm)	4.32	1.08
U_0 (m/s)	23.36	93.44
H_c (mm)	6.18	1.54
U_c (m/s)	6.02	24.11
Re_0	5,000	5,000
Ka_0	10.9	43.5
$Ka_{\widetilde{C}=0.5}$	2.6	32.0
Domain (x, y, z)	$12H_0 \times 24H_0 \times 3H_0$	$24H_0 \times 16H_0 \times 3H_0$
Grid size	$768 \times 586 \times 256$	$1536 \times 576 \times 256$

Ka_{cr}, mean shear is the dominant source of turbulent kinetic energy. For this mixture, the laminar flame thickness is $\delta_F = 0.435$ mm, and the laminar flame speed is $s_L = 1.195$ m/s, so the critical Karlovitz number is $Ka_{cr} = 6.7$.

The two flames investigated have Karlovitz number below (case K1) and above (case K2) the critical value. The bulk Reynolds number is kept constant, and the ratios U_0/U_c and H_0/H_c kept fixed, while the Karlovitz number is varied by modifying the turbulence strain rate U_0/H_0. Further characterization of the turbulence statistics of this configuration may be found in other works [57, 61].

To generate the DNS database, the Navier–Stokes equations are solved in the low Mach number limit using a semi-implicit iterative algorithm of Desjardins et al. [62], implemented in the code NGA. The species equations are solved with a monolithic scheme using an approximately factorized exact Jacobian [63].

The domain for case K1 has dimensions $12H_0 \times 24H_0 \times 3H_0$ in the streamwise (x), cross-stream (y), and spanwise (z) directions, respectively. The computational grid has $768 \times 586 \times 256$ points. Case K2 has a longer domain, $24H_0 \times 16H_0 \times 3H_0$, and a finer grid with $1536 \times 576 \times 256$ points. The boundary conditions are, in both cases, inflow on the $-x$ face, outflow on $+x$ face, free slip on $\pm y$ faces, and periodic in the z-direction.

The subdomain considered for the DMD analysis in case K1 has dimensions $7H_0 \times 4H_0 \times H_0$, or rather $390 \times 316 \times 86$ grid points, in the x-, y-, and z-directions, respectively. The subdomain contains about 10.6×10^6 grid points, so the snapshot dimension is $n \approx 3.3 \times 10^7$, since three variables (the three velocity components u, v, w) are considered. The number of snapshots considered is $m = 401$, and the space in time $\Delta t = 4\,\mu$s. Subsequently, for this DMD analysis, up to about 12.7 billion data values are used, or rather more than 300 GB.

In case K2, the subdomain considered has dimensions $8H_0 \times 4H_0 \times H_0$, or rather $454 \times 310 \times 86$ grid points, containing about 12.1×10^6 grid points. The amount of data for each snapshot is $n \approx 3.6 \times 10^7$, and for the complete analysis 14.6 billion values are used. The snapshots have been taken with an interval of $\Delta t = 0.325\,\mu$s.

This amount of data makes the standard DMD algorithm impossible to perform without the efficient parallel implementation of the snapshot algorithm [53, 56] presented in Sect. 8.3.3, which allows realization of each analysis in about 10,000 CPU-hours [52]. Before applying DMD, the variable values have been nondimensionalized with the values of the flow parameters in the central jet at the inlet. The total time window considered is on the order of two integral timescales, while the sampling frequency is one order of magnitude slower than the Kolmogorov timescale.

Figure 8.1 shows the instantaneous vorticity, OH mass fraction, and temperature fields for the x-y plane at the center of the domain considered for the analysis. The fields are fully developed and have reached a statistically stationary that permits temporal analysis of the dynamics with DMD without corruption from transient phenomena.

8.4.2 Results

The objective of this work is to identify through the DMD analysis the most relevant frequencies for various physical phenomena. The first subsection is devoted to the proof that the selection of the parameters of the DMD analysis satisfy the convergence criteria for both cases. In the second subsection, the most relevant DMD modes for streamwise velocity and temperature are shown, and a physical interpretation of their role is proposed.

8.4.2.1 Convergence of the Modes

The three criteria for DMD mode convergence are: decay of the residual (as defined by Schmid [2]), convergence of the eigenvalues (Ritz values), and decay of the mode amplitude [52]. In DMD analysis of DNS databases, the variation of the number of snapshots, the time spacing of the snapshots, and the choice of variables determines the convergence of the modes.

The residual in the reconstruction of the last snapshot and is influenced by both the number of snapshots, the time spacing of the snapshots. Although it decays to zero, the other criteria are more restrictive and have to be satisfied, as prescribed in the 'best practice" procedure introduced by Grenga et al. [52]. The time spanned by the snapshots, given by the product of the number of snapshots and the temporal spacing, is the most relevant factor for the convergence of the eigenvalues to the unit circle for the statistically stationary configuration considered: this must be longer than the longest integral timescale. The temporal spacing of the snapshot needs to be of order of the fastest Kolmogorov scale in order to obtain values of the amplitude of the modes that decay with increasing frequency. The total number of snapshots increases the spectral resolution of the modes. Therefore, only the three criteria combined provide a complete picture of mode convergence.

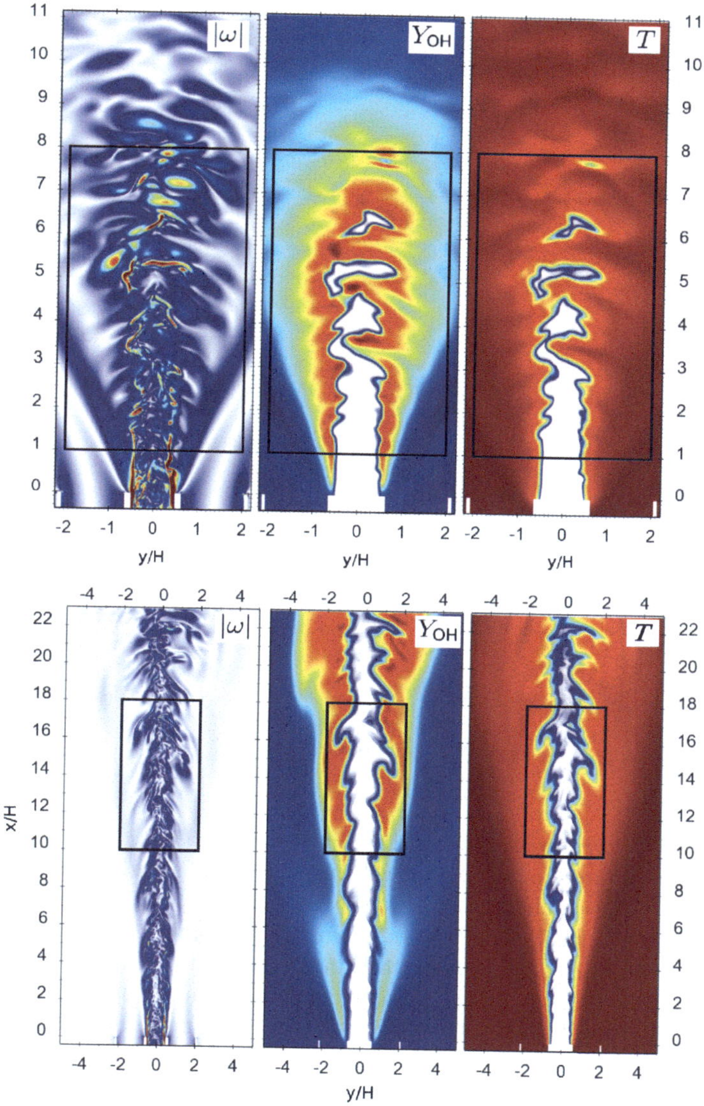

Fig. 8.1 Instantaneous images of vorticity, OH mass fraction, and temperature across a slice in the x-y plane for the case K1 (top) and K2 (bottom). The black boxes indicate the subdomains analyzed with DMD

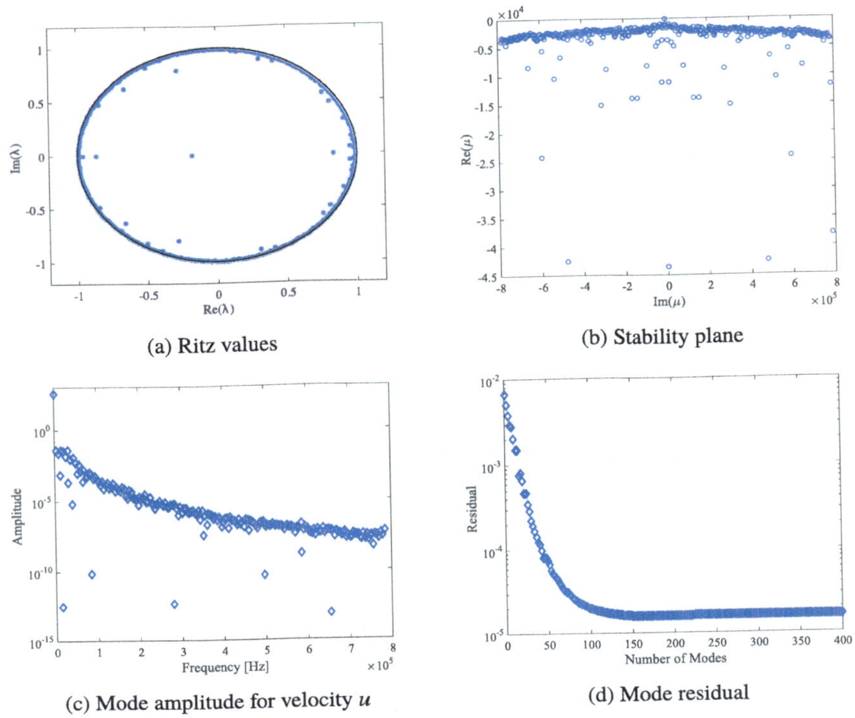

(a) Ritz values

(b) Stability plane

(c) Mode amplitude for velocity u

(d) Mode residual

Fig. 8.2 Convergence of the modes for case K1

The variables used to compute the DMD mode eigenvalues λ and eigenvectors T in Eq. 8.4 are the three components of velocity. The use of additional variables, such as temperature and species mass fractions, will not effect λ and T. The information contained in the velocity field well represents the dynamics of the system. Stated differently, the basis spanned by the DMD modes computed from velocity field data is sufficient for a comprehensive representation of the complex dynamics of the flow. The use of additional data in the analysis would not change the basis computed. Therefore, the field of any quantity of interest can be reconstructed on the basis determined from the velocity field. Additional details about this choice are provided in [52].

Figures 8.2a and 8.3a show the DMD mode eigenvalues, locations with respect to the unit circle $|\lambda_i| = 1$ for case K1 and K2, respectively. The unit circle represents a criterion for the stability of the modes. The eigenvalues lying outside the circle would identify explosive modes, while the ones lying inside identify dissipative modes, and the distance from the circle is a measure of the explosive/dissipative nature of the modes. For this statistically stationary flow, the expectation is that all modes should lie on the unit circle, which is the case as shown in the figures.

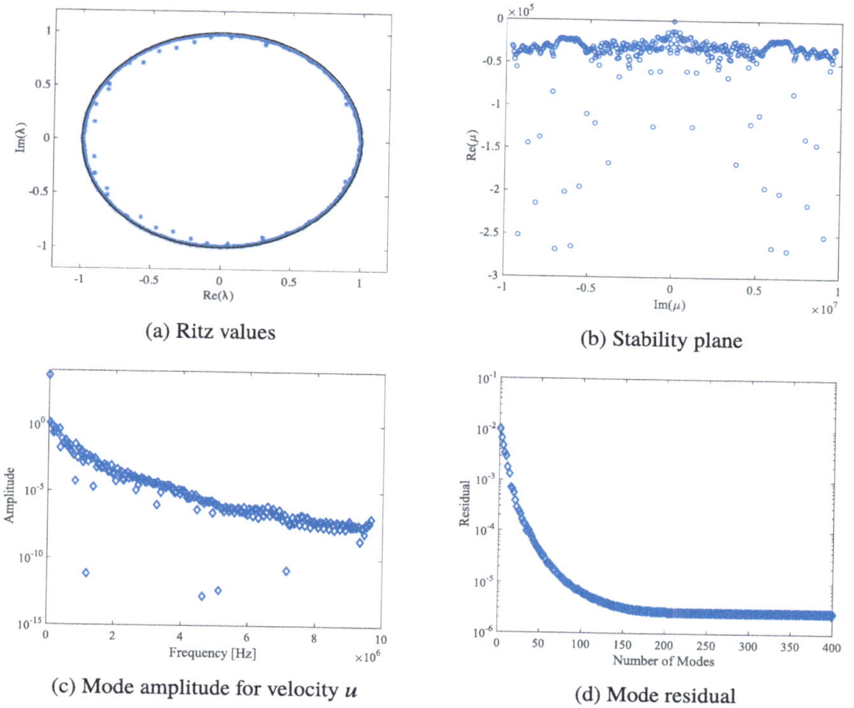

(a) Ritz values

(b) Stability plane

(c) Mode amplitude for velocity u

(d) Mode residual

Fig. 8.3 Convergence of the modes for case K2

The Ritz values are further transformed into the complex stability plane, shown in Figs. 8.2b and 8.3b, through the logarithmic transformation $\mu_i = \log(\lambda_i)/\Delta t$, where Δt is the sampling time step. The real part of μ_i is the exponential growth or decay rate of the ith mode, and the imaginary part represents the temporal frequency.

The eigenvalue corresponding to the first DMD mode is real and has real part nearly unity, so its value in the stability plane is zero. Figures 8.2b and 8.3b show that most of the modes have a relatively slow dissipation rate, and that the modes with the slowest frequencies are also the ones with slowest decay rate. The modes represented span the frequency range from 0 to 0.8 MHz for case K1 and to 10 MHz for case K2; however, according to the Nyquist criterion, only the modes having a frequency slower than half of the sampling frequency (or rather 125 kHz for case K1 and 1.5 MHz for case K2) represent actual physical phenomena, and the others can be classified as numerical noise.

The amplitude of each mode is evaluated as

$$|\phi_{k,j}|^2 = \frac{1}{\forall} \int \phi_{k,j}^2 dV, \tag{8.6}$$

where $\forall$ is the volume of the subdomain. Figures 8.2c and 8.3c show the amplitudes of the modes for the streamwise velocity versus the frequency. In both cases, there is a change of four orders of magnitude between the first mode and the next one. The first mode, which is stationary and has a frequency of zero, represents the mean value over the time interval considered. The amplitudes of the modes then decay with increasing frequency as is expected for broadband turbulent phenomena.

The values of each field for the last snapshot can be reconstructed from the DMD modes [2, 56], with an accuracy depending on both the number of snapshots and the time step used in the analysis. The normalized residual is defined as the normalized difference between the reconstruction of the last snapshot and the actual data:

$$r_{k,j} = \frac{\sum_{i=1}^{m-1} \phi_{k,i} \lambda^{j-1} - \mathbf{x}_{k,j}}{\mathbf{x}_{k,j}}. \tag{8.7}$$

The residual magnitude is evaluated as

$$|r_{k,j}|^2 = \frac{1}{\forall} \int r_{k,j}^2 dV, \tag{8.8}$$

where the square is an element-wise operation.

Figures 8.2d and 8.3d show the residual for the streamwise velocity versus the number of modes. The residual decreases as a larger number of modes and a smaller sampling time step are used for the reconstruction [52]. The choice of $m = 401$ for both cases and $\Delta t = 4\,\mu\text{s}$ and $0.325\,\mu\text{s}$ for case K1 and K2, respectively, allows the residual magnitude to converge to the minimum possible value.

8.4.2.2 Features of the Modes

The normalized amplitudes of the DMD modes of the streamwise velocity and temperature corresponding to the 15 lowest frequencies are shown in Fig. 8.4 for case K1. Note that each frequency corresponds a pair of complex conjugate modes, with the exception of mode 1 (the mean), each with the same amplitude. Therefore, the values represented in Fig. 8.4 correspond to the amplitudes of the 29 modes having the 15 lowest frequencies. These modes allow reconstruction of the velocity field with a residual (Eq. 8.8) of the order of 10^{-4}. Figure 8.4 shows the normalized amplitude for the velocity component u and the temperature for these modes (essentially a zoom-in of Fig. 8.2c) and highlights the value of the frequency of each mode and the amplitude decay rate with respect to frequency. The amplitude values are normalized by the amplitude of the first mode. As mentioned above, the amplitude decreases by four orders of magnitude after the first mode and then decays slowly. In other words, the energy content of the first mode, representing the mean flow, is four orders of magnitude larger that the following modes which represent the fluctuation. In addition, the energy content of fluctuations deceases with increasing frequency. The red

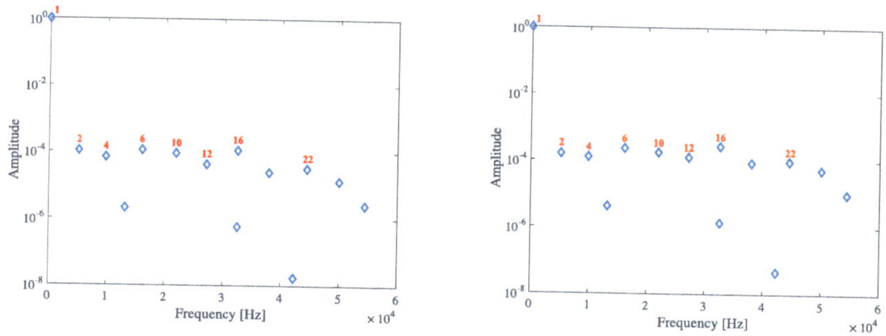

(a) Streamwise velocity

(b) Temperature

Fig. 8.4 Normalized amplitude of the most relevant modes for the streamwise velocity u and the temperature for case K1

numbers indicate the mode number for the modes with the largest amplitude that will be analyzed in detail.

The spatial structure of these modes is shown in Fig. 8.5 for the streamwise component of the velocity u for case K1. In this and subsequent figures, the modes are ordered by increasing frequency.

The mode with the largest amplitude is mode 1, Fig. 8.5a, which represents the mean of the field in the time window considered. This mode has a real eigenvalue with real part equal to one, $\lambda_1 = 1$, which implies that its growth/decay rate is zero, $\mathrm{Re}(\mu) = 0$, and the frequency also zero, $\mathrm{Im}(\mu) = 0$. Therefore, DMD is able to identify autonomously the mean of the field.

For case K1, the second mode ranked by amplitude is mode 16 represented in Fig. 8.5g. This mode, which has frequency 32 kHz, shows the small structures developing in the main jet and promoted by combustion, as discussed further below. The following modes, in order of amplitude, are 6 and 2 (Fig. 8.5d and b, respectively), which have a lower frequency and illustrate the largest structures in the flow, representing the interaction between the jet and the coflow. Mode 4 (Fig. 8.5c) has smaller amplitude compared to modes 2 and 6 but represents similar structures. Mode 10 and 12 (Fig. 8.5e and f, respectively) represent faster oscillating structures (about 25 kHz) at smaller scales than the lower frequency modes. The characteristic timescale of these modes is of the same order as the large-scale timescale (29 kHz at the inlet). Finally, mode 22 in Fig. 8.5, with a frequency of 44.5 kHz, is similar in structure to mode 16 and represents the development of structures in the jet promoted by combustion.

Figure 8.6 shows the DMD modes for the temperature field. The modes are obtained using Eq. 8.4 in which $\mathbf{x}_{k,i}$ is the ith snapshot of the temperature field, while λ_i and $[T]_{i,j}$ are obtained from the DMD analysis of the velocity fields.

As for the DMD modes for streamwise velocity, Fig. 8.6a represents the mean of the temperature field in the interval considered. There are several similarities in case K1 between the modes for temperature, Fig. 8.6, and velocity, Fig. 8.5. In comparison

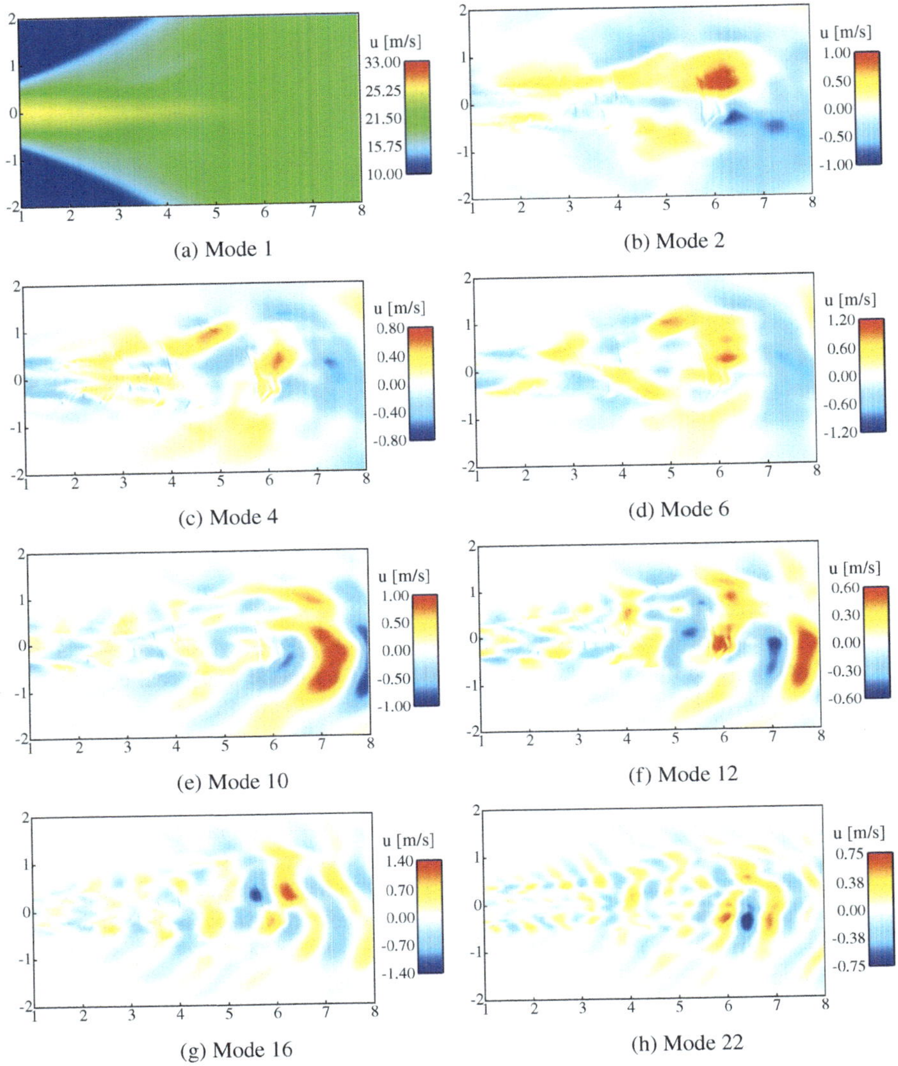

Fig. 8.5 Modes for the streamwise velocity u, case K1

the only change is that the amplitude of mode 10 becomes large than mode 2 (see Fig. 8.4). Mode 16, Fig. 8.6g, has the largest amplitude aside from the mean and is concentrated in the region of combustion. Given the structure of this mode, it seems to indicate the influence of combustion. Interestingly, this "combustion" mode is the largest in magnitude aside from the mean also for the streamwise velocity. This indicates a leading effect of combustion on the velocity dynamics, consistent with prior studies that have shown the dominant influence of combustion-induced dilatation on the turbulent kinetic energy [57, 61].

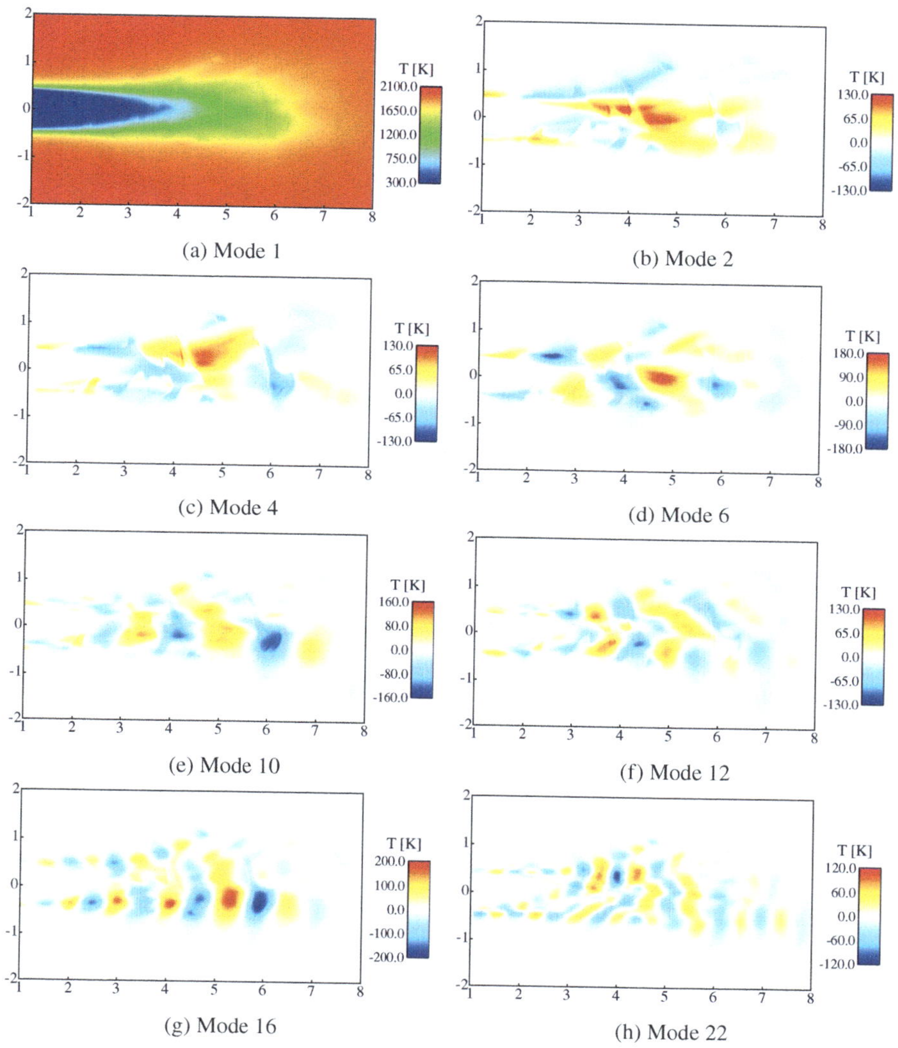

Fig. 8.6 Modes for the temperature, case K1

Figure 8.7, similar to Fig. 8.4, shows the normalized amplitudes for streamwise velocity and temperature for the first several modes for case K2 (essentially a zoom-in of Fig. 8.3c). The amplitude values are normalized with the amplitude of the first mode in order to highlight the relative importance of the modes and the decay rate. The red numbers indicate the mode number of the modes that are analyzed in detail since they have the largest amplitude.

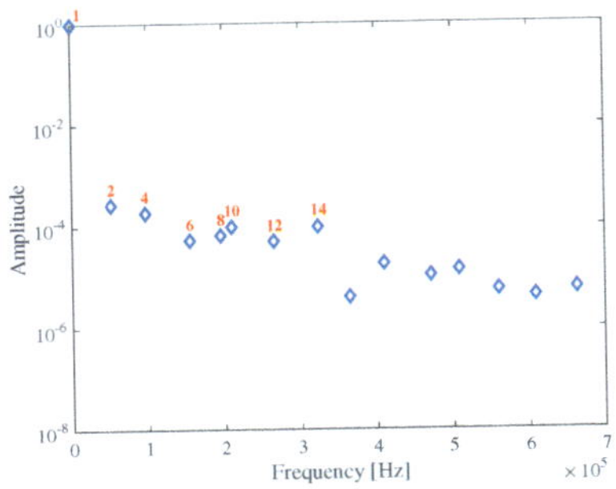

(a) Streamwise velocity

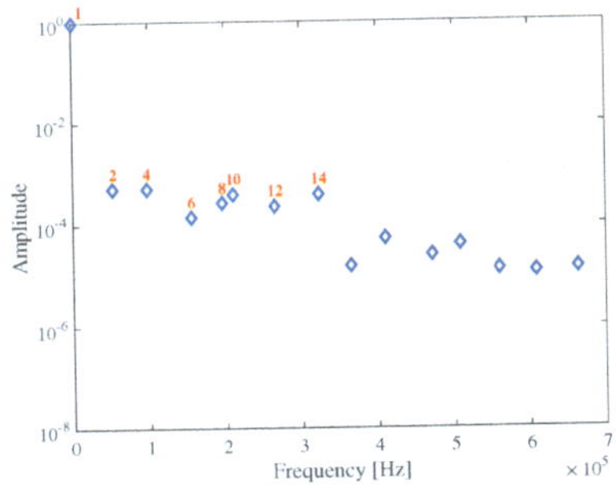

(b) Temperature

Fig. 8.7 Normalized amplitude of the most relevant modes for the streamwise velocity u and the temperature for the case K2

The modes for the streamwise component of velocity are shown in Fig. 8.8 for case K2. As for case K1, mode 1 has the largest amplitude (Fig. 8.8a) and represents the mean values of the fields in the time window considered.

For case K2, the rank of the modes by frequency also essentially follows the order by amplitude up until the last mode shown. All of these modes are successively smaller and faster flow structures that exhibit some similarity, for example, anti-symmetric and aligned with the streamwise direction. However, the highest frequency mode, Mode 14 (and perhaps to a lesser degree Modes 8 and 10), is qualitatively different with inclined structures with large magnitude extending into the coflow. These inclined structures are indicative of shear [64] and indicate its dominant influence on the velocity field at this higher Karlovitz number, as expected from previous studies [57, 61].

Figure 8.9 shows the DMD modes for the temperature field for case K2, obtained following the same procedure used for case K1.

As for the DMD modes for velocity, Fig. 8.9a represents the mean value of the temperature field in the interval considered. For case K2, the trends follow similarly to the velocity field. For the higher frequency modes, the inclined structures indicative of shear are even more evident. Additionally, compared to case K1, modes localized to the flame are less apparent.

For case K1, below the critical Karlovitz number, the combustion phenomena are more relevant than in the high-Karlovitz number case. Indeed, for case K1 the second most important mode after the mean is mode 16, which has a relatively high frequency and involves the combustion phenomena. Conversely, for case K2, the second and third most important modes after the mean are the slowest modes that represent the largest structures in the flow generated by the interaction between the jet and the coflow. These results are quite remarkable. Adding some physical interpretation, DMD is able to correctly identify the dominant physics in the two jet flames, notably the strong influence of combustion heat release on turbulence at low Karlovitz number and a reduced influence at high-Karlovitz number. No special conditioning of the data is required.

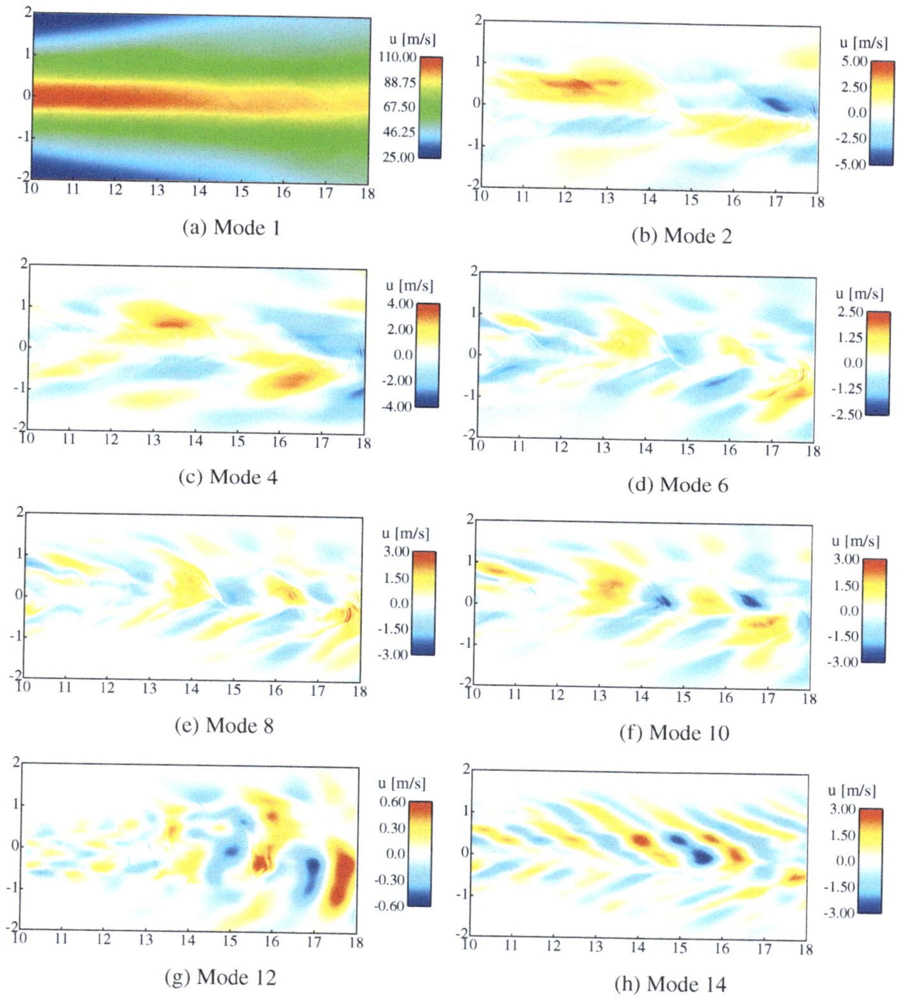

Fig. 8.8 Modes for the streamwise velocity u, case K2

8.5 Conclusion

DMD is a powerful method for analyzing dynamics of nonlinear systems using data generated either computationally or experimentally. The interest is the application of DMD to broadband turbulent combustion phenomena to identify relevant coherent structures and their frequencies.

Computational database from DNS, unlike experimental data, provides the complete range of spatial and temporal scales and also all fluid-dynamic and thermo-chemical quantities at any location and at any time in the three-dimensional domain.

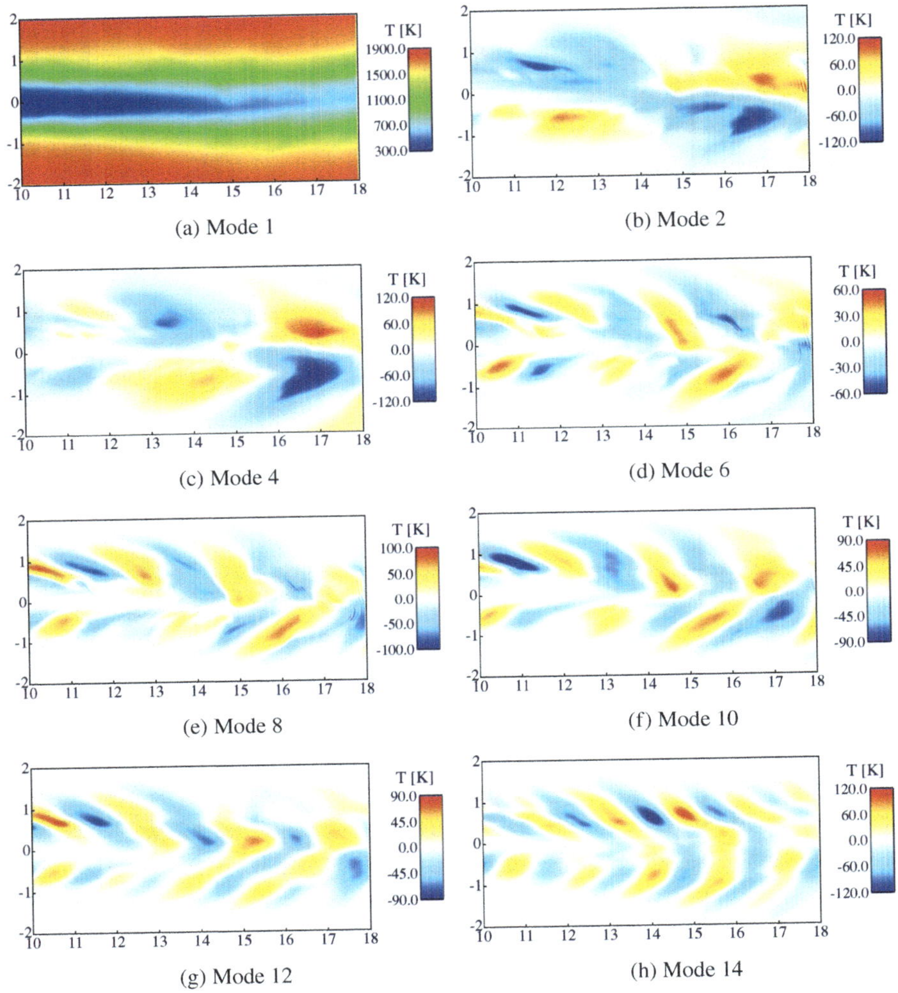

Fig. 8.9 Modes for the temperature, case K2

However, the amount of data generated makes the analysis impractical without using an algorithm specifically designed to minimize memory requirements and computational cost. The algorithm presented here, in conjunction with the additional criteria for DMD mode convergence, are the instruments for performing the DMD analysis efficiently and effectively.

This work also aims to shows the high potential of DMD as a data-driven tool to elucidate multi-physics, multi-scale interactions in turbulent combustion. The results show the capability of DMD for investigating a problem having broadband phenomena and for identifying a subset of modes, each of them evolving at specific frequency, which provide a major contribution to the flow field dynamics and characterize the physical phenomena. DMD has the capability to autonomously identify the relevant physical phenomena, highlighting how the influence of the heat release on turbulence changes with the Karlovitz number.

The analysis of the two cases having Karlovitz number below (case K1) and above (case K2) the critical value confirms that in the first case, the combustion phenomena perturb the evolution of the flow field, while in the second case the shear dominates.

Acknowledgements The authors gratefully acknowledge valuable support in the form of computational time on the TIGRESS high-performance computer center at Princeton University, which is jointly supported by the Princeton Institute for Computational Science and Engineering (PICSciE) and the Princeton University Office of Information Technology's Research Computing department.

References

1. P.J. Schmid, J.L. Sesterhenn, Bull. Am. Phys. Soc. (2008)
2. P.J. Schmid, J. Fluid Mech. **656** (2010)
3. C.W. Rowley, I. Mezić, S. Bagheri, P. Schlatter, D.S. Henningson, J. Fluid Mech. **641** (2009)
4. Y. Delorme, A.E. Kerlo, K. Anupindi, M.D. Rodefeld, S. Frankel, Fluid Dyn. Res. **46** (2014)
5. J. Xi, W. Zhao, PloS One **14**(1) (2019)
6. J.L. Proctor, P. Welkhoff, Int. Health **7** (2015)
7. L. Cui, W. Long, Phys. A: Stat. Mech. Its Appl. **461** (2016)
8. J. Yin, B. Liu, G. Zhu, Z. Xie, Sensors **18**(10) (2018)
9. J.L. Lumley, Atmospheric turbulence and radio wave propagation (1967)
10. J.L. Lumley, *Stochastic Tools in Turbulence* (Courier Corporation, North Chelmsford, 2007)
11. J.L. Lumley, *Transition and Turbulence* (Elsevier, Amsterdam, 1981)
12. P. Holmes, J. Lumley, G. Berkooz, *Cambridge Monographs on Mechanics* (Cambridge University Press, Cambridge, 1996)
13. K. Pearson, LIII. On lines and planes of closest fit to systems of points in space (1901)
14. H. Hotelling, J. Educ. Psychol. **24**(6) (1933)
15. M. Loeve, *Graduate Texts in Mathematics* (1978)
16. E.N. Lorenz, Empirical orthogonal functions and statistical weather prediction. Massachusetts Institute of Technology, Department of Meteorology, Cambridge, 1956
17. M. Ilak, C.W. Rowley, Phys. Fluids **20**(3) (2008)
18. J.R. Singler, Numer. Math. **121**(1) (2012)
19. B.O. Koopman, Proc. Natl. Acad. Sci. **17**(5) (1931)
20. K.K. Chen, J.H. Tu, C.W. Rowley, J. Nonlinear Sci. **22**(6) (2012)
21. Y. Saad, Linear Algebr. Its Appl. **34** (1980)
22. I. Mezić, A. Banaszuk, Phys. D: Nonlinear Phenom. **197**(1–2), 101 (2004)
23. I. Mezić, Nonlinear Dyn. **41**(1–3), 309 (2005)
24. C. Penland, Mon. Weather. Rev. **117**(10) (1989)
25. P.J. Goulart, A. Wynn, D. Pearson, in *51st IEEE Conference on Decision and Control* (2012)
26. J.N. Kutz, X. Fu, S.L. Brunton, SIAM J. Appl. Dyn. Syst. **15**(2) (2016)
27. B.R. Noack, W. Stankiewicz, M. Morzyński, P.J. Schmid, J. Fluid Mech. **809** (2016)
28. I. Mezić, Annu. Rev. Fluid Mech. **45** (2013)
29. P.J. Schmid, L. Li, M.P. Juniper, O. Pust, Theor. Comput. Fluid Dyn. **25**(1) (2011)
30. P.J. Schmid, Exp. Fluids **50**(4) (2011)
31. P.J. Schmid, D. Violato, F. Scarano, Exp. Fluids **52**(6) (2012)
32. D. Duke, D. Honnery, J. Soria, J. Fluid Mech. **691** (2012)

33. Q. Zhang, Y. Liu, S. Wang, J. Fluids Struct. **49** (2014)
34. S. Camarri, B.E. Fallenius, J.H. Fransson, J. Fluid Mech. **715** (2013)
35. S. Sarkar, S. Ganguly, A. Dalal, P. Saha, S. Chakraborty, Int. J. Heat Fluid Flow **44** (2013)
36. P.J. Schmid, K.E. Meyer, O. Pust, in *8th International Symposium on Particle Image Velocimetry-PIV09* (2009), 3
37. A. Seena, H.J. Sung, Int. J. Heat Fluid Flow **32**(6) (2011)
38. A. Seena, H.J. Sung, Int. J. Heat Fluid Flow **44** (2013)
39. L. Massa, R. Kumar, P. Ravindran, Phys. Fluids **24**(6) (2012)
40. R. Kumar, L. Massa, in *42nd AIAA Fluid Dynamics Conference and Exhibit* (2012)
41. T.W. Muld, G. Efraimsson, D.S. Henningson, Comput. Fluids **57** (2012)
42. V. Statnikov, T. Sayadi, M. Meinke, P. Schmid, W. Schröder, Phys. Fluids **27**(1) (2015)
43. F. Richecoeur, L. Hakim, A. Renaud, L. Zimmer (Center for Turbulence Research, Stanford University, 2012)
44. C. Souvick, M. Achintya, S. Swarnendu, in *N3L-International Summer School and Workshop on Non-Normal and Nonlinear* (2013)
45. S. Ghosal, in *ASME 2016 Dynamic Systems and Control Conference* (2016)
46. J.M. Quinlan, B.T. Zinn, in *50th AIAA/ASME/SAE/ASEE Joint Propulsion Conference* (2014)
47. C. Huang, W.E. Anderson, M.E. Harvazinski, V. Sankaran, AIAA J (2016)
48. A. Ghani, L. Gicquel, T. Poinsot, in *ASME Turbo Expo 2015: Turbine Technical Conference and Exposition* (2015)
49. E. Motheau, F. Nicoud, T. Poinsot, J. Fluid Mech. **749** (2014)
50. A. Abou-Taouk, S. Sadasivuni, D. Lörstad, B. Ghenadie, L.E. Eriksson, in *Proceedings of the 7th European Combustion Meeting* (2015)
51. A. Ghani, T. Poinsot, L. Gicquel, G. Staffelbach, Combust. Flame **162**(11) (2015)
52. T. Grenga, J. MacArt, M. Mueller, Combust. Theory Model. **22**(4) (2018)
53. J.H. Tu, C.W. Rowley, J. Comput. Phys. **231**(16) (2012)
54. L. Sirovich, Q. Appl. Math. **45**(3) (1987)
55. J.H. Tu, C.W. Rowley, D.M. Luchtenburg, S.L. Brunton, J.N. Kutz (2013), arXiv:1312.0041
56. B.A. Belson, J.H. Tu, C.W. Rowley, ACM Trans. Math. Softw. **40**(4) (2014)
57. J. MacArt, T. Grenga, M. Mueller, Combust. Flame **191** (2018)
58. S.G. Davis, A.V. Joshi, H. Wang, F. Egolfopoulos, Proc. Combust. Inst. **30**(1) (2005)
59. R.W. Bilger, Flow, Turbul. Combust. **72**(2) (2004)
60. S. Zhang, C.J. Rutland, Combust. Flame **102**(4) (1995)
61. J. MacArt, T. Grenga, M. Mueller, Proc. Combust. Inst. **37**(2) (2019)
62. O. Desjardins, G. Blanquart, G. Balarac, H. Pitsch, J. Comput. Phys. **227** (2008)
63. J.F. MacArt, M.E. Mueller, J. Comput. Phys. **326** (2016)
64. T. Sayadi, P.J. Schmid, Theor. Comput. Fluid Dyn. **30**(5), 415 (2016)

Chapter 9
Physics-Informed Data-Driven Prediction of Turbulent Reacting Flows with Lyapunov Analysis and Sequential Data Assimilation

Luca Magri and Nguyen Anh Khoa Doan

Abstract High-fidelity simulations of turbulent reacting flows enable scientific understanding of the physics and engineering design of practical systems. Whereas Direct Numerical Simulation (DNS) is the most suitable numerical tool to understand the physics, under-resolved and large-eddy simulations offer a good compromise between accuracy and computational effort in the prediction of engineering flows. This compromise speeds up the computations but reduces the space-and-time accuracy of the prediction. The objective of this chapter is to (i) evaluate the predictability horizon of turbulent simulations with chaos theory, and (ii) enable the space-and-time-accurate prediction of rare and transient events using a Bayesian statistical learning approach based on data assimilation. The methods are applied to DNS of Moderate or Intense Low-oxygen Dilution (MILD) combustion. The predictability provides an estimate of the time horizon within which the occurrence of ignition kernels and deflagrative modes, which are considered here as rare and transient events, can be accurately predicted. The accurate detection of ignition kernels and their evolution towards deflagrative structures are well captured on a coarse (under-resolved) grid when data is assimilated from a costly refined DNS. Physically, such an accurate prediction is important to understand the stabilization mechanism of MILD combustion. These techniques enable the space-and-time-accurate prediction of rare

L. Magri (✉)
Department of Engineering, University of Cambridge, Trumpington Street, Cambridge CB2 1PZ, UK
e-mail: lm547@cam.ac.uk

(visiting) Institute for Advanced Study, Technische Universität München, Lichtenbergstraße 2a, 85747 Garching, Germany

N. A. K. Doan
Fakultät für Maschinenwesen, Technische Universität München, Boltzmannstraße 15, 85747 Garching, Germany
e-mail: doan@tfd.mw.tum.de

Institute for Advanced Study, Technische Universität München, Lichtenbergstraße 2a, 85747 Garching, Germany

H. Pitsch and A. Attili (eds.), *Data Analysis for Direct Numerical Simulations of Turbulent Combustion*, https://doi.org/10.1007/978-3-030-44718-2_9

and transient events in turbulent flows by combining under-resolved simulations and experimental data, for example, from engine sensors. This opens up new possibilities for on-the-fly calibration of reduced-order models for turbulent reacting flows.

9.1 Introduction

Turbulent flows are chaotic flows. As such, the space-and-time accurate prediction of the solution is difficult to achieve because of the butterfly effect [29]: Two nearby initial conditions, which can differ by a very small amount, will practically diverge in time from each other with an exponential rate. This divergence rate is the dominant Lyapunov exponent. Whereas the statistics of turbulent flows are not significantly affected by tiny perturbations, the instantaneous solution is. (Having accurate predictions on the instantaneous solution is crucial for the prediction of rare and transient events.) For example, running the same code with the same initial conditions on a different number of processors should in principle provide two statistically equivalent solutions, but with completely different instantaneous fields after a few time steps. This fact is well-known in turbulent flows [5, 15, 37]. Albeit the butterfly effect seems to be a showstopper for the time-accurate prediction of turbulent flows, such a prediction can be greatly aided by physics-informed data-driven methods. In this chapter, first, we use dynamical systems theory to estimate the predictability of simulations with the Lyapunov time. Second, we improve the space-and-time-accurate prediction of rare and transient events with statistical learning by data assimilation. The turbulent reacting flow under investigation is relevant to Moderate or Intense Low-oxygen Dilution (MILD) combustion.

Dynamical systems theory provides the predictability of a chaotic simulation, which is the time scale after which the trajectories diverge due to the butterfly effect. There exist different approaches to characterize a chaotic solution [5, 15, 38]. On one hand, geometric approaches estimate the fractal dimension of the chaotic attractor, which provides an estimate of the active degrees of freedom of the chaotic dynamical system. An accurate measure is the Hausdorff dimension [19], which is often approximated by box counting, based on phase-space partitioning and correlation dimension based on time series analysis [24]. On the other hand, dynamical approaches estimate the entropy content of the solution, for example, via the Kolmogorov–Sinai entropy, and the separation rate of two nearby solutions via the Lyapunov exponents. The dominant Lyapunov exponent is a practical measure to compute the predictability of large-scale simulations because it (i) is easy to calculate [5] and (ii) does not depend on the initial conditions in ergodic processes [22]. In large-scale fluid-dynamics simulations, the Lyapunov exponent was calculated in channel and bluff-body flows[4], homogeneous isotropic turbulence [31, 32], reacting and non-reacting turbulent jets [32], a two-dimensional airfoil [20], backward-facing step [33], partially-premixed flames [23], to name only a few.

Once the predictability of the simulation is estimated, we assimilate data from a refined DNS on a coarse grid by statistical learning. Statistical learning by data assimilation finds the statistically optimal combination of model predictions and observations by combining concepts from control theory, probability theory and dynamic programming. The data assimilation technique used in this study is the ensemble Kalman filter [6, 17]. In the ensemble Kalman filter, a Monte Carlo approach is used to estimate the statistics at every time step [14], which makes it (i) a computationally efficient technique in terms of storage requirements, (ii) amenable to parallel computing and (iii) non-intrusive. Compared to other data assimilation techniques based on the Kalman filter, e.g. the extended Kalman filter, the ensemble Kalman filter is particularly suitable for highly nonlinear systems [18], such as turbulent diffusion flames [28] and premixed flames [42, 43], the latter of which display strongly nonlinear events, such as cusp formation and pinch-off. In non-reacting flows, the ensemble Kalman filter was applied to the prediction of aerodynamic flows [9, 10]. Data assimilation is practically useful to *make a qualitative reduced-order model quantitatively accurate*.

MILD combustion is a concept for future low-emission and high-energy efficiency combustion devices. This concept utilizes exhaust gas recirculation (EGR) to achieve a high reactant temperature and low-oxygen environment, which are crucial to obtain high energy efficiency and low pollutant emissions [8]. However, because of those particular turbulence and chemical conditions with recirculating gases carrying radicals, the physical understanding of MILD combustion is challenging. Indeed, in contrast to conventional combustion, reaction zones in MILD combustion have been shown both experimentally and from DNS to be volumetrically distributed [35], interacting with each other leading to a complex morphology [13, 30]. Furthermore, both ignition and deflagrative modes of combustion coexist, whose interaction contributes significantly to the global heat release dynamics [11]. However, despite the insight gained in these previous studies, the mechanism leading to the occurrence of the ignition kernels and the stabilization of deflagrative flames remains unknown, making their prediction a hard task. The objective of this chapter is to (i) calculate the predictability of the MILD combustion case under investigation by determining its Lyapunov exponent for different grids, and (ii) make an under-resolved DNS more predictive for the space-and-time-accurate evolution of chemical reactions, in particular, the occurrence of ignition kernels and their evolution into deflagrative structures. The second goal is achieved by statistical learning with Bayesian data assimilation, in particular, with the ensemble Kalman filter.

In Sect. 9.2, the fundamentals of Lyapunov analysis from chaos theory are presented to estimate the predictability time. A practical and non-intrusive algorithm for the calculation of the (dominant) Lyapunov exponent and corresponding covariant Lyapunov vector is described. In Sect. 9.3, the ensemble Kalman filter is explained with a Bayesian approach. The algorithm is shown step by step for parallel computing. In Sect. 9.4, the MILD combustion simulation is explained along with the DNS setup. In Sect. 9.5, the predictability and ensemble Kalman filtered DNS are computed for MILD combustion. The chapter ends with conclusions and future directions.

9.2 Lyapunov Analysis from Chaos Theory

This section introduces the key concepts and results of Oseledets' theorem [34], which lay out the fundamentals of Lyapunov analysis [21]. The turbulent reacting flow problem is governed by partial differential equations, i.e. the Navier–Stokes equations, mass and energy conservation, and equations for the chemistry. On spatial discretization, the turbulent reacting problem can be cast as a nonlinear dynamical system

$$\begin{cases} \dot{\boldsymbol{q}}(t) = \boldsymbol{F}(\boldsymbol{q}(t), \boldsymbol{\theta}) \\ \boldsymbol{q}(0) = \boldsymbol{q}_0 \end{cases}, \tag{9.1}$$

where $\dot{()} \equiv d()/dt$; $\boldsymbol{q} \in \mathbb{R}^{N_{dof}}$ is the state vector (e.g. pressure, velocity, species, etc., at each discrete location), where N_{dof} denotes the degrees of freedom from numerical discretization of the spatial derivatives; $\boldsymbol{\theta}$ is the vector containing the parameters of the system; and $\boldsymbol{F} : \mathbb{R}^{N_{dof}} \rightarrow \mathbb{R}^{N_{dof}}$ is a nonlinear function, which encapsulates the discretized boundary conditions. The subscript 0 denotes the initial condition. In this chapter, the parameters are constant, therefore, the dependency on $\boldsymbol{\theta}$ will be dropped unless necessary for clarity. In Lyapunov analysis, we calculate the evolution of infinitesimal perturbations. The solution is split, accordingly, as

$$\boldsymbol{q}(t) = \bar{\boldsymbol{q}}(t) + \boldsymbol{q}'(t), \tag{9.2}$$

where $\bar{\boldsymbol{q}}(t)$ is the original (unperturbed) solution of (9.1), and $\boldsymbol{q}'(t)$ is the infinitesimal perturbation such that $||\boldsymbol{q}'(t)|| \sim O(\epsilon)$, where $\epsilon \rightarrow 0$. The perturbation is governed by the tangent equation

$$\begin{cases} \dot{\boldsymbol{q}}' = \boldsymbol{J}(t)\boldsymbol{q}' \\ \boldsymbol{q}'(0) = \boldsymbol{q}'_0 \end{cases}, \tag{9.3}$$

where $\boldsymbol{J}(t) \equiv \frac{d\boldsymbol{F}}{d\boldsymbol{q}}\big|_{\bar{\boldsymbol{q}}(t)}$ is the Jacobian. In a turbulent flow, the Jacobian is time-dependent because the unperturbed solution is chaotic. The formal evolution of the perturbation, $\boldsymbol{q}'$, is given by the tangent propagator, which maps $\boldsymbol{q}'$ from time t to time $t + \tilde{t}$, as

$$\boldsymbol{q}'(t + \tilde{t}) = \boldsymbol{M}(t, \tilde{t})\boldsymbol{q}'(t), \tag{9.4}$$

where $\dot{\boldsymbol{M}} = \boldsymbol{J}(\tilde{t})\boldsymbol{M}$ with $\boldsymbol{M}(t, 0) = \boldsymbol{I}$ and $\boldsymbol{I}$ being the identity matrix. From Eq. (9.3), the norm of an infinitesimal perturbation, $\boldsymbol{q}'_0$, asymptotically evolves as an exponential [34]

$$||\boldsymbol{q}'(\tilde{t})|| = ||\boldsymbol{q}'_0||e^{\lambda(\boldsymbol{q}'_0, \bar{\boldsymbol{q}})\tilde{t}} \qquad \tilde{t} \rightarrow \infty. \tag{9.5}$$

The characteristic Lyapunov exponent is defined as

$$\lambda(\boldsymbol{q}'_0, \bar{\boldsymbol{q}}) = \lim_{\tilde{t}\to\infty} \frac{1}{\tilde{t}} \log \frac{||\boldsymbol{M}(0,\tilde{t})\boldsymbol{q}'_0||}{||\boldsymbol{q}'_0||} = \lim_{\tilde{t}\to\infty} \frac{1}{\tilde{t}} \log \frac{||\boldsymbol{q}'(\tilde{t})||}{||\boldsymbol{q}'_0||}. \tag{9.6}$$

Physically, the Lyapunov exponent is the average exponential expansion rate of an infinitesimal volume of the phase space, which moves along the attractor. A chaotic system, hence a turbulent flow, has at least one positive (unstable) Lyapunov exponent. In other words, a turbulent flow is characterized by an unstable tangent space. Furthermore, in a turbulent flow, there exists a zero Lyapunov exponent, which is associated with the neutrally stable direction of the unperturbed solution, $\bar{\boldsymbol{q}}(t)$. As mathematically and numerically shown by [25, 26], (i) if the attractor is a steady solution, the Lyapunov exponents are equal to the real part of the eigenvalues of the Jacobian at the fixed point; and (ii) if the attractor is a periodic solution, the Lyapunov exponents are equal to the real part of the Floquet exponents. The Lyapunov exponent is norm-independent, however, some norms are more physically significant than others. Oseledets' theorem [34] proves that there exist N Lyapunov exponents $\lambda_1 \geq \lambda_2 \geq \cdots \geq \lambda_{N_{dof}}$, which are constants of the chaotic attractor. To each Lyapunov exponent, λ_j, there exists an associated covariant Lyapunov vector, $\boldsymbol{\phi}_j$, which is the solution of

$$\frac{d\boldsymbol{\phi}_j}{d\tilde{t}} = \boldsymbol{J}\boldsymbol{\phi}_j - \lambda_j\boldsymbol{\phi}_j. \tag{9.7}$$

In this chapter, we do not consider degenerate Lyapunov exponents, which, however, can appear in thermoacoustic oscillations [25, 26]. Equation (9.7) shows that a covariant Lyapunov vector is evolved by the tangent system and has bounded norm because the second term on the right-hand side suppresses the exponential growth. A tangent vector, $\boldsymbol{q}'_0$, can be decomposed in a basis of covariant Lyapunov vectors, $\{\boldsymbol{\phi}_1, \ldots, \boldsymbol{\phi}_{N_{dof}}\}$. Assuming ergodicity [34, 38], the Lyapunov exponents are independent of the initial conditions, i.e. $\lambda(\boldsymbol{q}'_0, \bar{\boldsymbol{q}}) = \lambda_j$. Figure 9.1 illustrates pictorially the significance of Lyapunov exponents and covariant Lyapunov vectors. Three covariant Lyapunov vectors are shown at two different instants, each associated with a different Lyapunov exponent, which can be positive (unstable, red colour), zero (neutrally stable, orange colour) and negative (stable, green colour).

9.2.1 Non-intrusive Computation of the Dominant Lyapunov Exponent and Covariant Lyapunov Vector

We focus on the dominant Lyapunov exponent and covariant Lyapunov vector. The dominant Lyapunov exponent will be denoted λ, i.e. the subscript 1 is dropped. Obtaining accurate estimates of the dominant Lyapunov exponent and covariant Lyapunov vector is straightforward even in large-scale simulations. A non-intrusive method is based on the calculation of the *separation* trajectory, also known as the *error* trajectory. The separation trajectory is the difference between two nearby trajectories (which can be Eulerian fields), which originate from two close initial conditions.

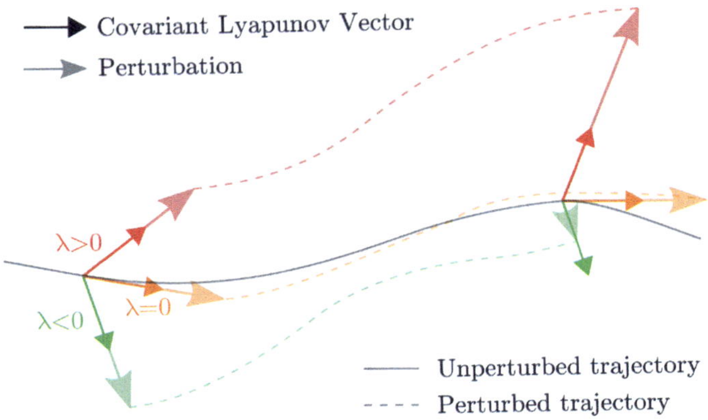

Fig. 9.1 Schematic of the significance of unstable (red), neutral (orange) and stable (green) covariant Lyapunov vectors and exponents [25]

Because it is almost sure for the separation trajectory to have a component—even tiny—in the direction of the dominant covariant Lyapunov vector, the separation trajectory will almost surely diverge along the covariant vector with an exponential divergence rate provided by the Lyapunov exponent. This is why the dominant Lyapunov exponent and covariant Lyapunov vector are of paramount importance in chaotic flows. They can be calculated as described in the following practical and non-intrusive algorithm.

1. **Statistically converged solution.** Run a numerical simulation (9.1) until statistical convergence is reached ($\bar{\boldsymbol{q}}(t)$).
2. **Reset time**, $t = t_0$.
3. **Perturb**. At $t = t_0$, evaluate the perturbed solution $\boldsymbol{q}^*$ as

$$\boldsymbol{q}^*(t_0) = \bar{\boldsymbol{q}}(t_0) + \boldsymbol{\xi}||\bar{\boldsymbol{q}}(t_0)||, \tag{9.8}$$

   where $\boldsymbol{\xi}$ is a random vector with a norm on the order of ϵ, typically in the range 10^{-9}–10^{-3}, and $||\boldsymbol{q}(t_0)||$ is a norm of order 1.
4. **Separation trajectory.** Advance both solutions, $\bar{\boldsymbol{q}}(t_0)$ and $\boldsymbol{q}^*(t_0)$, to some time t_f and evaluate the separation trajectory

$$\Delta\boldsymbol{q}(t) = \boldsymbol{q}^*(t) - \bar{\boldsymbol{q}}(t) \qquad t_0 \leq t \leq t_f. \tag{9.9}$$

5. **Identification of the linear region** $t_1 \leq t \leq t_2$ where $\ln(\|\Delta\boldsymbol{q}(t)\|)$ grows linearly. t_f in item 4 must be larger than t_2. If the linear region of the separation trajectory looks noisy, i.e. the local Lyapunov exponent markedly fluctuates, an ensemble simulation is recommended.

6. **Covariant Lyapunov vector.** This is the separation trajectory in the linear region, i.e. $\Delta \boldsymbol{q}(t)$ $t_1 \leq t \leq t_2$.
7. **Lyapunov exponent.** Because of (9.5), the Lyapunov exponent is the slope of the linear region, which can be obtained by linear regression

$$\lambda = \frac{1}{t_2 - t_1} \ln \left(\frac{\|\Delta \boldsymbol{q}(t_2)\|}{\|\Delta \boldsymbol{q}(t_1)\|} \right). \tag{9.10}$$

9.2.1.1 Predictability

Several definitions of predictability are offered in the literature [5]. However, for the purpose of this chapter, the *predictability* is defined as the *Lyapunov time*, which is the inverse of the Lyapunov exponent

$$t_p \equiv \frac{1}{\lambda}. \tag{9.11}$$

From Eq. (9.10), the predictability is the time that a norm of the separation trajectory takes to get amplified by $e \approx 2.718$. Physically, the predictability provides a time scale for the divergence of two nearby trajectories due to the chaotic nature of turbulent flows. Nastac et al. [32] showed that the Lyapunov time scale is slightly larger than the Kolmogorov time scale for both forced-homogeneous isotropic turbulence and turbulent jets. The predictability is an important factor for the time-accurate calculation of rare and transient events, such as extinction and re-ignition. Because these rare and transient events occur within a small time scale, the numerical simulation and grid should accurately resolve the dynamics within the predictability. In turbulent (reacting and non-reacting) large-eddy and direct numerical simulations, as the grid resolution approaches the smallest physical scales, the Lyapunov exponent, hence predictability, reaches a plateau [32]. Therefore, the Lyapunov exponent was proposed as a metric to assess the quality of a large-eddy simulation [32]: If grid a has the same Lyapunov exponent as grid b, the grid with fewer degrees of freedom can be used to predict rare and transient deterministic events.

9.3 Data Assimilation from Bayesian Inference

Data assimilation is a method to make a model more predictive given data from external observations. The physical model provides a prediction on the solution, which is called *forecast*, which is *updated* with a statistical learning algorithm to provide a more accurate state, which is called *analysis*. Data assimilation is here treated as a problem in Bayesian inference, where existing knowledge is quantified in the form of a probability distribution over candidate solutions. When external data becomes available, the probability distribution is updated, effectively combining the existing knowledge with the data. To introduce data assimilation as a Bayesian problem, we consider the nonlinear dynamical system (9.1) at discrete times denoted by the subscript k. The evolution of the state is governed by the deterministic physical model $\boldsymbol{F}$, its parameters $\boldsymbol{\theta}$ and the initial condition $\boldsymbol{q}_0$. The data assimilation problem is cast as a probabilistic state- space model

$$\boldsymbol{q}_k = \boldsymbol{q}_{k-1} + \int_{t_{k-1}}^{t_k} \boldsymbol{F}(\boldsymbol{q}(t), \boldsymbol{\theta})\,\mathrm{d}t \sim p(\boldsymbol{q}_k \mid \boldsymbol{q}_{k-1}, \boldsymbol{\theta}, \boldsymbol{F}), \tag{9.12}$$

$$\boldsymbol{y}_k = \boldsymbol{M}(\boldsymbol{q}_k) \sim p(\boldsymbol{y}_k \mid \boldsymbol{q}_k). \tag{9.13}$$

The states $\boldsymbol{q}_k$ and the observations $\boldsymbol{y}_k$ are considered realizations inside their respective probabilistic state spaces (symbol $\sim$), whose degrees of belief are provided by the conditional probability distributions $p(\boldsymbol{q}_k \mid \boldsymbol{q}_{k-1}, \boldsymbol{\theta}, \boldsymbol{F})$ and $p(\boldsymbol{y}_k \mid \boldsymbol{q}_k)$, respectively. We assume that (i) the physical model is described by a Markov chain; and (ii) observations are conditionally independent in time, which means that the probability of an observation depends only on the current state. The objective of data assimilation is to find the joint probability distribution $p(\boldsymbol{q}_{0:N}, \boldsymbol{y}_{1:N}, \boldsymbol{\theta}, \boldsymbol{F})$, which provides a complete statistical description because all the probability distributions of interest ensue from it. (The symbol : signifies that all states are considered, e.g. $\boldsymbol{q}_{0:N}$ signifies that the state $\boldsymbol{q}$ is considered from time step 0 to N inclusive.) First, in *state estimation*, the objective is to compute $p(\boldsymbol{q}_{0:N} \mid \boldsymbol{y}_{1:N}, \boldsymbol{\theta}, \boldsymbol{F})$, which provides the belief in a series of states given a physical model and parameters. Second, in *parameter estimation*, the objective is to compute $p(\boldsymbol{\theta} \mid \boldsymbol{y}_{1:N}, \boldsymbol{F})$, which provides the belief in a set of parameters given a physical model. Parameter estimation can be combined with state estimation, where the parameters are concatenated in the state vector such that they are subject to the same probabilistic state space (Eq. (9.12)) [43]. Third, in *model comparison*, the objective is to compute $p(\boldsymbol{F} \mid \boldsymbol{y}_{1:N})$, which provides the model that is more likely to be predictive as compared to other models. Practically, the joint probability distribution is rarely computed because the probabilistic state space, which spans multiple time steps, is high dimensional [2]. Therefore, state estimation computes the conditional probability distribution at a single time step. Conditional probability distributions that are relevant to state estimation are (i) filtering, (ii) smoothing and (iii) prediction. Filtering estimates our belief in the current state given observations in the past and present, i.e. $p(\boldsymbol{q}_k \mid \boldsymbol{y}_{1:k}, \boldsymbol{\theta}, \boldsymbol{F})$. Smoothing estimates our belief in the current state given observations in the past, future and present, i.e. $p(\boldsymbol{q}_k \mid \boldsymbol{y}_{1:N}, \boldsymbol{\theta}, \boldsymbol{F})$. Prediction estimates our belief in the future state given the observations in the past and present, i.e. $p(\boldsymbol{q}_{N+1} \mid \boldsymbol{y}_{1:N}, \boldsymbol{\theta}, \boldsymbol{F})$. In Sect. 9.5.1, we will perform state estimation without smoothing through an ensemble Kalman filter. For smoothing, parameter estimation and model performance in premixed flames, the reader may refer to [42, 43].

9.3.1 Bayesian Filter and Ensemble Kalman Filter

To introduce the ensemble Kalman filter with a Bayesian approach, we first explain the Bayesian filter. First, the state is predicted by the Chapman–Kolmogorov equation (prediction step)

$$p(\boldsymbol{q}_k \mid \boldsymbol{y}_{1:k-1}, \boldsymbol{\theta}, \boldsymbol{F}) = \int p(\boldsymbol{q}_k \mid \boldsymbol{q}_{k-1}, \boldsymbol{\theta}, \boldsymbol{F}) p(\boldsymbol{q}_{k-1} \mid \boldsymbol{y}_{1:k-1}, \boldsymbol{\theta}, \boldsymbol{F}) \, \mathrm{d}\boldsymbol{q}_{k-1}. \quad (9.14)$$

Typically, the Chapman–Kolmogorov equation, which involves a high-dimensional integral, is solved by a Monte Carlo simulation. Second, the state is updated by Bayes' rule (update step)

$$p(\boldsymbol{q}_k \mid \boldsymbol{y}_{1:k}, \boldsymbol{\theta}, \boldsymbol{F}) = \frac{p(\boldsymbol{y}_k \mid \boldsymbol{q}_k) p(\boldsymbol{q}_k \mid \boldsymbol{y}_{1:k-1}, \boldsymbol{\theta}, \boldsymbol{F})}{p(\boldsymbol{y}_k \mid \boldsymbol{y}_{1:k-1}, \boldsymbol{\theta}, \boldsymbol{F})}, \quad (9.15)$$

where the left-hand side is the posterior, the first term in the numerator of the right-hand side is the likelihood and the second term is the prior. The Bayesian filter is sequential in nature, which reduces the complexity of data assimilation and enables on-the-fly statistical learning. In the Kalman filter, the prior and the likelihood in the update step of the Bayesian filter are assumed to be normal, i.e. $p(\boldsymbol{q}_k \mid \boldsymbol{y}_{1:k-1}, \boldsymbol{\theta}, \boldsymbol{F}) = \mathcal{N}\left(\boldsymbol{q}_k \mid \boldsymbol{\psi}^f, \boldsymbol{C}^f_{\psi\psi}\right)$ and $p(\boldsymbol{y}_k \mid \boldsymbol{q}_k) = \mathcal{N}\left(\boldsymbol{y}_k \mid \boldsymbol{M}\boldsymbol{q}_k, \boldsymbol{C}_{\epsilon\epsilon}\right)$, where $\mathcal{N}$ is a normal distribution with mean in the first argument and covariance matrix in the second argument. The measurement operator, $\boldsymbol{M}$, is assumed linear in this chapter. The mean of the prior is denoted by $\boldsymbol{\psi}^f$, the covariance matrix by $\boldsymbol{C}^f_{\psi\psi}$, and the covariance matrix of the likelihood, also known as the observation error, by $\boldsymbol{C}_{\epsilon\epsilon}$. The Kalman filter [27] finds the filtered distribution as follows:

$$p(\boldsymbol{q}_k \mid \boldsymbol{y}_{1:k}, \boldsymbol{\theta}, \boldsymbol{F}) = \mathcal{N}\left(\boldsymbol{q}_k \mid \boldsymbol{\psi}^a, \boldsymbol{C}^a_{\psi\psi}\right), \quad (9.16)$$

$$\boldsymbol{\psi}^a = \boldsymbol{\psi}^f + \underbrace{\left(\boldsymbol{M}\boldsymbol{C}^f_{\psi\psi}\right)^T \left[\boldsymbol{C}_{\epsilon\epsilon} + \boldsymbol{M}\boldsymbol{C}^f_{\psi\psi}\boldsymbol{M}^T\right]^{-1}}_{\text{Kalman gain}} \underbrace{\left(\boldsymbol{y} - \boldsymbol{M}\boldsymbol{\psi}^f\right)}_{\text{Innovation}}, \quad (9.17)$$

$$\boldsymbol{C}^a_{\psi\psi} = \boldsymbol{C}^f_{\psi\psi} - \left(\boldsymbol{M}\boldsymbol{C}^f_{\psi\psi}\right)^T \left[\boldsymbol{C}_{\epsilon\epsilon} + \boldsymbol{M}\boldsymbol{C}^f_{\psi\psi}\boldsymbol{M}^T\right]^{-1} \left(\boldsymbol{M}\boldsymbol{C}^f_{\psi\psi}\right), \quad (9.18)$$

where the superscript f denotes *forecast* (everything related to the prediction), and the superscript a denotes *analysis* (everything related to the update). If the physical model is highly nonlinear, such as turbulent flows, the prediction of the covariance matrix $\boldsymbol{C}^f_{\psi\psi}$ for the Kalman filter is computed by Monte Carlo sampling. This is called the ensemble Kalman filter [17]. Instead of a mean $\boldsymbol{\psi}$ and a covariance matrix $\boldsymbol{C}_{\psi\psi}$, a distribution is represented by a sample $\boldsymbol{\psi}^j$, $j = 1, 2, \ldots, n$ where n is the number of samples. During the prediction step, the ensemble members $\boldsymbol{\psi}^j$ evolve in time independently. Before the update step, the statistics are estimated from the sample as follows:

$$\overline{\boldsymbol{\psi}} \approx \frac{1}{n}\sum_{j=1}^{n}\boldsymbol{\psi}^j \quad , \quad \boldsymbol{C}_{\psi\psi} \approx \frac{1}{n-1}\sum_{i=1}^{n}(\boldsymbol{\psi}^i - \overline{\boldsymbol{\psi}}) \otimes (\boldsymbol{\psi}^i - \overline{\boldsymbol{\psi}}), \tag{9.19}$$

where $\otimes$ is the dyadic product. The sample covariance matrix $\boldsymbol{C}_{\psi\psi}$ involves division by $n-1$ instead of n to avoid a sample bias. There exist several implementations of the ensemble Kalman filter, which differ in the update step. In the most straightforward implementation of the ensemble Kalman filter [17], each ensemble member is individually updated (Eq. (9.17)). In so doing, the observations must be randomly perturbed to guarantee a statistically consistent analysis scheme [6]. This is the approach used in this chapter. Alternatively, to avoid the introduction of randomly generated numbers in the observations, the square root filter can be used [40, 43], where the mean and the deviations of the ensemble members are updated. While no spurious errors due to the random perturbations of the observations are introduced, the square root filter requires a singular value decomposition.

9.3.1.1 Numerical Implementation of the Ensemble Kalman Filter

The Kalman gain $\boldsymbol{K} \equiv \boldsymbol{C}^f_{\psi\psi}\boldsymbol{M}^T\left[\boldsymbol{C}_{\epsilon\epsilon} + \boldsymbol{M}\boldsymbol{C}^f_{\psi\psi}\boldsymbol{M}^T\right]^{-1}$ is not explicitly computed because it is more efficient to perform a number of matrix-vector computations from right to left. From Eq. (9.17), the i-th sample of the analysis is conveniently expressed as

$$\boldsymbol{\psi}^{a,i} = \boldsymbol{\psi}^{f,i} + \underbrace{\frac{1}{n-1}\sum_{j=1}^{n}\left(\boldsymbol{\psi}^{f,j} - \overline{\boldsymbol{\psi}^f}\right)\left[\boldsymbol{M}\left(\boldsymbol{\psi}^{f,j} - \overline{\boldsymbol{\psi}^f}\right)\right]^T}_{\boldsymbol{C}_{\psi\psi}\boldsymbol{M}^T} \boldsymbol{b}^i, \tag{9.20}$$

where $\boldsymbol{K}\boldsymbol{b}^i = \boldsymbol{y}^i - \boldsymbol{M}\boldsymbol{\psi}^{f,i}$. The parallel algorithm is summarized in the following gray box using the indices i and j as in (9.20).

1. Compute $\boldsymbol{y}^i - \boldsymbol{M}\boldsymbol{\psi}^{f,i}$ on each CPU i;
2. Compute $\overline{\boldsymbol{\psi}^f}$ `MPI_Allreduce` to make it available on all CPUs;
3. Compute $\boldsymbol{\psi}^{f,j} - \overline{\boldsymbol{\psi}^f}$ on each CPU j;
4. Compute $\boldsymbol{M}\boldsymbol{C}_{\psi\psi}\boldsymbol{M}^T$ with $O(mn^2)$ operations as

$$\boldsymbol{M}\boldsymbol{C}_{\boldsymbol{\psi}\boldsymbol{\psi}}\boldsymbol{M}^T = \frac{1}{n-1}\sum_{j=1}^{n}\left[\boldsymbol{M}\left(\boldsymbol{\psi}^{f,j} - \bar{\boldsymbol{\psi}}\right)\right] \otimes \left[\boldsymbol{M}\left(\boldsymbol{\psi}^{f,j} - \bar{\boldsymbol{\psi}}\right)\right]; \tag{9.21}$$

5. Compute $\boldsymbol{C}_{\epsilon\epsilon} + \boldsymbol{M}\boldsymbol{C}^f_{\boldsymbol{\psi}\boldsymbol{\psi}}\boldsymbol{M}^T$;
6. Solve $\left(\boldsymbol{C}_{\epsilon\epsilon} + \boldsymbol{M}\boldsymbol{C}^f_{\boldsymbol{\psi}\boldsymbol{\psi}}\boldsymbol{M}^T\right)\boldsymbol{b}^i = \boldsymbol{y}^i - \boldsymbol{M}\boldsymbol{\psi}^{f,i}$ for $\boldsymbol{b}_i$ on each CPU i (e.g. with LAPACK), and collect all $\boldsymbol{b}^i$ on all CPUs (e.g. MPI_Allgatherv);
7. Compute the scalars $\left[\boldsymbol{M}\left(\boldsymbol{\psi}^{f,j} - \overline{\boldsymbol{\psi}^f}\right)\right]^T \boldsymbol{b}^i$ on each CPU j;
8. Compute $\boldsymbol{C}^f_{\boldsymbol{\psi}\boldsymbol{\psi}}\boldsymbol{M}^T\boldsymbol{b}_i$ and evaluate Eq. (9.20) on each CPU i (MPI_Reduce).

The ensemble covariance matrix $\boldsymbol{C}^f_{\boldsymbol{\psi}\boldsymbol{\psi}}$ of the model prediction has rank $n-1$, and so does the m-by-m matrix $\boldsymbol{M}\boldsymbol{C}^f_{\boldsymbol{\psi}\boldsymbol{\psi}}\boldsymbol{M}^T$. Depending on the covariance matrix $\boldsymbol{C}_{\epsilon\epsilon}$ of the measurement, the m-by-m matrix $\boldsymbol{C}_{\epsilon\epsilon} + \boldsymbol{M}\boldsymbol{C}^f_{\boldsymbol{\psi}\boldsymbol{\psi}}\boldsymbol{M}^T$ may be singular. Therefore, if $n \le m$, the matrix $\boldsymbol{M}\boldsymbol{C}^f_{\boldsymbol{\psi}\boldsymbol{\psi}}\boldsymbol{M}^T$ is rank deficient. In the absence of measurement errors, solving the linear system in item 6 is a minimum-norm least-squares problem. An alternative approach is to use the square root filter (Sect. 9.3.1). Note that a finite ensemble size can cause the covariance to be underestimated, which can lead to covariance collapse [39]. There are a few tricks to avoid this, such as a simple covariance inflation [1, 41]

$$\boldsymbol{\psi}^{a,i} = \overline{\boldsymbol{\psi}^a} + \beta(\boldsymbol{\psi}^{a,i} - \overline{\boldsymbol{\psi}^a}) + \boldsymbol{\alpha}, \tag{9.22}$$

where β is the multiplicative covariance inflation, which could be a scalar or a matrix, and $\boldsymbol{\alpha}$ is the additive covariance inflation, which is a vector.

9.4 Direct Numerical Simulation of MILD Combustion

In this chapter, the analysis methods described in Sects. 9.2 and 9.3 are applied to Direct Numerical Simulation (DNS) of Moderate or Intense Low-oxygen Dilution (MILD) combustion. The DNS data considered for this analysis is generated following the methodology presented in [13]. The configuration is a non-premixed methane/diluted air MILD combustion with internal EGR at atmospheric pressure. The DNS domain is a cube of 1 cm^3 with inflow/non-reflecting Navier–Stokes Characteristic Boundary Conditions (NSCBC) [36] outflow boundary conditions in the x-direction and periodic boundary conditions in the y and z directions. As to the generation of the initial conditions, in a first preprocessing stage, initial fields of turbulence, mixture fraction, Z, reaction progress variable c, and species mass fractions, Y_α, are generated. To do so, first, laminar premixed flames under MILD combustion conditions are computed for various values of Z and the scalar mass fractions are

Table 9.1 DNS initial conditions. Λ_0 is the integral length scale, η is the Kolmogorov time scale and u' is the initial root-mean square value of the velocity fluctuations

Grid	Λ_0/ℓ_Z	$\langle X_{O_2}\rangle$	$X_{O_2}^{max}$	ℓ_Z/ℓ_c	$\langle Z\rangle$	Z_{st}	σ_Z	$\langle c\rangle$	σ_c	u' [m/s]	η [μs]
512^3	0.60	0.0270	0.035	1.30	0.008	0.010	0.0084	0.56	0.26	16.67	47.7

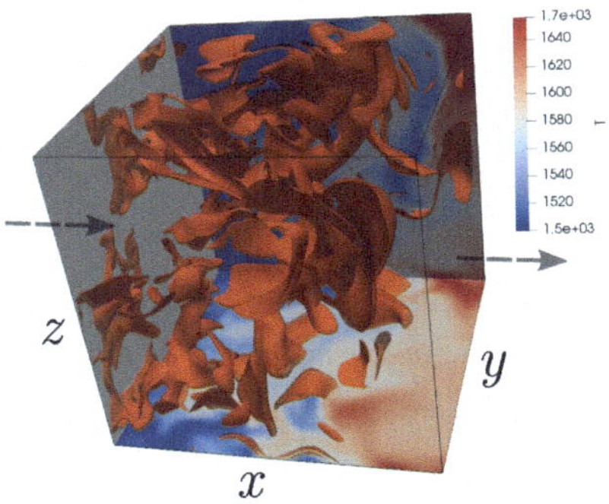

Fig. 9.2 Isosurface of heat release rate field in DNS of MILD combustion (threshold at $\dot{Q} = 1.75 \cdot 10^8$ [W/m^3]) with temperature field (in [K]) on the bottom and side surfaces

tabulated as a function of Z and c. Subsequently, initial 3D reaction progress variable and mixture fraction fields are generated with prescribed means, $\langle Z\rangle$ and $\langle c\rangle$, and length scales, ℓ_Z and ℓ_c using the method of [16]. Z is defined using Bilger's definition [3] and the progress variable is based on the fuel mass fraction. The scalar mass fractions from the laminar flames calculations are then mapped onto these 3D fields. Additionally, an initial decaying homogeneous isotropic turbulence field is generated with the required turbulence characteristics. This turbulence field is then combined with the scalar fields previously generated, which interact in a non-reacting simulation for about one large-eddy turnover time to ensure that the velocity-scalar correlations have sufficiently evolved. The fields obtained are then used as the initial and inflowing conditions for the second stage, which is the actual reacting DNS of MILD combustion. Further details on this methodology can be found in [13].

In this chapter, DNS similar to the AZ1 case of [13] with a 512^3 mesh is performed. The initial characteristics of this DNS are summarized in Table 9.1. Additionally, the oxidizer considered is composed of 3.5% of O_2, 13.4% of H_2O, 6.7% of CO_2 and 76.4% of N_2. The initial turbulence Reynolds numbers based on the integral length scale and Taylor microscale are, respectively, $Re_t = 96$ and $Re_\lambda = 34.73$. The chemical mechanism used here is an extended version of the Smooke methane/air mechanism as detailed in [13]. We removed OH* chemistry to allow the time step to be larger, i.e. $\Delta t = 10$ ns. A study of the chemical markers, such as OH*, can be found in [12].

A typical heat release rate field is presented in Fig. 9.2. There exist multiple isosurfaces of heat release rate that are not confined to thin reaction zones, where both ignition and deflagration combustion modes can be found [11].

In addition to the fine DNS with a 512^3 mesh, DNS on coarser grids of 96^3, 128^3 and 256^3 points are performed. These DNS are initialized using the solution of the fine DNS after it has reached a steady state. The DNS is run with the code SENGA2 [7], which solves the fully compressible reacting Navier–Stokes equations with a tenth-order finite difference scheme for spatial discretization and a third-order low-storage Runge–Kutta scheme for time integration on CSD3 (Cambridge, UK), a Tier-2 cluster.

9.5 Results

9.5.1 *Predictability of DNS of MILD Combustion*

Following the procedure described in Sect. 9.2, the predictability of the DNS of MILD combustion is assessed on coarse DNS grids with, respectively, 96^3, 128^3 and 256^3 grid points. Each simulation is initialized from the last snapshot of a statistically converged DNS on the 512^3 grid. Subsequently, the inlet velocity of the coarse simulations are perturbed, with $\epsilon = 10^{-3}$ and the separation trajectory of each simulation is presented in Fig. 9.3.

The values of the Lyapunov exponent, extracted by linear regression, highlighted using dashed lines in Fig. 9.3, are summarized in Table 9.2. It is seen that the Lyapunov exponent increases between 96^3 and 128^3 and saturates for 256^3. This indicates that

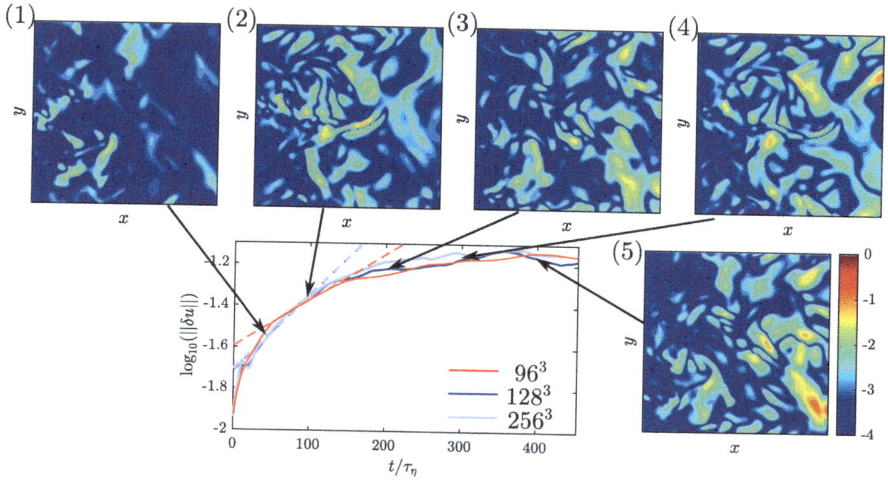

Fig. 9.3 Separation based on velocity as a function of time with meshes 96^3 (red), 128^3 (blue) and 256^3 (violet). The dashed lines indicate the slope of the linear region of the separation trajectories. Additional panels show the mid x–y plane of the norm of the separation trajectory, $\log_{10}(||\delta u||)$, for the mesh 128^3. In the linear region, this is the norm of the covariant Lyapunov velocity vector

Table 9.2 Lyapunov exponents and predictability of the MILD combustion configuration

	96^3	128^3	256^3
$\lambda\tau_\eta$	0.00509	0.00835	0.00826
$1/(\lambda\tau_\eta)$	195.46	119.76	121.06
$t_p = 1/\lambda$ [μs]	562.2	342.7	346.4

most of the chaotic dynamics is captured by the 128^3 grid. Furthermore, the separation frames highlight that the separation occurs in the upstream region of the domain, where the decaying isotropic turbulence is being injected into the domain (see panels (1) and (2) in Fig. 9.3). In that region, the separation originates chiefly because of how the perturbed and unperturbed velocity fields are evolving into different turbulence fields, upstream of the main heat-releasing regions.

In addition to the first highlighted linear region, there appears a second seemingly linear region for $t/\tau_\eta = 200$ to 300. This suggests that another phenomenon is responsible for the separation of the simulations at a later time. Indeed, large separation downstream of reaction zones appear and contribute to the increase in separation between the simulations (see panels (3) and (4) in Fig. 9.3). This separation is due to the diverging heat release rate fields, which yield different velocity fields downstream of the reaction zones. We conjecture that the change in slope is linked to the change in timescale between the regions upstream and downstream of the reaction zones, i.e. the downstream region has a much smaller turbulence level due to the heat release. Finally, it should be noted that the particular DNS cases have a quite limited Reynolds number and have high temperature with large viscosity. As a result, the Lyapunov exponents of MILD combustion are smaller than those of the forced-homogeneous isotropic turbulence of [32]. Combustion makes a turbulent flow more predictable from a dynamical systems point of view.

9.5.2 *State Estimation of DNS of MILD Combustion*

The ensemble Kalman filter described in Sect. 9.3 is applied here to improve the prediction of the heat release rate evolution in the DNS of MILD combustion. In the present problem, the measurements are taken from the finely resolved DNS (on the 512^3 mesh) and the forecasts are obtained by running an ensemble of 10 coarse DNS, each with a mesh of 128^3 nodes. The initial velocity field for each coarse DNS is perturbed by a factor $\xi||u||$, where ξ follows a Gaussian distribution centred around 10^{-3} with standard deviation of 10^{-4}. The observations from the fine DNS, which mimic observations from an experimental rig, are located in a cube of length 3 mm centred around the point (3.149, 3.149, 3.149) [mm] as illustrated in Fig. 9.4. It was observed from the fine DNS that this location corresponds to a region where both ignition and flame propagation occur, which are difficult to predict accurately in space

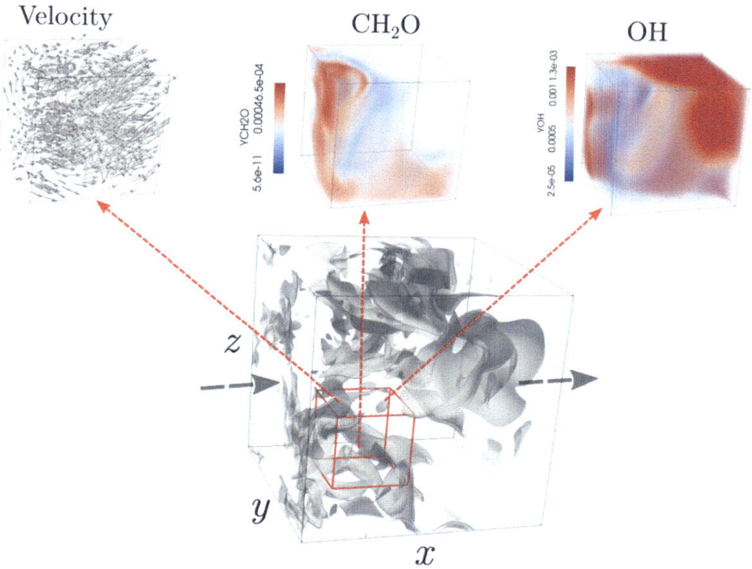

Fig. 9.4 Observation box (in red), where the measured quantities (velocity field, CH_2O and OH) are taken

and time. Therefore, this region is a spot-on location to apply data assimilation to appreciate its performance. In the observation box, sampling locations are uniformly distributed with 21 samples in each direction for a total of 9261 sampling locations. At each of these locations, the values of velocity (u, v and w), Y_{OH} and Y_{CH_2O} are collected from the fine DNS to mimic experimental measurements from Particle Image Velocimetry (PIV) or Planar Laser-Induced Fluorescence (PLIF). The state vector, $\boldsymbol{\psi}$, is taken to be the same as the experimental measurements, hence, the observation matrix $\boldsymbol{M}$ is the identity matrix. Furthermore, the observation error is taken to be the standard deviation of the observed quantity in a smaller cube of length $78\,\mu$m, which corresponds to 1/40 of the entire observation box centred around each measurement location from the fine DNS. This corresponds to passing from the fine grid of 512^3 nodes to 128^3 nodes on the coarse DNS. Data assimilation is performed every $40\,\mu$s, which corresponds to approximately one-tenth of the average ignition delay time for the mixture under study. The evolution of the logarithm of the Root-Mean Squared (RMS) error (RMSE), computed as the trace of the ensemble covariance matrix, RMSE $= \text{tr}(\boldsymbol{C}^f_{\psi\psi})$ where $\boldsymbol{\psi}$ is just composed of the measured states, is shown in Fig. 9.5a. In the beginning, the RMS error is small because nearly identical initial conditions are used for all cases. Because of the chaotic nature of the turbulence flow, the RMS subsequently grows. At each data assimilation event, the uncertainty is markedly decreased by approximately half a decade. Another metric to assess how more predictive the filtered solution is with respect to the unfiltered solution is the evolution of the normalized error of the heat release rate (Fig. 9.5b)

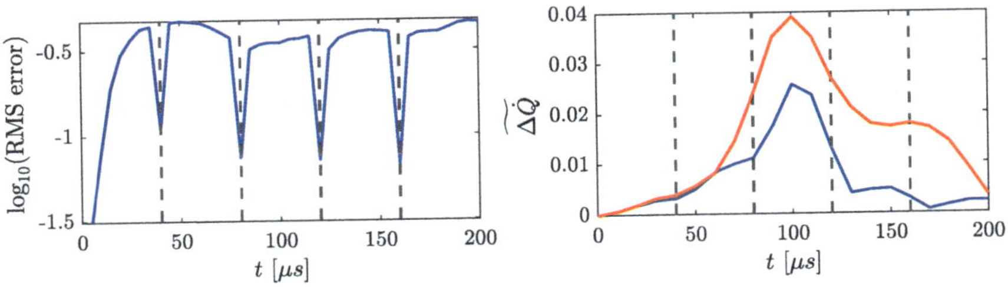

Fig. 9.5 Evolution of the (left) RMSE and (right) normalized error of $\dot{Q}$ from the coarse DNS (red) and the average evolution from the ensemble Kalman filter (blue). Dashed vertical lines indicate when the data from the fine DNS is assimilated

$$\widetilde{\Delta \dot{Q}} = \frac{\sqrt{\int_V (\dot{Q}_m - \dot{Q}_c)^2 dV}}{\int_V \dot{Q}_m dV} \tag{9.23}$$

which is integrated into the observation box, V, where $\dot{Q}_m$ is the heat release rate computed on the fine DNS and $\dot{Q}_c$ is either the heat release rate from the coarse DNS (red line in Fig. 9.5b) or the one from the mean evolution obtained from the Kalman Filter procedure (blue line in Fig. 9.5b). It can be observed that the Ensemble Kalman Filter procedure allows for an improvement on the prediction of the evolution of the heat release rate with an overall lower error. In addition to these global statistics, Fig. 9.6 illustrates the evolution of the heat release rate in the observation box during the time period shown in Fig. 9.5 for the fine DNS with a mesh of 512^3, the coarse DNS with a mesh of 128^3 and the mean field from the Ensemble Kalman filter method, denoted EnsKal DNS in the following. In frames (1) and (2), the morphology of the reaction zones is approximately the same for all simulations as they are evolving from the same initial field (or the slightly perturbed field in the case of the Ensemble Kalman filter). The coarse features of the reaction zones for the coarse and EnsKal DNS come from the under-resolved DNS, which cannot capture all the detailed morphological features of the fine DNS. In frame (3), which is after the first data assimilation event, a specific ignition kernel is highlighted. While both the coarse and EnsKal DNS show that they capture that feature, its evolution in time (frame (4)) is fuller and more quantitatively accurate in the EnsKal DNS. This indicates that the EnsKal DNS has assimilated the correct consumption of species, in particular of CH_2O, which acts as a precursor, to predict the heat release morphology. Additionally, in frame (5), one can observe another ignition kernel, highlighted in the blue box, whose shape has been better captured by EnsKal DNS. From frame (5) to (6), there are major reaction-zone interactions, which roughly correspond to the peak in error at $t = 100\,\mu s$ in Fig. 9.5b. From frame (5) to (7), one can observe the evolution of the previous reaction zone, which has merged with neighbouring reaction zones to form a deflagrative structure. While the merging is captured by all the simulations, the morphology of the EnsKal DNS more accurately represents the morphology of the fine DNS. This is thanks to the successful assimilation of the velocity

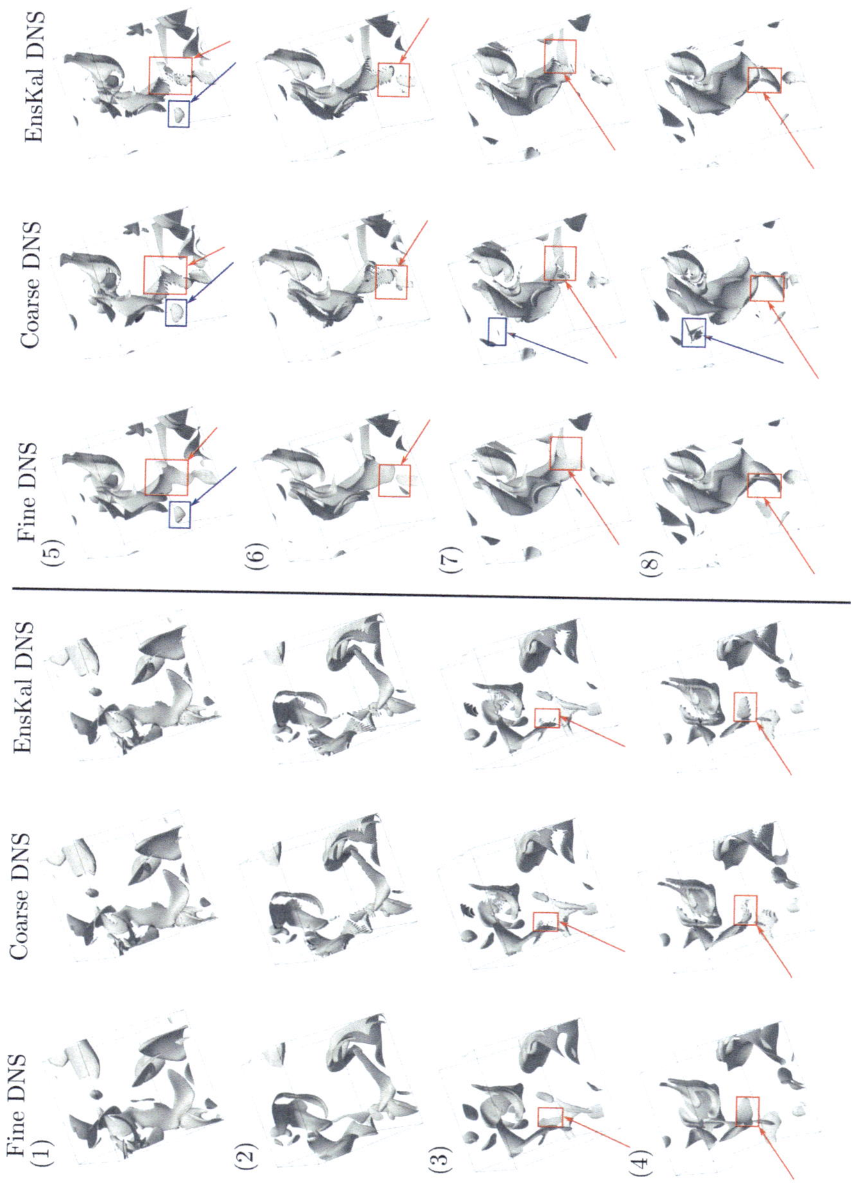

Fig. 9.6 Evolution of the isosurface of $\dot{Q} = 1.75 \cdot 10^8$ [W/m^3] in the observation box during the simulation (left: fine DNS, middle: coarse DNS, right: average of the ensemble). Red and blue boxes highlight some of the key features. Panels 1–8 are, respectively, for $t =$ 0, 35, 60, 80, 105, 120, 140, 155 μs in Fig. 9.5b

field, which is an essential physical quantity to assimilate for the propagation of the various interacting reaction zones. In conclusion, the turbulent MILD combustion case presented highlights how data assimilation improves space-and-time-accurate prediction of both macroscopic quantities (volumetric heat release rate, Fig. 9.5b) and local quantities (heat release morphology, Fig. 9.6).

9.6 Conclusions and Future Directions

The butterfly effect limits the time-accurate prediction of turbulent flows, which are chaotic flows. Tools from chaos theory and statistical learning are, thus, called upon. The inverse of the dominant Lyapunov exponent provides an estimate of the predictability of the flow, which is a characteristic time horizon within which the butterfly effect makes two nearby trajectories diverge. The ensemble Kalman filter enables statistical learning of the state given predictions from a physical model and observations from a more refined simulation or laboratory experiments. We use Lyapunov theory and the ensemble Kalman filter on a turbulent reacting problem relevant to Moderate or Intense Low-oxygen Dilution (MILD) combustion. In MILD combustion, it is paramount to accurately capture ignition kernels and their evolutions towards deflagration because this interplay stabilizes the combustion process. The occurrence and dynamics of ignition kernels is seemingly random (albeit deterministic) due to this turbulent mixture containing high-temperature spots and recirculating radicals. Assessing the predictability of simulations helps to determine whether they can reliably capture such transient events, for example, in terms of the mesh size requirements. By assimilating data from a refined simulation, the ensemble Kalman filter greatly improves the space-and-time-accurate prediction in an under-resolved simulation by detecting the occurrence of ignition kernels, the stabilization of the deflagrative mode and intricate changes in the flow morphology. The physics-informed data-driven methods presented open up new possibilities for the space-and-time-accurate prediction of rare and transient events in turbulent flows by combining an under-resolved simulation, or a reduced-order model, with data from experiments. As an example, this can be exploited to learn and calibrate subgrid-scale models for large-eddy simulation.

Acknowledgements The authors acknowledge the support of the Technical University of Munich - Institute for Advanced Study, funded by the German Excellence Initiative and the European Union Seventh Framework Programme under grant agreement no. 291763. L. Magri also acknowledges the Royal Academy of Engineering Research Fellowships Scheme. This work was performed using resources provided by the Cambridge Service for Data Driven Discovery (CSD3) operated by the University of Cambridge Research Computing Service (www.csd3.cam.ac.uk), provided by Dell EMC and Intel using Tier-2 funding from the Engineering and Physical Sciences Research Council (capital grant EP/P020259/1), and DiRAC funding from the Science and Technology Facilities Council (www.dirac.ac.uk). The authors acknowledge Prof. Cant for providing the code SENGA2. Francisco Huhn and Hans Yu are gratefully acknowledged for their invaluable inputs on Lyapunov analysis (Sect. 9.2) and data assimilation (Sect. 9.3), respectively.

References

1. J.L. Anderson, S.L. Anderson, A Monte Carlo implementation of the nonlinear filtering problem to produce ensemble assimilations and forecasts. Mon. Weather Rev. **127**(12), 2741–2758 (2002)
2. R. Bellman, *Dynamic Programming* (Dover Publications, New York, 2003)
3. R.W. Bilger, S.H. Starner, R.J. Kee, On reduced mechanisms for methane-air combustion in nonpremixed flames. Combust. Flame **80**, 135–149 (1990)
4. P.J. Blonigan, P. Fernandez, S.M. Murman, Q. Wang, G. Rigas, L. Magri, Towards a chaotic adjoint for LES, Center for Turbulence Research, Summer Program (2016)
5. G. Boffetta, M. Cencini, M. Falcioni, A. Vulpiani, Predictability: a way to characterize complexity. Phys. Rep. **356**, 367–474 (2002)
6. G. Burgers, P.J. van Leeuwen, G. Evensen, Analysis scheme in the ensemble Kalman filter. Mon. Weather Rev. **126**(6), 1719–1724 (1998)
7. R. S. Cant. SENGA2 User Guide (CUED-THERMO-2012/04, 2nd edn.). Technical report, University of Cambridge (2013)
8. A. Cavaliere, M. de Joannon, MILD combustion. Prog. Energy Combust. Sci. **30**, 329–366 (2004)
9. A.F.C. da Silva, T. Colonius, Ensemble-based state estimator for aerodynamic flows. AIAA J. **56**(7), 2568–2578 (2018)
10. D. Darakananda, A.F.D.C. da Silva, T. Colonius, J.D. Eldredge, Data-assimilated low-order vortex modeling of separated flows. Phys. Rev. Fluids **3**(12), 1–24 (2018)
11. N.A.K. Doan, N. Swaminathan, Autoignition and flame propagation in non-premixed MILD combustion. Combust. Flame **201**, 234–243 (2019)
12. N.A.K. Doan, N. Swaminathan, Analysis of markers for combustion mode and heat release in MILD combustion using DNS data. Combust. Sci. Technol. **191**(5–6), 1059–1078 (2019)
13. N.A.K. Doan, N. Swaminathan, Y. Minamoto, DNS of MILD combustion with mixture fraction variations. Combust. Flame **189**, 173–189 (2018)
14. A. Doucet, N. Freitas, N. Gordon (eds.), *Sequential Monte Carlo Methods in Practice* (Springer, New York, 2001)
15. J.P. Eckmann, D. Ruelle, Ergodic theory of chaos and strange attractors. Rev. Mod. Phys. **57**, 617–656 (1985)
16. V. Eswaran, S.B. Pope, Direct numerical simulations of the turbulent mixing of a passive scalar. Phys. Fluids **31**(3), 506–520 (1988)
17. G. Evensen, Sequential data assimilation with a nonlinear quasi-geostrophic model using Monte Carlo methods to forecast error statistics. J. Geophys. Res. **99**(C5), 10143 (1994)
18. G. Evensen, *Data Assimilation - The Ensemble Kalman Filter* (Springer, Berlin, 2009)
19. J.D. Farmer, E. Ott, J.A. Yorke, The dimension of chaotic attractors. Phys. D **7**(1–3), 153–180 (1983)
20. P. Fernandez, Q. Wang, Lyapunov spectrum of the separated flow around the NACA 0012 airfoil and its dependence on numerical discretization. J. Comput. Phys. **350**, 453–469 (2017)
21. F. Ginelli, H. Chate, R. Livi, A. Politi, Covariant Lyapunov vectors. J. Phys. A: Math. Theor. **46**(25), 254005 (2013)
22. I. Goldhirsch, P.-L. Sulem, S.A. Orszag, Stability and Lyapunov stability of dynamical systems: a differential approach and a numerical method. Phys. D: Nonlinear Phenom. **27**(3), 311–337 (1987)
23. M. Hassanaly, V. Raman, Ensemble-LES analysis of perturbation response of turbulent partially-premixed flames. Proc. Combust. Inst. **37**(2), 2249–2257 (2019)
24. R.C. Hilborn, *Chaos and Nonlinear Dynamics* (Oxford University Press, Oxford, 1994)
25. F. Huhn, L. Magri, Stability, sensitivity and optimisation of chaotic acoustic oscillations. J. Fluid Mech. **882**, A24 (2020)
26. F. Huhn, L. Magri, Optimisation of chaotically perturbed acoustic limit cycles. Nonlinear Dyn. (2020). https://doi.org/10.1007/s11071-020-05582-x

27. R.E. Kalman, A new approach to linear filtering and prediction problems. J. Basic Eng. **82**(1), 35 (1960)
28. J.W. Labahn, H. Wu, B. Coriton, J.H. Frank, M. Ihme, Data assimilation using high-speed measurements and LES to examine local extinction events in turbulent flames. Proc. Combust. Inst. **37**(2), 2259–2266 (2019)
29. E.N. Lorenz, Deterministic nonperiodic flow. J. Atmos. Sci. **20**(2), 130–141 (1963)
30. Y. Minamoto, N. Swaminathan, R.S. Cant, T. Leung, Morphological and statistical features of reaction zones in MILD and premixed combustion. Combust. Flame **161**(11), 2801–2814 (2014)
31. P. Mohan, N. Fitzsimmons, R.D. Moser, Scaling of Lyapunov exponents in homogeneous isotropic turbulence. Phys. Rev. Fluids **2**, 114606 (2017)
32. G. Nastac, J. Labahn, L. Magri, M. Ihme, Lyapunov exponent as a metric for assessing the dynamic content and predictability of large-eddy simulations. Phys. Rev. Fluids **2**(9), 094606 (2017)
33. A. Ni, Q. Wang, Sensitivity analysis on chaotic dynamical systems by non-intrusive least squares shadowing (NILSS). J. Comput. Phys. **347**, 56–77 (2017)
34. V.I. Oseledets, A multiplicative ergodic theorem: Lyapunov characteristic numbers for dynamical systems. Trans. Mosc. Math. Soc. **19**, 197–231 (1968)
35. I.B. Özdemir, N. Peters, Characteristics of the reaction zone in a combustor operating at MILD combustion. Exp. Fluids **30**, 683–695 (2001)
36. T. Poinsot, Boundary conditions for direct simulations of compressible viscous flows. J. Comput. Phys. **101**, 104–129 (1992)
37. S.B. Pope, Ten questions concerning the large-eddy simulation of turbulent flows. New J. Phys. **6** (2004)
38. D. Ruelle, Ergodic theory of differentiable dynamical systems. Publications mathematiques de l'IHES **50**(1), 27–58 (1979)
39. P.J. van Leeuwen, Comment on "Data assimilation using an ensemble Kalman filter technique". Mon. Weather Rev. **127**(6), 1374–1377 (2002)
40. J.S. Whitaker, T.M. Hamill, Ensemble data assimilation without perturbed observations. Mon. Weather Rev. **130**(7), 1913–1924 (2002)
41. J.S. Whitaker, T.M. Hamill, Evaluating methods to account for system errors in ensemble data assimilation. Mon. Weather Rev. **140**(9), 3078–3089 (2012)
42. H. Yu, T. Jaravel, M. Juniper, M. Ihme, L. Magri, Data assimilation and optimal calibration in nonlinear models of flame dynamics. J. Eng. Gas Turb. Power **141**(2), 121010 (2019)
43. H. Yu, M.P. Juniper, L. Magri, Combined state and parameter estimation in level-set methods. J. Comput. Phys. **399**, 108950 (2019)

Chapter 10
Data-Based Modeling for the Crank Angle Resolved CI Combustion Process

Jan Schilliger, Nils Keller, Severin Hänggi, Thivaharan Albin, and Christopher Onder

Abstract For new combustion control concepts such as Combustion Rate Shaping, a crank angle resolved model of the compression ignition (CI) combustion process is necessary. The complex CI combustion process involving fuel injection, turbulent flow, and chemical reactions has to be reproduced. However, to be suitable for control, it has to be computationally efficient at the same time. To allow for learning-based control, the model should be able to adapt to the current measurement data. This paper proposes two algorithms that model the CI combustion dynamics by learning a crank angle resolved model from past heat release rate (HRR) measurement data. They are characterized by short learning and evaluation times, low calibration effort, and high adaptability. Both approaches approximate the total HRR as the linear superposition of the HRRs of individual fuel packages. The first algorithm approximates the HRR of a single fuel package as a Vibe function and identifies the parameters by solving a nonlinear program having the squared difference between the measured HRR and the superposition as cost. The second algorithm approximates the individual packages' HRRs as Gaussian distributions and estimates the parameters by solving a nonlinear program with the Kullback–Leibler divergence between the measurement and the superposition as cost function using the expectation–maximization algorithm. Both algorithms are validated using test bench measurement data.

Keywords Combustion rate shaping · Mixture models · Expectation maximization · Sequential quadratic programming · Heat release rate modeling

---

J. Schilliger and N. Keller contributed equally to the publication.

J. Schilliger · N. Keller (✉) · S. Hänggi · T. Albin · C. Onder
Institute for Dynamic Systems and Control, ETH Zürich, Sonneggstrasse 3, 8092 Zürich, Switzerland
e-mail: nilkelle@ethz.ch

J. Schilliger
e-mail: janschi@ethz.ch

H. Pitsch and A. Attili (eds.), *Data Analysis for Direct Numerical Simulations of Turbulent Combustion*, https://doi.org/10.1007/978-3-030-44718-2_10

10.1 Introduction

The main goal of the current development of powertrains with compression ignition engines is set on the reduction of CO_2 emissions while minimizing other pollutant emissions at the same time. A major limitation of improving the performance of CI engines in terms of pollution, noise, and CO_2 emissions are the well-known NO_x-soot trade-off and soot-noise trade-off [9, 13, 21]. Different methods are employed to overcome these trade-offs. For instance, exhaust gas aftertreatment systems are applied. These can be effective, but have their own drawbacks, such as expensive material cost. A different approach is the direct reduction of engine-out pollutant emissions. This is usually preferable to using abatement systems, since no additional cost during vehicle lifetime is incurred. A promising approach are new combustion processes along with innovative control concepts.

The conventional approach for CI combustion control can be classified as a cycle-to-cycle feedback control, which is based on cycle-integral parameters. For characterization of the combustion process several surrogate parameters are defined, which represent the cycle-integral behavior. The most prominent ones are the indicated mean effective pressure, as correlator for load and the combustion average, i.e. the time point of 50% burnt fuel, as correlator for the fuel efficiency. However, for an increasing number of fuel injections this conventional approach reaches its limitations, as in the case of many injection events, the cycle-integral parameters are no longer suitable to represent the performance parameters, such as efficiency and pollutant emissions. A new control concept to overcome these drawbacks is the so-called Combustion Rate Shaping (CRS).

10.1.1 Model-Based Combustion Rate Shaping

CRS describes the control of the combustion on a crank angle-based resolution based on a very flexible fuel injection. The measured value is typically the in-cylinder pressure, which can be measured with an in-cylinder pressure sensor. Note that the transformation from in-cylinder pressure to heat release is bijective and thus heat release rate and in-cylinder pressure are equivalent. Controlling the angle resolved combustion process leads to maximum degrees of freedom to control the combustion process. As a consequence, the combustion process can be operated in an optimal manner, thus reducing pollutant emissions and increasing efficiency. Modern injectors allow for a large number of injection events, thus offering the needed flexibility for actuation, such that CRS is viable.

There are two different application possibilities for CRS. CRS can be used as a feedback-control algorithm on a cycle-to-cycle basis. It allows to adapt the injection profile during operation, e.g., in a vehicle. The feedback algorithm allows to reject arising disturbances, e.g., due to aging. The second possibility is the application of CRS to find the best suitable injection profile on an engine test bench. The results can be used to design lookup tables, which are applied during operation in an open-loop manner [11].

In both cases, the task of the controller is to calculate an appropriate injection profile. The injection profile is characterized by the start of injection and duration of injection for each injection event. Due to the many degrees of freedom of such an injection profile, its computation is very demanding. The combustion process is characterized by strong cross-couplings, e.g., the first injection influences the second, and a highly nonlinear behavior. As a consequence, conventional control strategies such as PID control are not suitable anymore. Instead, the complex system dynamics have to be accounted for. Furthermore, heuristic tuning is limited due to the curse of dimensionality. In [22], an iterative learning control CRS approach with low structure complexity is used to generate a continuous injection rate profile. To further exploit the degrees of freedom offered by CRS, model-based control approaches are promising. Model-based control approaches allow to circumvent the curse of dimensionality, take into account the difficult combustion dynamics, and offer a wide range of operation [1].

Within model-based control, the process model plays a central role. As already mentioned, CI combustion is hard to model as the involved processes are highly nonlinear. This includes the diesel injection, the fuel spray atomization and vaporization, the in-cylinder turbulence, and the oxidation, to name a few. Precise combustion models require extensive CFD simulations, which have run-times many orders of magnitude larger than the duration of an engine cycle. These run-times are not suitable for use within CRS. Simplified physics-based or data-based models can be employed to trade some accuracy for faster computation, therefore making model-based control feasible.

A CRS controller based on an invertible physics-based model with real-time capabilities is shown to deliver good results in [11, 12]. A similar concept is employed in a flex-fuel setting in [16], where it is shown to perform well under a changing blend of fuels. Reduced physics-based models can be computationally less expensive and have good extrapolation capabilities. However, they are usually high-dimensional, leading to long calibration times. As a consequence, online adaptation of the parameters based on measurements can be very expensive for these nonlinear models with many parameters.

Data-based models also allow for reduced computation times and high accuracy. They usually lack in terms of the extrapolation capabilities but they can be parametrized in a computationally very efficient manner. As a result, data-based modeling is a promising approach for use within learning-based control algorithms.

10.2 Data-Based Modeling Based on Linear Superposition

For the proposed data-based modeling approach, the combustion process is represented by a low-dimensional set of parameters, which are learned from past measurement data. These parameters are not derived based on physical arguments, but rather to best-fit present data. The purely data-based approach offers flexibility and lightweight learning with high accuracy and robustness over a wide operating range, since the model is updated using efficient algorithms on a cycle-to-cycle basis.

In the following, the model structure is introduced. The continuous injection profile is denoted as $u(\varphi)$ and the continuous heat release rate as $y(\varphi)$, respectively. To simplify the derivations, u and y are subsequently discretized and u has units of energy. The injected fuel energy (subsequently, "injected fuel" or "injection profile" always refers to energy) can be converted to injected fuel mass by dividing by the lower heating value of the fuel. Let $u(\varphi)$ be discretized into N_u samples, i.e., $\boldsymbol{u} = [u_1 \; \dots \; u_{N_u}]^{\mathsf{T}}$. Similarly, $y(\varphi)$ is discretized into N_y samples, i.e., $\boldsymbol{y} = [y_1 \; \dots \; y_{N_y}]^{\mathsf{T}}$. The model maps an input vector $\boldsymbol{u}$ onto an output vector $\boldsymbol{y}$. Let k index the discrete crank angle instances. Furthermore, the u_i is the energy stored in the fuel injected at crank angle instance i. Because only positive fuel flow is possible, $u_i \geq 0$ holds. Also, all fuel injected is equal to the total heat released in a cycle $Q_{f,tot}$, i.e.,

$$Q_{f,tot} = \sum_{i=1}^{N_u} u_i. \tag{10.1}$$

Assuming that all injected fuel mass is burned and that each injected fuel package has a distinct heat release rate associated to the package, the heat release rate $\boldsymbol{y}$ is assumed to be a linear combination of the distinct heat release rates associated with the packages u_i, denoted as basis functions $f(\theta_i)$, where θ_i denotes a set of parameters, i.e.,

$$y_k = \sum_{i=1}^{N_u} u_i f(k|\theta_i). \tag{10.2}$$

This is a reasonable assumption as any injected fuel package has a fixed energy content that usually completely oxidizes. Note that no further assumption has been made on the basis function. In a model derived from physical principles, the basis functions $f(k|\theta_i)$ are replaced with a general function $F(\boldsymbol{u}, k)$ for all injected packages that directly takes into account the cross-couplings. From (10.2), it is visible that the basis functions $f(\theta_i)$ represent the impulse response associated with package i, which is linearly scaled by u_i, and thus $\boldsymbol{y}$ can be described as a linear time-variant system. The objective of CRS is to minimize the error between a reference trajectory $\boldsymbol{y}^{ref}$ and the heat release rate, i.e.,

$$\boldsymbol{e}_{crs} = \boldsymbol{y}^{ref} - \boldsymbol{y}^{meas}. \tag{10.3}$$

For this purpose, model-based control is used. The plant/model mismatch is denoted by

$$\boldsymbol{e} = \boldsymbol{y}^{meas} - \boldsymbol{y}^{pred}, \tag{10.4}$$

where y_k^{meas} is the measured heat release rate, and y_k^{pred} is the predicted heat release rate, both at instance k. The goal is to minimize $\boldsymbol{e}$ to learn the underlying model. In the following, two algorithms to minimize $\boldsymbol{e}$ are introduced.

10.3 Vibe Functions and Nonlinear Programming

Heat release $Q_f(\varphi)$ and heat release rate (HRR) for Diesel combustion are often modeled as Vibe functions. The Vibe function is parametrized by a shape factor m, the start of combustion φ_{soc}, and the combustion duration $\Delta\varphi_{cd}$. According to Vibe [19], the normalized heat release rate of an injected fuel package can be written as

$$\zeta(\varphi|m, \Delta\varphi_{cd}, \varphi_{soc}) = \frac{d}{d\varphi}\frac{Q_f(\varphi|m, \Delta\varphi_{cd}, \varphi_{soc})}{Q_{f,tot}} = \theta(\varphi - \varphi_{soc})a\frac{(m+1)}{\Delta\varphi_{cd}}\varphi_{rel}^{m}e^{-a\varphi_{rel}^{m+1}}, \tag{10.5}$$

where

$$\varphi_{rel} = \frac{\varphi - \varphi_{soc}}{\varphi_{cd}} \tag{10.6}$$

and

$$\theta(x) = \begin{cases} 1 & x \geq 0 \\ 0 & x < 0 \end{cases} \tag{10.7}$$

which is the Heaviside function, which does not have a continuous first-order derivative. It can be approximated by a sigmoid function when a continuous derivative is required. In (10.5), a is an efficiency parameter and assuming that 99.9% of the injected fuel has been oxidized until the end of combustion, it is set to $a = 6.908$.

Fig. 10.1 shows the principle of the Vibe function mixture model. With (10.2) and using Vibe functions as basis functions to describe the heat release rates of a single fuel package, y_k^{pred} may be written as

$$y_k^{pred} = \sum_{i\in\mathcal{I}} u_i\zeta(\varphi|m_i, \Delta\varphi_{cd,i}, \varphi_{soc,i}). \tag{10.8}$$

The superposition of the Vibe base functions $\zeta(m_i, \Delta\varphi_{cd,i}, \varphi_{soc,i})$ that are parametrized by the parameters m_i, $\Delta\varphi_{cd,i}$, $\varphi_{soc,i}$ for each fuel package i. Together, these make up $\boldsymbol{y}^{pred}$ which is also shown in Fig. 10.1. Using the Vibe function as introduced in (10.5), the goal is to minimize the mean squared error (MSE), i.e.,

$$\min \frac{1}{N_y}\boldsymbol{e}^{\top}\boldsymbol{e} \tag{10.9}$$

which is equivalent to

$$\min_{\boldsymbol{m}, \boldsymbol{\Delta\varphi}_{cd}, \boldsymbol{\varphi}_{soc}} \|\boldsymbol{y}^{meas} - \boldsymbol{y}^{pred}(\boldsymbol{m}, \boldsymbol{\Delta\varphi}_{cd}, \boldsymbol{\varphi}_{soc})\|_2^2 \tag{10.10}$$

where the predicted heat release rate is a linear combination of the heat release rates of the single packages

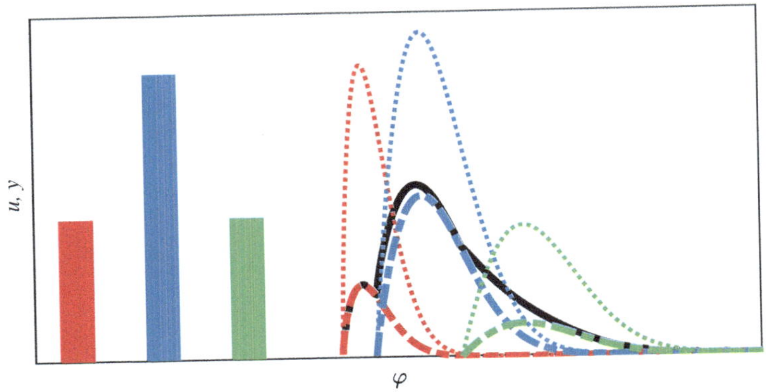

Fig. 10.1 The principle of a Vibe function mixture model with three mixture components ($N_u = 3$). Referring to (10.8), the histogram bars represent u_i, the dotted curves $\zeta(m_i, \Delta\varphi_{cd,i}, \varphi_{soc,i})$, the dashed curves $u_i\zeta(m_i, \Delta\varphi_{cd,i}, \varphi_{soc,i})$, and the black curve y^{pred}

$$\boldsymbol{y}^{pred}(\boldsymbol{m}, \boldsymbol{\Delta\varphi}_{cd}, \boldsymbol{\varphi}_{soc}) = \sum_{i=1}^{N_u} u_i\zeta(m_i, \Delta\varphi_{cd,i}, \varphi_{soc,i}), \tag{10.11}$$

where $\boldsymbol{m}, \boldsymbol{\Delta\varphi}_{cd}, \boldsymbol{\varphi}_{soc} \in \mathbb{R}^{N_u}$ are vectors consisting of the parameters associated with the Vibe functions of the different fuel packages, i.e., m_i, $\Delta\varphi_{cd,i}$, and $\varphi_{soc,i}$ are the parameters of ζ. Note that the scaling factor $\frac{1}{N_y}$ from (10.9) is omitted, as it does not affect the optimal solution.

The predicted output $\boldsymbol{y}^{pred}$ is parametrized by $\boldsymbol{m}$, $\boldsymbol{\Delta\varphi}_{cd}$, and $\boldsymbol{\varphi}_{soc}$. These parameters are usually within certain bounds under normal operating conditions. Thus, by constraining the parameters to realistic values, the performance of the resulting non-convex NLP can be improved, as unrealistic solutions are rejected [10]. Furthermore, it holds that the start of combustion φ_{soc} is sequential with the injection, i.e., $\varphi_{soc,i} \leq \varphi_{soc,i+1}$. The complete optimization problem can then be written as

$$\min_{\boldsymbol{\varphi}_{soc}\boldsymbol{\Delta\varphi}_{cd}, \boldsymbol{m}} \quad \|\boldsymbol{y}^{meas} - \boldsymbol{y}^{pred}(\boldsymbol{m}, \boldsymbol{\Delta\varphi}_{cd}, \boldsymbol{\varphi}_{soc})\|_2^2 \tag{10.12a}$$

$$\text{s.t.} \quad \boldsymbol{\varphi}_{soc,min} \leq \boldsymbol{\varphi}_{soc} \leq \boldsymbol{\varphi}_{soc,max}, \tag{10.12b}$$

$$\boldsymbol{\Delta\varphi}_{cd,min} \leq \boldsymbol{\Delta\varphi}_{cd} \leq \boldsymbol{\Delta\varphi}_{cd,max}, \tag{10.12c}$$

$$\boldsymbol{m}_{min} \leq \boldsymbol{m} \leq \boldsymbol{m}_{max}, \tag{10.12d}$$

$$\varphi_{soc,i} \leq \varphi_{soc,i+1}, \tag{10.12e}$$

$$\varphi_{soi,i} \leq \varphi_{soc,i} \tag{10.12f}$$

where (10.12e) is a constraint to preserve the order of injection. It is intuitive that the packets should not "overtake" each other, but it is not clear how physically accurate this intuition actually is and it is not trivial to encode this constraint. This can be done as seen in (10.12e), but other methods could also be possible. One could restrict the HRR peaks as such: Let $\tilde{\varphi}_{peak,i}$ denote an approximation of the real peak HRR $\varphi_{peak,i}$ of the respective ith Vibe function. The distance from $\varphi_{soc,i}$ to the peak $\varphi_{peak,i}$ is a function of m_i and $\Delta\varphi_{cd,i}$, but no closed form exists. However, the function can be approximated, for example, using a low-order polynomial. Therefore, (10.12e) can be formulated in various ways to directly or indirectly and softly constrain that the injection order is preserved.

The constraint (10.12f) ensures that the combustion of fuel packet i does not begin before it has been injected, where $\varphi_{soi,i}$ denotes the location where fuel packet i is injected.

To simplify the optimization problem, we concatenate all optimization variables in a single optimization variable $\boldsymbol{x} = \left[\boldsymbol{\varphi}_{soc}^{\mathsf{T}}\ \boldsymbol{m}^{\mathsf{T}}\ \boldsymbol{\Delta\varphi}_{cd}^{\mathsf{T}}\right]^{\mathsf{T}}$ and all inequality constraints are captured in a single affine inequality constraint

$$\boldsymbol{b} \leq \boldsymbol{A}\boldsymbol{x}. \tag{10.13}$$

Thus, the resulting optimization problem is

$$\begin{aligned} \min_{\boldsymbol{x}} \quad & \|\boldsymbol{y}^{meas} - \boldsymbol{y}^{pred}(\boldsymbol{x})\|_2^2 \\ \text{s.t.} \quad & \boldsymbol{b} \leq \boldsymbol{A}\boldsymbol{x} \end{aligned} \tag{10.14}$$

The resulting optimization problem can be solved efficiently, for example, using sequential quadratic programming (SQP) [1]. The main idea of SQP is to iteratively solve a quadratic approximation of the optimization problem, i.e., at the jth iteration we have

$$\begin{aligned} \min_{\boldsymbol{s}_x} \quad & \frac{1}{2}\boldsymbol{s}_x{}^{\mathsf{T}}\nabla_{xx}^2 L(\boldsymbol{x}_j, \boldsymbol{\mu}_j)\boldsymbol{s}_x + \boldsymbol{s}_x{}^{\mathsf{T}}\nabla J(\boldsymbol{x}_j) \\ \text{s.t.} \quad & \boldsymbol{b} - \boldsymbol{A}(\boldsymbol{x}_j + \boldsymbol{s}_x) \leq \boldsymbol{0} \end{aligned} \tag{10.15}$$

where $L(\boldsymbol{x}_j, \boldsymbol{\mu}_j)$ is the Lagrangian of the original optimization problem at iteration j and $J(\boldsymbol{x}_j)$ the respective cost function of the original optimization problem, with

$$\boldsymbol{x}_{j+1} = \boldsymbol{x}_j + \boldsymbol{s}_x$$

and

$$\boldsymbol{\mu}_{j+1} = \boldsymbol{\mu}_{QP},$$

where $\boldsymbol{\mu}_{QP}$ are the Lagrange multipliers associated with the quadratic optimization problem. The SQP algorithm is terminated when

$$\|\nabla L(\boldsymbol{x}, \boldsymbol{\mu})\|_2^2 < \epsilon,$$

which indicates that a local minimum has been found. The parameter ϵ is the convergence tolerance and should be chosen small enough. Another possibility is to use a nonlinear solver, i.e., IPOPT [20] to solve the nonlinear optimization problem (10.12).

The NLP is initialized with shape factor m_i, combustion duration $\varphi_{cd,i}$, and the start of combustion $\varphi_{soc,i}$ for the different i being linearly spaced on intervals whose boundaries are obtained by experiment.

A big advantage of using SQP to solving the NLP is the fact that the computational effort of solving a single SQP step is predictable. This enables its use also in a cycle-to-cycle feedback controller. Instead of letting the SQP fully converge, it is also possible to only execute a certain number of SQP steps (e.g., just one) and then use the parameters to build the combustion model, which is then used to compute the injection profile for the next cycle. Based upon the measurements of the next cycle and the predictions, the next SQP steps are executed, where the parameters are initialized with the parameters of the preceding cycle.

10.4 Gaussian Mixture Models and the EM Algorithm

Section 10.3 introduces the Vibe function as a good approximation of the heat release rate of an injected fuel package. In this section, the Vibe function is replaced by a Gaussian function, which is uniquely parametrized by the mean μ and variance σ^2. Thus the predicted heat release rate is a linear superposition of Gaussian distributions with different parameters mean μ and variance σ^2. The Gaussian distribution is denoted by

$$\mathcal{N}(k|\mu, \sigma^2) = \frac{1}{\sqrt{2\pi\sigma^2}} e^{-\frac{(k-\mu)^2}{2\sigma^2}}. \tag{10.16}$$

The resulting superposition consisting of N_u Gaussian distributions is then

$$y_k^{pred} = \sum_{i=1}^{N_u} u_i \mathcal{N}(k|\mu_i, \sigma_i^2). \tag{10.17}$$

By definition, the Gaussian function acts as normalized rate and thus has the unit s^{-1}. The similarity to (10.8) is readily visible. An example what such a superposition looks like is given in Fig. 10.2. The use of Gaussian distributions instead of Vibe functions is interesting because it only has two parameters per fuel package and it has some nice mathematical properties. Namely, it can be solved much quicker using an iterative algorithm with closed-form solutions, therefore, circumventing the NLP used for Vibe functions. Note the following two important properties:

1. $\frac{\sum_{k=1}^{N_y} y_k}{Q_{f,tot}} = 1$
2. $y_k \geq 0, \forall k$.

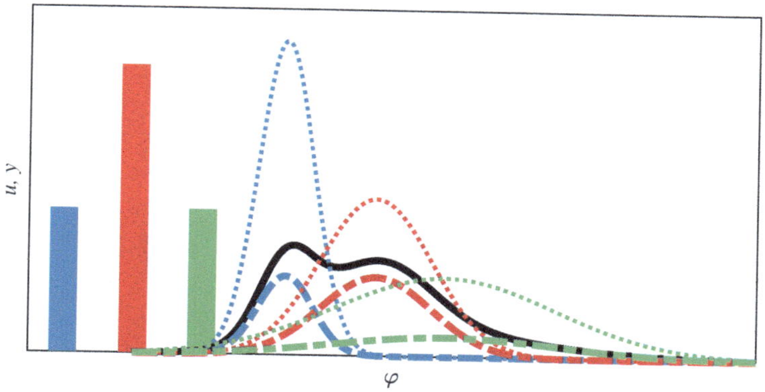

Fig. 10.2 The principle of a Gaussian Mixture Model with three mixture components ($N_u = 3$). Referring to (10.17), the histogram bars represent u_i, the dotted curves $\mathcal{N}(\cdot|\mu_i, \sigma_i^2)$, the dashed curves $u_i\mathcal{N}(\cdot|\mu_i, \sigma_i^2)$, and the black curve y^{pred}

These are the two properties that characterize a probability distribution. Thus

$$\bar{y}_k^{meas} = \frac{y_k^{meas}}{Q_{f,tot}} \quad \text{and} \quad \bar{y}_k^{pred} = \frac{\sum u_i \mathcal{N}(k|\mu_i, \sigma_i^2)}{Q_{f,tot}} \tag{10.18}$$

can be seen as probability distributions. Remember, we want to fit the parameters to the measured data. It is possible to use tools from probability theory, since these functions fulfill the criteria for probability distributions. As will be visible shortly, the mathematics used in the following are invariant to linear scaling. Thus, $\boldsymbol{y}^{meas}$, and $\boldsymbol{y}^{pred}$ are used instead of $\bar{\boldsymbol{y}}^{meas}$ and $\bar{\boldsymbol{y}}^{pred}$ to facilitate notation.

An often-used measure to judge the overlap of two appropriately defined probability distributions $P(x)$ and $Q(x)$ is the Kullback–Leibler divergence, defined as

$$D_{KL}(P\|Q) = -\sum_{x\in\mathcal{X}} P(x)\log\left(\frac{Q(x)}{P(x)}\right) = -\sum_{x\in\mathcal{X}} \left[P(x)\log\left(Q(x)\right) - P(x)\log\left(P(x)\right)\right]. \tag{10.19}$$

The Kullback–Leibler divergence is nonnegative, and $D_{KL} = 0$ holds only if $P(x) = Q(x)$ almost everywhere [5]. Because of these properties, the Kullback–Leibler divergence is often used to find probability distributions with high overlap. As shown above, the heat release rate curves have the same properties as a probability distribution. Thus, to find the parameters μ_i, σ_i, the following optimization problem is set up:

$$\min_{\boldsymbol{\mu}, \boldsymbol{\sigma}} \quad D_{KL}(\boldsymbol{y}^{meas} \| \boldsymbol{y}^{pred}(\boldsymbol{\mu}, \boldsymbol{\sigma})) \tag{10.20}$$

Since $\sum_{k=1}^{N_y} y_k^{meas}\log(y_k^{meas})$ is a constant term, (10.20) reduces to

$$\min_{\boldsymbol{\mu}, \boldsymbol{\sigma}} \quad -\sum_{k=1}^{N_y} y_k^{meas} \log \left(y_k^{pred}(\boldsymbol{\mu}, \boldsymbol{\sigma}) \right) \tag{10.21}$$

which corresponds to

$$\min_{\boldsymbol{\mu}, \boldsymbol{\sigma}} \quad -\sum_{k=1}^{N_y} y_k^{meas} \log \left(\sum_{i=1}^{N_u} u_i \mathcal{N}(k|\mu_i, \sigma_i^2) \right). \tag{10.22}$$

Unfortunately, there exists no closed-form solution to (10.22). However, (10.22) is equivalent to a clustering problem using Gaussian Mixture Models (GMM) with fixed weights for the Gaussian distributions. It is possible to solve this optimization problem using the well-known Expectation-Maximization (EM) algorithm [5]. But first, the so-called responsibility γ is introduced as

$$\gamma_{ik} = \frac{u_i \mathcal{N}(k|\mu_i, \Sigma_i)}{\sum_{j=1}^{N_u} u_j \mathcal{N}(k|\mu_j, \Sigma_j)}. \tag{10.23}$$

The responsibility γ_{ik} is the fraction of the heat release rate at time k stemming from injected package i.

The EM algorithm from can be adapted to this problem by simply omitting the update steps for the weights of the different Gaussian distributions. This is possible, because the inputs u_k are assumed to be known, and thus are not optimization variables. Then, the EM to iteratively solve (10.22) is composed of the following steps [5]:

1. Initialize the means μ_i and covariances σ_i and evaluate the initial value of the cost function of (10.22).
2. **E step**: Evaluate the responsibilities using the current parameter values

$$\gamma_{ik} = \frac{u_i \mathcal{N}(k|\mu_i, \sigma_i)}{\sum_{j=1}^{N_u} u_j \mathcal{N}(k|\mu_j, \sigma_j)}. \tag{10.24}$$

3. **M step**: Re-estimate the parameters using the current responsibilities

$$\mu_i^{\text{new}} = \frac{1}{C_i} \sum_{k=1}^{N_y} \gamma_{ik} k \tag{10.25}$$

$$\sigma_i^{\text{new}} = \frac{1}{C_i} \sum_{k=1}^{N_y} \gamma_{ik} (k - \mu_i^{\text{new}})^2, \tag{10.26}$$

where

$$C_i = \sum_{k=1}^{N_y} \gamma_{ik}$$

can be seen as the total heat released by injected fuel package i. Based on the argument that earlier injected packages oxidize earlier, the constraint $\mu_i < \mu_{i+1}$ is enforced by reinitializing any μ_i that do not fulfill the constraint, starting from step 1.

4. Evaluate the cost function

$$-\sum_{k=1}^{N_y} y_k^{meas} \log \left(\sum_{i=1}^{N_u} u_i \mathcal{N}(k|\mu_i^{\text{new}}, (\sigma_i^{\text{new}})^2) \right) \tag{10.27}$$

and check for convergence of either the parameter vectors $\boldsymbol{\mu}$ and $\boldsymbol{\sigma}$, or the cost. If the convergence criterion is not satisfied, return to step 2.

The EM algorithm for GMM faces the problem that the optimization problem in (10.22) is not lower bounded. Specifically, if there exist i and k such that $\mu_i = k$, then for $\sigma_i \to 0$, the cost function in (10.22) can get arbitrarily close to $-\infty$. This problem is well known and has been treated in several publications, e.g., [4, 5] or [3]. The most often-used solution is to set a lower limit for the σ_i. If the lower bound for the σ_i is violated for any i, then these μ_i and σ_i are randomly reinitialized.

Two initialization routines are investigated, the first being drawing the μ_i and σ_i randomly from within reasonable bounds, e.g., from a calibrated probability distribution. The second algorithm splits the measured HRR into N_u segments, where the ith segment integrates to u_i. For each of these segments, a Gaussian distribution is fitted using a simple maximum-likelihood approach. The μ_i and σ_i found by this algorithm are then used to initialize the EM algorithm. This initialization routing is referred to as decoupled initialization in Sect. 10.5.

10.5 Validation of the Fitted Model

The performance of the algorithms is validated against measurements using the measured HRR and injection rate signals as inputs to the learning algorithms. The HRR and injection rate are measured with a sampling angle of $\varphi_s = 0.1$ °CA. The measurements were conducted on a four-cylinder diesel engine test bench at ETH Zürich with 2.0 L displacement. The engine is turbocharged and features exhaust gas recirculation (EGR). A total of 18 measurements at distinct operating points with two to four discrete injection events were conducted at 2000 rpm, with EGR mass rates ranging from 5% to 20%. The injection rate is resampled to have $N_u = 60$ sampling points between approximately −18 °CAaTDC and 21 °CAaTDC, resulting in a sampling angle of $\varphi_{s,u} \approx 0.65$ °CA. Note that in this section, all trajectories have been normalized, such that the sum of all samples of a trajectory equals one. The calculations are conducted on a personal computer equipped with an Intel Core i7-7700 (3.60 GHz, 8 MB) processor.

10.5.1 Nonlinear Programming for Vibe Model

The algorithm is initialized as described in Sect. 10.3. A representative measurement is shown throughout this section. Both SQP and IPOPT are capable of solving the optimization. Since the used SQP routine is implemented by the authors without too much effort to maximize the efficiency, it is generally slower than the highly optimized IPOPT routines. The implementation is done using the CasADi framework [2] in MATLAB. The resulting fit after performing 102 and 389 IPOPT iterations is shown in Fig. 10.3, which correspond to an MSE of about 7×10^{-8} and 1×10^{-9}, respectively. In Fig. 10.4, the evolution of the value of the cost function is plotted over the number of solver iterations. Not shown is the error during the solver's presolve routine, where, starting from an initial guess, a feasible point is searched through a series of auxiliary problems involving barrier functions. The initial guess in this case is not inside the feasible region, and the solver performs eight presolve steps, during which the value of the objective can be increasing. These presolve steps are not counted as solver iterations. The solution found by the solver after the presolve steps is also shown in Fig. 10.3. Because the reduction in the cost is greatest at the beginning, these optimization algorithms are often terminated before complete convergence is reached. An error that is robustly reached is about 1×10^{-8}. Both algorithms, SQP and IPOPT, have run-times on the order of tens of seconds to convergence. Most of this time can be attributed to inefficient implementation, in particular assembling the necessary CasADi objects, and it is reasonable to assume that the algorithms could be made fast enough to be applicable to real-time control, since in particular SQP strategies are commonly used in real-time nonlinear model predictive control (NMPC) [1]. The so-called real-time iteration scheme that is often used in real-time NMPC even only performs a single SQP step [6].

10.5.2 Expectation–Maximization for Gaussian Mixture Model

Two major aspects that influence the performance of the EM algorithm are the convergence threshold and the initialization. Both also have a strong influence on the run-time. Another parameter that has a major influence on the run-time, but also on the quality of the fit is the chosen discretization of the injection signal. These three aspects are subsequently investigated.

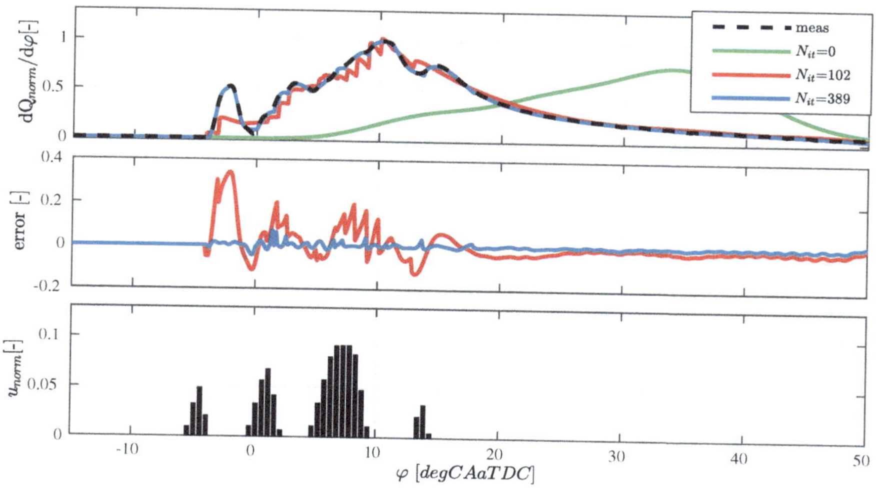

Fig. 10.3 The Vibe function model fitted by solving an NLP approximately for a four pulse injection compared to the measured HRR. Shown is the solution after the presolve steps of the solver and the solution after different numbers of solver iterations

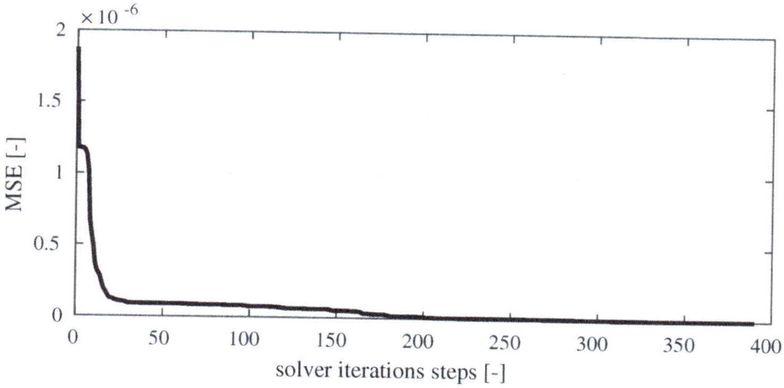

Fig. 10.4 The evolution of the cost function over the number of performed solver iterations

To investigate when to terminate the EM algorithm, the MSE is plotted over the performed iterations for a representative experiment in Fig. 10.5 on the left. The convergence is fast in the first few iterations and begins to stall after about ten iterations. In Fig. 10.5 on the right the decadic logarithm of the MSE is plotted to make the discontinuities that occur more visible. These stem from degenerations of one or more mixture components, which are then randomly reinitialized, leading to an increase in the MSE. The figure shows that these discontinuities do not prohibit the EM algorithm with random reinitializations from converging. Experience shows that a robust convergence tolerance ϵ for near-complete convergence can be chosen as $\epsilon_1 = 1 \times 10^{-8}$. If, however, the computational effort must be minimized, choosing

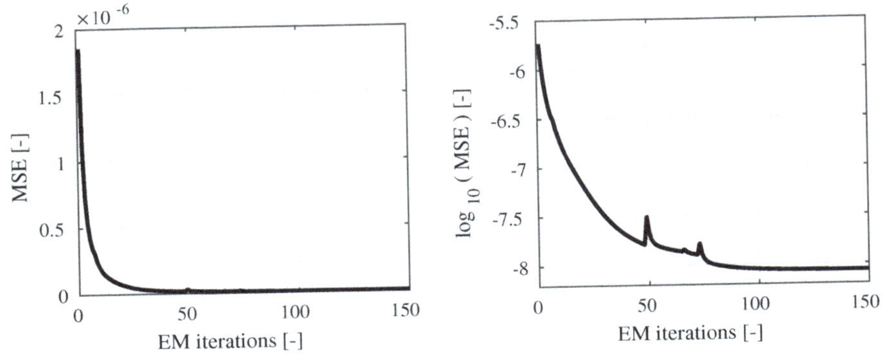

Fig. 10.5 The evolution of the MSE (left) and the decadic logarithm of the MSE (right) of the EM algorithm for GMM for a representative experiment

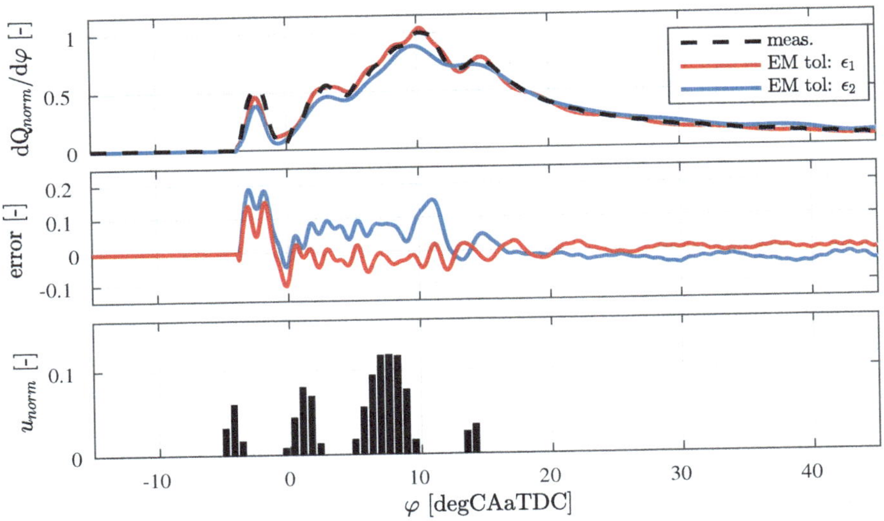

Fig. 10.6 The GMM model fitted with the EM algorithm for a four pulse injection compared to the measured HRR for two different convergence tolerances

$\epsilon_2 = 4 \times 10^{-8}$, leading to incomplete convergence, is a reasonable trade-off between accuracy and computation time.

The effect on the quality of the fit is shown in Fig. 10.6. The algorithm using ϵ_1 as termination condition approximates the measured HRR better, however the fit of the algorithm using ϵ_2 is reasonably good as well.

All measurements converged below the specified tolerances. Table 10.1 shows number of iterations and computation time averaged over the measurements.

Using the decoupled initialization reduces both the number of iterations until convergence and the total computation time for both the fully converged (denoted ϵ_1)

Table 10.1 The average total run-times and average number of iterations of the EM algorithm until convergence for two tolerances, two initialization algorithms for $N_u = 60$. Standard deviations are shown in parentheses

Initialization	$N_u = 60$			
	Run-time [s]		Iterations [-]	
	ϵ_1	ϵ_2	ϵ_1	ϵ_2
Random	1.16 (0.29)	0.55 (0.36)	80.8 (21.5)	36.2 (19.9)
Decoupled	0.29 (0.21)	0.02 (0.01)	20.1 (15)	1.5 (0.5)

Table 10.2 The average total run-times and average number of iterations of the EM algorithm until convergence for two tolerances, two input discretizations with decoupled initialization. Standard deviations are shown in parentheses

N_u	Decoupled Initialization			
	Run-time [s]		Iterations [-]	
	ϵ_1	ϵ_2	ϵ_1	ϵ_2
60	0.29 (0.21)	0.02 (0.01)	20.1 (15)	1.5 (0.5)
120	0.11 (0.02)	0.04 (0.01)	3.7 (0.4)	1.5 (0.5)

as well as the partially converged (denoted ϵ_2) scenario. Note that the implementation is not optimized at all, but still the decoupled-ϵ_2 scenario already terminates in only 70 ms, showing how efficient the algorithm is. The average computation time per iteration is similar for all settings at about 15 ms.

To see the influence of $\varphi_{s,u}$, the number of support points of u is doubled from $N_u = 60$ to $N_u = 120$. The comparison is shown in Table 10.2 only for decoupled initialization. The average computation time per iteration is about 29 ms, which is about double of what was measured for $N_u = 60$. Increasing N_u leads to increased computation times for the partially converged ϵ_2 setting, but decreased time for the fully converged ϵ_1 setting. The initial convergence per iteration is greater for $N_u = 60$, however it is slower towards the plateau region of the MSE.

10.5.3 Comparison

It is evident that the EM algorithm for the GMM model computes faster than the NLP for Vibe basis functions. Additionally, the fit measured by the MSE is of comparable quality. The Vibe function is motivated by experiment and has been reported to be a good approximation for the HRR of diesel combustion. However, to obtain a model that can be learned to produce a good fit to a measured HRR, using Gaussian function instead of Vibe functions seems to come at no drawbacks but with computational benefits. The conducted experiments do not show any advantage of using the Vibe basis function model over the GMM model.

10.6 Conclusion and Outlook

In this paper, two novel data-based learning algorithms for modeling of the crank angle resolved CI combustion process are presented. The intended purpose of the approaches is the learning-based combustion rate shaping. The advantages of the presented algorithms compared to other state-of-the-art models used in CRS, is the faster calibration time. This allows to use the approach even for cycle-to-cycle learning of the combustion model. Additionally, the presented algorithms are more robust to noise compared to other learning-based algorithms, as they make use of all the available data. In both cases, the data-based model is given by a superposition of basis functions.

The first approach uses the Vibe function as the basis function. The parameter estimation is based on a nonlinear optimization problem, where the objective is to minimize the squared distance between the measured heat release rate and the data-based model. It successfully converges to the measured heat release rate. While the algorithm is still too slow for cycle-to-cycle control, with run-times in the order of seconds it is applicable to reduce calibration times on test benches. Furthermore, for application of this algorithm, it is possible to optimize the implementation and speed up the computation time.

In the second algorithm, Gaussian distribution function is used as basis function, which only has two parameters, the mean and the variance. Thus, fewer parameters need to be learned. Furthermore, the specific cost function is chosen as the Kullback–Leibler divergence. The resulting optimization problem can be solved iteratively, using an adapted version of the expectation maximization algorithm. Since for each iteration closed-form solutions are available, the algorithm is much faster. Besides achieving a significant speedup compared to random initialization, the decoupled initialization routine offers another advantage: it eliminates the need for heuristic tuning of (hyper)parameters that underlie the random initialization routine, and can therefore improve the robustness of the algorithm. Tailoring the algorithm, e.g., using fast lookup tables to evaluate the Gaussian distribution, the algorithm is applicable for online cycle-to-cycle control.

The Vibe function is motivated by the experiment and has been reported to be a good approximation of the HRR of CI combustion. However, both algorithms lead to models of equal quality, while the GMM model can be obtained with much lower computational effort. Thus, approximating the Vibe function as a Gaussian distribution leads to good fits and increased computation speeds.

Further research should focus on testing these algorithms in closed-loop combustion rate shaping to assess the quality of these learning algorithms to deal with a control framework.

Acknowledgements The research was performed as part of the Research Unit (Forschergruppe) FOR 2401 "Optimization based Multiscale Control for Low Temperature Combustion Engines," which is funded by the German Research Association (Deutsche Forschungsgemeinschaft, DFG). The support is gratefully acknowledged.

References

1. T. Albin, *Nonlinear Model Predictive Control of Combustion Engines: From Fundamentals to Applications* (Springer, 2020)
2. J.A.E. Andersson, J. Gillis, G. Horn, J.B. Rawlings, M. Diehl, CasADi – A software framework for nonlinear optimization and optimal control. Mathematical Programming Computation (In Press, 2018)
3. J.P. Baudry, G. Celeux, Em for mixtures. Stat. Comput. **25**(4), 713–726 (2015)
4. C. Biernacki, S. Chrétien, Degeneracy in the maximum likelihood estimation of univariate gaussian mixtures with em. Stat. Probabil. Lett. **61**(4), 373–382 (2003)
5. C.M. Bishop, *Pattern Recognition and Machine Learning* (Springer, 2006)
6. M. Diehl, H.G. Bock, J.P. Schlöder, A real-time iteration scheme for nonlinear optimization in optimal feedback control. SIAM J. Control. Optim. (2005)
7. P.G. Dowell, S. Akehurst, R.D. Burke, An improved rate of heat release model for modern high-speed diesel engines. J. Eng. Gas Turbines Power **139**(9), (2017)
8. J. Galindo, J. Lujan, J. Serrano, L. Hernández, Combustion simulation of turbocharger hsdi diesel engines during transient operation using neural networks. Appl. Therm. Eng. **25**(5–6), 877–898 (2005)
9. L. Guzzella, C. Onder, *Introduction to Modeling and Control of INTERNAL COMBUSTION ENGINE SYSTEMS* (Springer Science & Business Media, 2009)
10. R. Isermann, *Engine Modeling and Control* (Springer, Berlin, 2014)
11. C. Jörg, J. Schaub, D. Neumann, S. Pischinger, Diesel combustion control via rate shaping. MTZ Worldwide **79**(4), 16–21 (2018)
12. C. Jãrg, Development of a combustion rate shaping controller for transient engine operation on a direct injection compression ignition engine. Ph.D. thesis, RWTH Aachen (2017)
13. T. Li, H. Ogawa, Analysis of the trade-off between soot and nitrogen oxides in diesel-like combustion by chemical kinetic calculation. SAE Int. J. Engines **5**(2), 94–101 (2012)
14. L. Ljung, System identification, *Wiley Encyclopedia of Electrical and Electronics Engineering* (1999) pp. 1–19
15. X. Luo, M. Donkers, B. de Jager, F. Willems, Systematic design of multivariable fuel injection controllers for advanced diesel combustion. *IEEE Transactions on Control Systems Technology* (2018)
16. D. Neumann, S. Pischinger, J. Schaub, B. Heuser, Flex-fuel-kompatibilität durch digitale verbrennungsratenregelung. MTZ - Motortechnische Zeitschrift **80**(5), 30–39 (2019)
17. D. Rether, M. Grill, A. Schmid, M. Bargende, Quasi-dimensional modeling of ci-combustion with multiple pilot-and post injections. SAE Int. J. Engines **3**(1), 12–27 (2010)
18. J. Syed, R.U. Baig, S. Algarni, Y.S. Murthy, M. Masood, M. Inamurrahman, Artificial neural network modeling of a hydrogen dual fueled diesel engine characteristics: an experiment approach. Int. J. Hydrog. Energy **42**(21), 14750–14774 (2017)
19. I.I. Vibe, Semi-empirical expression for combustion rate in engines, *Proceedings of Conference on Piston Engines, USSR Academy of sciences* (Moscow, 1956), pp. 185–191
20. A. Wächter, L.T. Biegler, On the implementation of an interior-point filter line-search algorithm for large-scale nonlinear programming. Math. Program. **106**(1), 25–57 (2006)
21. S.P. Wenzel, Modellierung der ruß-und nox-emissionen des dieselmotors (2006)
22. R. Zweigel, F. Thelen, D. Abel, T. Albin, T, Iterative learning approach for diesel combustion control using injection rate shaping, in *2015 European Control Conference (ECC)* (2015), pp. 3168–3173

Chapter 11
From Discrete and Iterative Deconvolution Operators to Machine Learning for Premixed Turbulent Combustion Modeling

P. Domingo, Z. Nikolaou, A. Seltz, and L. Vervisch

Abstract Following the rapid and continuous progress of computing power, allowing for increasing the mesh resolution in large eddy simulation (LES), new modeling strategies appear which are based on a direct treatment of the now well resolved, but still not fully resolved scalar signals. Along this line, deconvolution or inverse filtering, either based on discrete or iterative operators, is first discussed. Recent results obtained from a direct numerical simulation (DNS) database and LES of a premixed turbulent jet flame are presented. The analysis confirms the potential of deconvolution to approximate the unclosed non-linear terms and the SGS fluxes. Then, the introduction of machine learning in turbulent combustion modeling is illustrated in the context of convolutional neural networks.

P. Domingo · A. Seltz · L. Vervisch (✉)
CORIA – CNRS, Normandie Université INSA de Rouen,
76801 Saint-Etienne-du-Rouvray, France
e-mail: vervisch@coria.fr

P. Domingo
e-mail: domingo@coria.fr

A. Seltz
e-mail: seltz@coria.fr

A. Seltz
Safran Aircraft Engines Site de Villaroche Rond-Point René Ravaud-Réau,
77550 Moissy Cramayel, France

Z. Nikolaou
Computation-based Science and Technology Research Centre (CaSToRC),
The Cyprus Institute, 2121 Nicosia, Cyprus
e-mail: z.nicolaou@cyi.ac.cy

H. Pitsch and A. Attili (eds.), *Data Analysis for Direct Numerical Simulations of Turbulent Combustion*, https://doi.org/10.1007/978-3-030-44718-2_11

11.1 Introduction

Virtual prototyping has become an essential ingredient in the development and optimization of combustion systems, such as furnaces, boilers, internal combustion engines, and gas turbines [1–5]. Even relying on high-performance computing, the spatial resolutions reached by the grids do not allow for fully resolving the smallest scales of the flow motions and scalar signals when simulating real combustion chambers. The introduction of modeling is therefore mandatory to deal with unresolved sub-grid scale (SGS) fluctuations affecting the transport of mass, momentum, and energy. In addition, chemical reactions' mechanisms are strongly non-linear and specific tools must also be developed to compute burning rates over grids which are coarse at the scale of the thin reaction zones observed in flames [6–9].

In large eddy simulation (LES), a spatial filtering operation is applied to the first principle equations of aerothermochemistry, and any quantity $\overline{\phi}(\underline{x}, t)$ resolved by the grid reads

$$\overline{\phi}(\underline{x}, t) = \int_{-\infty}^{+\infty} \phi(\underline{x}', t) G_{\Delta}(\underline{x}' - x) dx' , \tag{11.1}$$

where $G_{\Delta}(\underline{x}' - x)$ denotes a filtering operation of characteristic size Δ and $\phi(\underline{x}, t)$ is the fully resolved signal, i.e., the solution that would be observed in a direct numerical simulation (DNS) resolving all the scales [10].

Because of unresolved SGS effects, the filtered value of any non-linear term $\overline{T(\phi)}$ differs from $T(\overline{\phi})$. So far, closures were developed in the literature in the form of mathematical expressions, providing estimation of the sub-grid scale effects from the knowledge of the quantities resolved by the grid, then $\overline{T(\phi)} = \mathcal{T}(\overline{\phi})$, where $\mathcal{T}(\overline{\phi})$ denotes the physical model used to express $\overline{T(\phi)}$ from the resolved field $\overline{\phi}$.

The application to turbulent flames of a different paradigm for modeling, which is based on a direct data analysis, is discussed in this chapter. Two main points are addressed:

- First, two methods of signal reconstruction based, respectively, on approximate inverse filtering and iterative deconvolution are discussed. In these two approaches, a tentative true signal $\phi^{\star}(\underline{x}, t) = \mathcal{L}_{\Delta}^{-1}\left[\overline{\phi}(\underline{x}, t)\right]$ is approximated from $\overline{\phi}(\underline{x}, t)$, the resolved field, to compute the non-linear terms which are then explicitly filtered,

$$\overline{T(\phi)}(\underline{x}, t) = \overline{T\left(\mathcal{L}_{\Delta}^{-1}\left[\overline{\phi}(\underline{x}, t)\right]\right)} . \tag{11.2}$$

- Second, a neural network is trained to relate the filtered non-linear terms (i.e., $\overline{T(\phi)}$) to three-dimensional distributions of resolved quantities surrounding the considered grid point (i.e., $T(\overline{\phi})$ or any function of $\overline{\phi}$). A mapping function $\mathcal{H}$ is then obtained so that

$$\overline{T(\phi)}(\underline{x}, t) = \mathcal{H}\left[T(\overline{\phi}(\underline{x}_1, t)), \ldots, T(\overline{\phi}(\underline{x}_N, t)\right] , \tag{11.3}$$

with $\underline{x}_j$ the N points selected around $\underline{x}$ to build an input image for a neural network previously trained.

Both approaches are evaluated a priori from different DNS databases, for the prediction of unresolved SGS convective and diffusive fluxes and for the modeling of filtered chemical sources.

11.2 Flame Filtering and Signal Reconstruction

11.2.1 Filtered Scalar Balance Equation

The budget equations for scalars describing flames (chemical species mass fractions, energy, temperature) may be cast in the generic form,

$$\frac{\partial \rho \phi}{\partial t} + \nabla \cdot (\rho \mathbf{u} \phi) = \nabla \cdot (\rho D \phi) + \dot{\omega} \,, \tag{11.4}$$

where ρ is the density, $\mathbf{u}$ the velocity vector, D a molecular diffusion coefficient, and $\dot{\omega}$ a chemical source. Applying the filtering operation (11.1) leads to

$$\frac{\partial \overline{\rho} \tilde{\phi}}{\partial t} + \nabla \cdot (\overline{\rho} \tilde{\mathbf{u}} \tilde{\phi}) = \nabla \cdot (\overline{\rho} D(\tilde{\phi}) \nabla \tilde{\phi}) + \overline{\dot{\omega}} + \nabla \cdot \tau \,, \tag{11.5}$$

where $\tilde{\phi} = \overline{\rho \phi}/\overline{\rho}$ and

$$\tau = \underbrace{\left(\overline{\rho D \nabla \phi} - \overline{\rho} D(\tilde{\phi}) \nabla \tilde{\phi}\right)}_{\text{SGS diffusion}} - \underbrace{\left(\overline{\rho \mathbf{u} \phi} - \overline{\rho} \tilde{\mathbf{u}} \tilde{\phi}\right)}_{\text{SGS convection}}, \tag{11.6}$$

are two unresolved SGS fluxes. The molecular diffusion coefficient usually depends non-linearly on the flow thermochemical properties and the notation $D(\tilde{\phi})$ denotes D computed from the resolved fields.

11.2.2 Signal Deconvolution

Two methodologies for signal reconstructions are now discussed after introducing a relation useful to determine the filter size which should actually be applied according to the mesh resolution and the number of points required within the flame.

11.2.2.1 Relation Between Flame Thickness, Filter Size, and Mesh Size

For a mesh of characteristic size h and a laminar premixed flame thickness δ_L, the relation between the number of points within the filtered signal and the filter size Δ may be approximated as [11]

$$\frac{h}{\Delta} = \frac{1}{n}\sqrt{\frac{\pi}{6} + \frac{\delta_L^2}{\Delta^2}}\,, \tag{11.7}$$

where a Gaussian filtering operation has been assumed

$$G(\underline{x}) = (6/(\pi\Delta^2))^{3/2}\exp(-6\underline{x}\cdot\underline{x}/\Delta^2)\,, \tag{11.8}$$

with a flame front modeled with an error function. This relation shows that for a LES with $\Delta = 5\delta_L$, the mesh size should verify $h = 0.75\Delta/n$. Since $n > 1$, h is therefore expected smaller than the filter size applied to the scalar fields.

11.2.2.2 Approximate Deconvolution and Explicite Filtering (ADEF)

Most filtering operations may be approximated as a diffusive process formulated implicitly

$$\overline{\phi}(\underline{x}, t) = \phi(\underline{x}, t) + d\nabla^2\overline{\phi}(\underline{x}, t)\,, \tag{11.9}$$

with $d = \Delta^2/24$ for a Gaussian filter [12]. From this relation, an approximate deconvolution operator is readily obtained in which all derivatives are resolved over the coarse mesh ($h < \Delta$):

$$\phi^\star(\underline{x}, t) = \mathcal{L}_\Delta^{-1}\left[\overline{\phi}(\underline{x}, t)\right] = \overline{\phi}(\underline{x}, t) - d\nabla^2\overline{\phi}(\underline{x}, t)\,. \tag{11.10}$$

To illustrate these filtering/deconvolution operations, Fig. 11.1 shows $T(x)$ (line), the temperature signal through a one-dimensional unstrained stoichiometric methane/air premixed flame, along with $\overline{T}(x)$ (dashed line) and $\tilde{T}(x)$ (circle). The deconvoluted signal $T^\star(x)$ (square) perfectly matches $T(x)$. Moreover, the solving of the budget equations of the one-dimensional filtered flame with all the non-linear terms computed from approximate deconvolution and filtering, returns the expected $\tilde{T}(x)$ (triangle). This was done here for $\Delta/\delta_L = 8$ with a second order interpolation of the scalar signals inside computing cells ($h = 0.5\Delta$) [13]. Flame deconvolution with an appropriate sub-grid interpolation or a regularization procedure thus appears as a robust tool [14, 15]. Sometimes deconvolution is also associated to a scale similarity hypothesis [16].

Approximate deconvolution is easily generalized to three-dimensional turbulent flames. Figure 11.2 shows snapshots of the simulation of a stoichiometric methane/air turbulent bunsen flame studied experimentally [17]. These simulations are performed

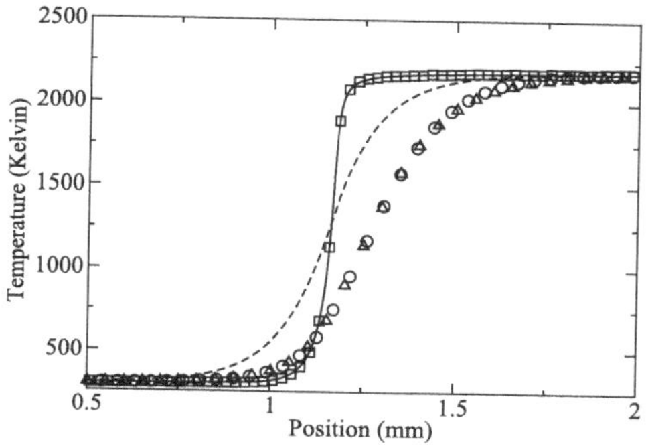

Fig. 11.1 Temperature profiles across a one-dimensional premixed flame. Line: $T(x)$; Square: $T^\star(x)$ from deconvolution (Eq. (11.10)); Dash: $\overline{T}(x)$; Circle: $\tilde{T}(x)$. Solving balance equation for filtered 1D flame from deconvolution Triangle: $\tilde{T}(x)$. $\Delta/\delta_{fl} = 8$. Reprinted with permission [13]

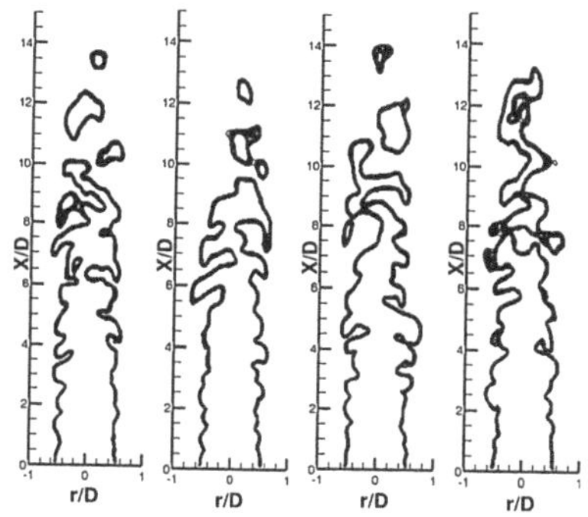

Fig. 11.2 Time sequence of iso-filtered source of progress variable, $\Delta t = 1.8$ ms. Reprinted with permission [13]

with the flow solver SiTCom [18], considering the Navier–Stokes equations in their fully compressible form together with a balance equation for $\tilde{c}$, a filtered progress variable ($c = 0$ in fresh gases and $c = 1$ in burnt gases [6]). Tabulated detailed chemistry [19, 20] is coupled to deconvolution of this progress variable with Eq. (11.10) to simulate the premixed turbulent jet flame. The tabulation is based on the GRI-3.0 [21] detailed chemistry, with a progress variable defined from CO, CO_2, H_2O, and NO_x as in [22]. The thermochemical quantities are then read from a lookup table $\phi = \phi_T(c)$ and the filtered quantities may be written from deconvolution and explicit filtering,

$$\overline{\rho}\tilde{\phi} = \overline{\rho^\star \phi_T(c^\star)} = \overline{\mathcal{L}_\Delta^{-1}[\overline{\rho}]\phi_T\left(\frac{\mathcal{L}_\Delta^{-1}[\overline{\rho}\tilde{c}]}{\mathcal{L}_\Delta^{-1}[\overline{\rho}]}\right)}\,. \tag{11.11}$$

The convective terms are discretized with a fourth-order centered skew-symmetric-like scheme [23] and the diffusive terms with a fourth-order centered scheme. Time is advanced explicitly with a third-order Runge–Kutta method and NSCBC boundary conditions [24] are imposed at inlet and outlet, with the measured profiles with synthetic turbulence [25] prescribed at inlet.

The time sequence ($\Delta t = 1.8$ ms) of iso-burning rate (Fig. 11.2) illustrates the high level of resolved flame wrinkling in this jet flame simulation featuring $u'/S_L = 5$,

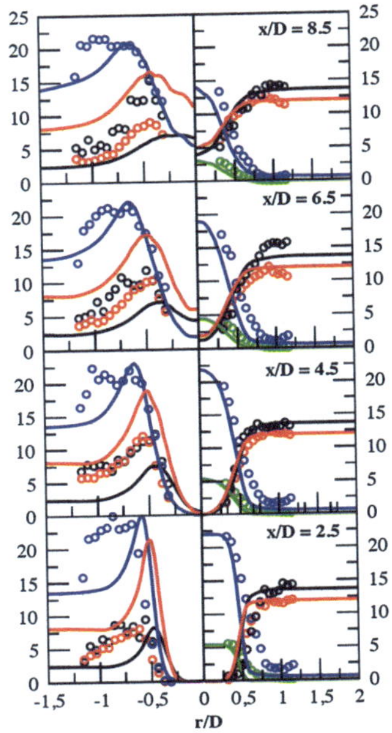

Fig. 11.3 Mass fractions in %. Symbol measurements [17]. Line: LES. In each subfigure: Left: Red: $CO \times 10$. Black: $H_2 \times 100$. Blue: $OH \times 75$. Right: Blue: O_2. Green: CH_4. Red: H_2O. Black: CO_2. Reprinted with permission [26]

with u' the characteristic velocity fluctuation and S_L the laminar flame speed. The formation of pockets at the flame tip is also observed. The mesh resolution[1] is of the order of 170 μm. The approximate deconvolution and explicit filtering approach coupled with tabulated chemistry allows for simulating the flame and predicting species mass fractions without resorting to any adjustment of model parameters. Results are given in Fig. 11.3. The experimentalist reported a measurement error range between 8% and 15% for major species concentration and up to 25% for minor species [19, 20]. The major species O_2, CH_4, H_2O, and CO_2 agree with measurements within this range (Fig. 11.3-right). The chemistry tabulation limited to a progress variable as input does not include strain rate as a parameter in these simulations, explaining the higher departure for the minor species CO, H_2 and OH.

11.2.2.3 Iterative Deconvolution and Explicit Filtering (IDEF)

As an alternative to approximate deconvolution based on discretized inverse filtering, the Van Cittert algorithm is another well-established iterative deconvolution algorithm [27, 28]. The calculation of derivatives is avoided by expressing the deconvoluted scalar field from

---

[1] A resolution of 50 μm would be necessary to fully resolve the flame (i.e., DNS) with tabulated chemistry and between 10 μm and 80 μm to resolve the Kolmogorov scale.

$$\phi^{\star n+1} = \phi^{\star n} + b(\overline{\phi} - \overline{\phi^{\star n}}) \,. \tag{11.12}$$

The initial condition of the iteration is $\phi^{\star 0} = \overline{\phi}$ and b is defined so that the deconvoluted variable stays within specific bounds [11]. If ϕ_l and ϕ_h are the low and high bounds, on option consists in writing

$$b = \begin{cases} \dfrac{\phi^{\star n} - \phi_l}{\phi_a - \phi_l} & \text{if } \phi \leq \phi_a \\[2ex] \dfrac{\phi^{\star n} - \phi_h}{\phi_a - \phi_h} & \text{otherwise} \end{cases} \tag{11.13}$$

with the average value $\phi_a = (\phi_l + \phi_h)/2$. The number of iterations is monitored through a volume averaged error $e_v = \langle |\overline{\phi} - \overline{\phi^\star}| \rangle$, so that iterations can proceed as long as e_v decreases, in practice for well-resolved premixed flames ($\Delta < 10\delta_L$), the iterations never exceed a few decades.

Correlations between conditionally averaged progress variable obtained from (11.12) compared to DNS values are shown in Fig. 11.4. The data are from a DNS of a methane/air planar premixed flame interacting with homogeneous turbulence [29]. In this DNS, $u'/S_L = 14$, the turbulent Reynolds number is $\mathrm{Re}_T = 230$, the Damköhler number $\mathrm{Da} = 1.17$, and the Karlovitz number $\mathrm{Ka} = 13$.

These three-dimensional direct simulations have been performed using the SENGA2 code [30], which solves the compressible reacting flow of Navier–Stokes equations using a 10th-order finite difference scheme for interior points, and a fourth-order Runge–Kutta scheme for the time-stepping. The DNS-filtered variables were sampled onto a LES ($h/\Delta = 0.25$) mesh using high-order Lagrange polynomials, values which are then used for the deconvolution [11]. The strong correlation between the prediction and the DNS reference (Fig. 11.4) confirms the potential of the iterative procedure for signal reconstruction. The time-averaged Pearson correlation coefficients are above 0.9850 for the three filter sizes ($\Delta = \delta_L$, $2\delta_L$, and $3\delta_L$) considered.

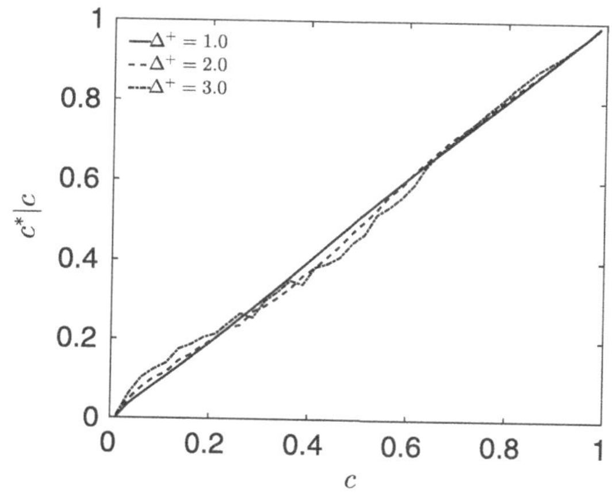

Fig. 11.4 Conditionally averaged progress variable as obtained from IDEF based on the actual value as obtained from the DNS. $\Delta^+ = \Delta/\delta_L$ Reprinted with permission [11]

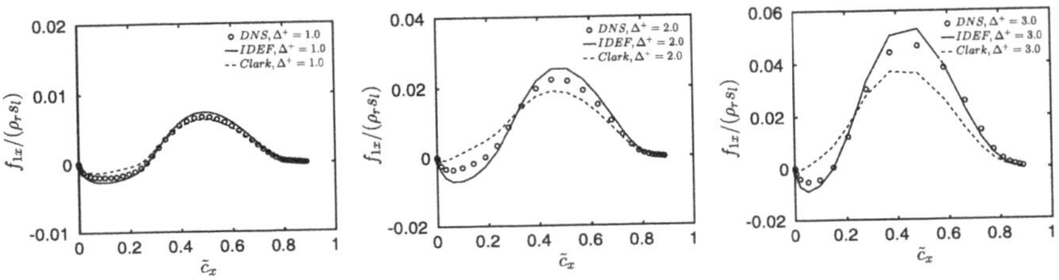

(a) $u'/S_L = 3.18$, $\mathrm{Re}_T = 53$, Da = 5.19, Ka = 1.39.

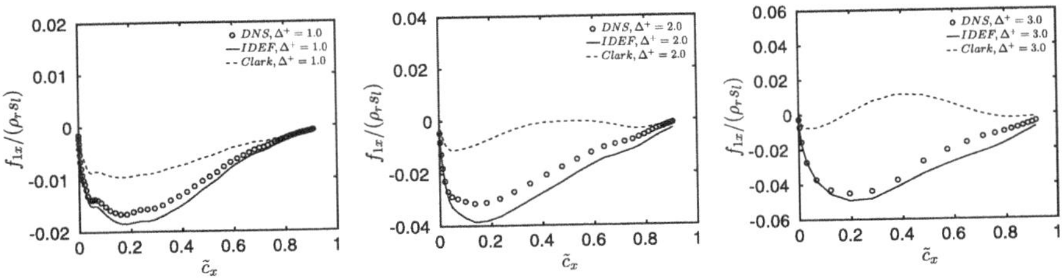

(b) $u'/S_L = 9$, $\mathrm{Re}_T = 150$, Da = 1.8, Ka = 6.62.

Fig. 11.5 SGS convective flux in the direction normal to the mean flame front averaged in the transverse plane, plotted versus a progress variable ($c_x = 0$ in fresh gases and $c_x = 1$ in fully burnt gases). o: DNS. –: IDEF. – –: Clark model [32]. Left: $\Delta/\delta_L = 1$, Middle: 2, Right: 3. Reprinted with permission [31]

On this basis, the iterative deconvolution was further applied to compute unresolved SGS fluxes, as the convective one [31]

$$\tau_u = \overline{\rho \mathbf{u} \phi} - \overline{\rho} \tilde{\mathbf{u}} \tilde{\phi} = \overline{(\rho \mathbf{u})^\star \frac{(\rho \phi)^\star}{\rho^\star}} - \overline{\rho} \tilde{\mathbf{u}} \tilde{\phi} \,. \tag{11.14}$$

Evaluation against the same turbulent planar flame DNS configuration [29] of this SGS flux projected in the direction normal to the mean flame front is shown in Fig. 11.5. In this figure, comparisons against the well-established Clark model [32] are also added; this physical model reads

$$f_{1x} = \tau_u \cdot \mathbf{e}_1 = \overline{\rho} \frac{\Delta^2}{12} \frac{\partial \tilde{u}_1}{\partial x_j} \frac{\partial \tilde{\phi}}{\partial x_j} \,. \tag{11.15}$$

As expected the transition from counter-gradient SGS turbulent transport to the usual gradient turbulent flux is found in the data by varying the turbulence intensity [33]. Indeed, $f_{1x} > 0$ (counter-gradient transport) is seen in Fig. 11.5(a) for the progress variable larger than 0.3, and $f_{1x} < 0$ (gradient transport) is observed in Fig. 11.5(b). The transport through the flame which was mainly driven by gas expansion for $u'/S_L = 3.18$ is rather governed by turbulent mixing for $u'/S_L = 9$. Interestingly, the deconvolution procedure perfectly reproduces this transition, while the physical model given by the relation (11.15) misses this very basic, but essential, property of SGS turbulent transport in premixed turbulent flame.

Following the same line, iterative deconvolution was also found effective to model the SGS stress tensor in premixed turbulent V-flame [34].

11.3 Machine Learning for Turbulent Combustion Modeling

11.3.1 DNS Embedded in LES Database

An additional DNS database was obtained in the configuration of the premixed turbulent bunsen flame discussed above in Sect. 11.2.2.2 [17]. This DNS is located downstream of the well-resolved LES of the first sections of the piloted premixed stoichiometric fuel-air round jet (Fig. 11.6); LES that provides turbulent flame conditions for the DNS inlet plane [26, 35, 36]. In this simulation, both LES and DNS are run simultaneously.

This dual simulation was achieved by embedding, inside the LES mesh, a zone where the resolution is sufficiently high to fully resolve the thin reaction zones and the smallest turbulent scales. The nozzle diameter is D = 12 mm and the jet Reynolds number is 24,000 (bulk nozzle velocity of 30 $m{\cdot}s^{-1}$ and turbulent kinetic energy of 3.82 $m^2{\cdot}s^{-2}$). The pilot is composed of fully burnt gases at $T_b = 2200$ K. The LES mesh contains 171 million nodes for a domain 16D×8D×8D, leading to a resolution of the order of 150 μm. The resolution in the DNS zone is uniformed at 50 μm, to secure a full resolution of the flow and flame scales for the chosen operating conditions [26].

Both LES and DNS rely on chemistry tabulation with a stoichiometric premixed flamelet with fresh gases at $T_o = 300$ K (GRI-3.0 mechanism [21] and progress variable defined as in [22]), without any SGS modeling in the DNS part. The flame thermal thickness based on the progress variable field is of the order of $\delta_L \approx 400$ μm. The mesh in the DNS zone is composed of 28.58 million nodes (243×343×343), over a physical domain of 12 mm×18 mm×18 mm. The DNS starts at $4.5D$ downstream of the nozzle, and at $5.5D$ the mesh is coarsened again to progressively evolve toward LES resolution (Fig. 11.6). Modeling based on a progress variable presumed pdf approach is used for the chemical sources and species in the LES zones [37] and the SGS momentum fluxes are approximated with the Vreman model [38].

As for the jet flame LES of Sect. 11.2.2.2, these simulations are performed using the flow solver SiTCom [18]. A Gaussian filtering operation (Eq. (11.8)), with filter size $\Delta = 0.3$ mm $= 0.75\ \delta_L$, $\Delta = 0.6$ mm $= 1.50\ \delta_L$, and $\Delta = 0.9$ mm $= 2.25\ \delta_L$, is applied to the DNS variables in order to generate a priori LES- filtered quantities. Thus the resolution of the a priori fields varies from well resolved to coarse LES from the reaction zone point of view ($\Delta = 0.9$ mm is 18 times larger than the DNS grid resolution).

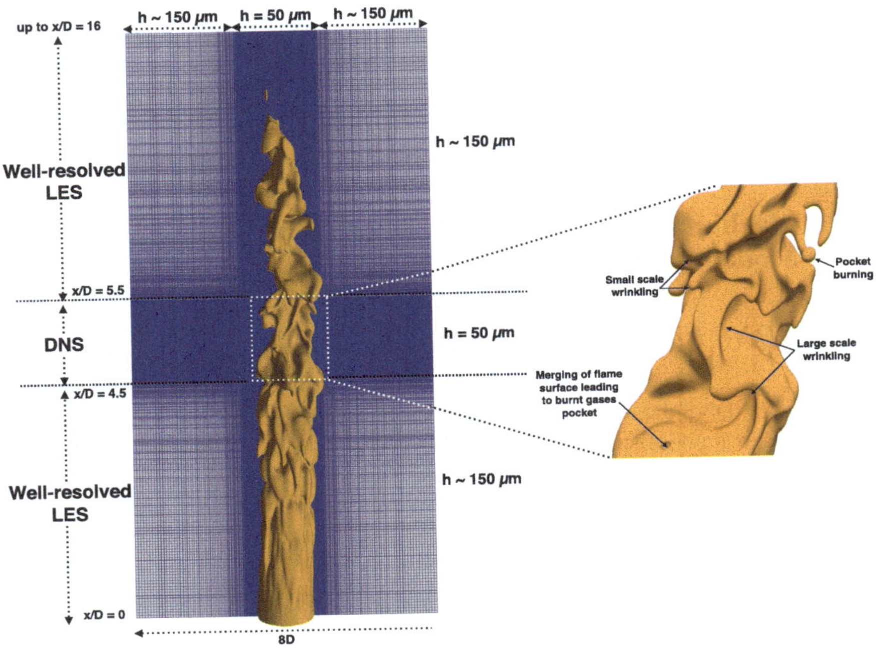

Fig. 11.6 LES-DNS snapshot of the jet flame simulation [26]. Mesh and iso-progress variable $c = 0.8$. h: resolution. Right: zoom of iso-$c = 0.8$ in the DNS zone (different angle view). Reprinted with permission [26]

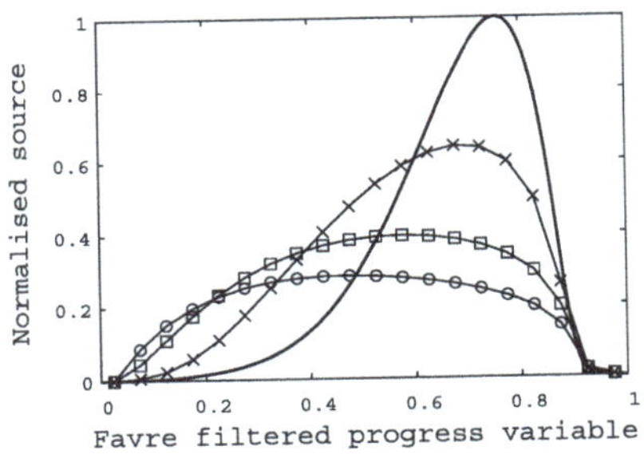

Fig. 11.7 Thick-line: $\dot{\omega}^+(\tilde{c})$ vs $\tilde{c}$ as obtained from a 1D laminar flame (tabulated chemistry). $\langle \overline{\dot{\omega}^+} \mid \tilde{c} \rangle$ from DNS vs $\tilde{c}$ for filter sizes $\times$: 0.3 mm, $\square$: 0.6 mm, $\circ$: 0.9 mm, ($\delta_L = 0.4$ mm). Reprinted with permission [39]

In the DNS zone, the flame surface is highly convoluted (Fig. 11.6-right), with flame wrinkling covering a large spectrum of length scales, including complex flame topological properties such as pockets of fresh or burnt gases.

Figure 11.7 shows $\langle \overline{\dot{\omega}^+} \mid \tilde{c} \rangle$, the statistical mean over the DNS domain of the normalized filtered progress variable source, conditioned on values of $\tilde{c}$. The subscript '+' denotes source terms normalized by their maximum value in the reference freely propagating laminar premixed flame. The result obtained using a 1D laminar flame, $\dot{\omega}^+(\tilde{c})$, is also shown as a solid line. The maximum of $\langle \overline{\dot{\omega}^+} \mid \tilde{c} \rangle$ decreases while increasing the filter size and thus while increasing unresolved fluctuations [6]. Fol-

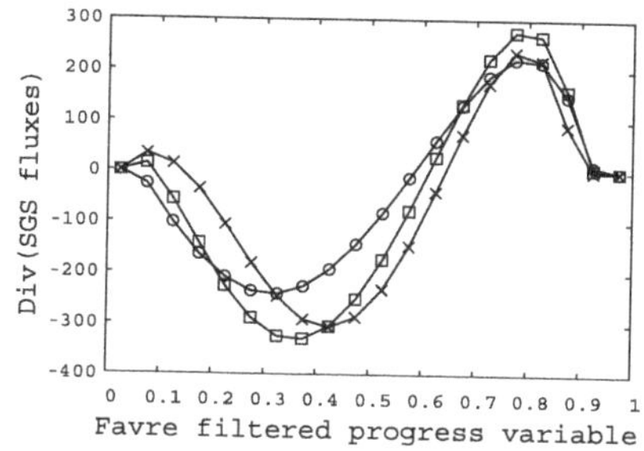

Fig. 11.8 $\langle \nabla \cdot \tau \mid \tilde{c} \rangle$ vs $\tilde{c}$. Filter size ×: 0.3 mm, □: 0.6 mm, o: 0.9 mm. Reprinted with permission [39]

lowing the thickening of the filtered flame front, the response of this conditional filtered source term also spreads in progress variable space for increasing filter sizes, up to $\Delta = 0.9\,\text{mm}$. Figure 11.7 clearly demonstrates that $\overline{\dot{\omega}^+} \neq \dot{\omega}^+(\tilde{c})$ and not only the amplitude, but also the shape of the burning rate versus the progress variable, vary with the filter size. A simple re-scaling by a factor would therefore not be sufficient to properly model $\overline{\dot{\omega}}$ from $\dot{\omega}(\tilde{c})$. Similarly, Fig. 11.8 shows the conditional mean of the sum of the SGS convective and diffusive fluxes (Eq. (11.6)), which also requires modeling.

11.3.2 Machine Learning for Turbulent Combustion Modeling

The underlying idea discussed in this section consists of finding a mapping function $\mathcal{G}$, so that the filtered chemical source $\overline{\dot{\omega}}(\underline{x}, t)$ is constructed from the Three-dimensional distribution surrounding $\underline{x}$ of the chemical sources computed from the resolved field, i.e.,

$$\overline{\dot{\omega}}(\underline{x}, t) = \mathcal{G}\left[\dot{\omega}(\tilde{c}(\underline{x}_1, t)), \ldots, \dot{\omega}(\tilde{c}(\underline{x}_N, t))\right] , \tag{11.16}$$

where N is the number of LES mesh cells involved in the mapping.

Similarly, a second mapping function $\mathcal{F}$ is sought to express $\nabla \cdot \tau$, the divergence of the SGS fluxes (Eq. 11.6), from the three-dimensional distribution of resolved diffusive budgets $\nabla \cdot (\overline{\rho} D_c(\tilde{c}) \nabla \tilde{c})$ surrounding the $\underline{x}$ location,

$$\nabla \cdot \tau(\underline{x}, t) = \mathcal{F}\left[\nabla \cdot \left(\overline{\rho} D_c(\tilde{c}) \nabla \tilde{c}(\underline{x}_1, t)\right), \cdots, \nabla \cdot \left(\overline{\rho} D_c(\tilde{c}) \nabla \tilde{c}(\underline{x}_N, t)\right)\right] . \tag{11.17}$$

These two mapping function are built from convolutional neural networks (CNN), which conveniently allow for interpolating over a large number of reference values. Typically, if a set of N_L reference values of filtered chemical sources $\overline{\dot{\omega}}[i]$ and divergence of the SGS fluxes $\nabla \cdot \tau[i]$ are known and are associated through the mappings to well-defined properties of the resolved field (here sources and diffusive budgets computed from the resolved scalar fields), well-trained neural networks will return $\Omega_i(\underline{x}, t)$ and $T_i(\underline{x}, t)$ as the weights to be affected to the ith values contributing to the filtered quantities, i.e.,

$$\overline{\dot{\omega}}(\underline{x}, t) = \sum_{i=1}^{N_L} \Omega_i(\underline{x}, t) \times \overline{\dot{\omega}}[i] \,, \tag{11.18}$$

$$\nabla \cdot \tau(\underline{x}, t) = \sum_{i=1}^{N_L} T_i(\underline{x}, t) \times \nabla \cdot \tau[i] \,. \tag{11.19}$$

The $\overline{\dot{\omega}}[i]$, $\nabla \cdot \tau[i]$, Ω_i, and T_i are calibrated by training the neural network from the DNS database. For a given filter size Δ, a three-dimensional test box of size $(2h)^3$ is built around each of the $M = 28.58$ million DNS nodes, where h is defined from Eq. (11.7). Figure 11.9 illustrates this process. The test box is centered at $\underline{x}$ and contains $N = 27$ points which hold the three-dimensional distributions of $\dot{\omega}(\tilde{c}(\underline{x}_j, t))$ and $\nabla \cdot (\overline{\rho} D_c(\tilde{c}) \nabla \tilde{c})\, (\underline{x}_j, t)$, for $j = 1, \ldots, N$. These data are stored and constitute the 'images' that will be processed by the CNN. From the DNS, the 'labels' of each ith image are $\overline{\dot{\omega}}[i] = \overline{\dot{\omega}}(\underline{x}, t)$ and $\nabla \cdot \tau[i] = \nabla \cdot \tau(\underline{x}, t)$ for $i = 1, \ldots, N_L$. In the present work $N_L = 2000$.

Two networks of similar structures (same number of layers, convolution kernels, etc.) are trained, one for the chemical source and one for the SGS fluxes. For each value of Δ, the following procedure is applied:

- First, 1000 images with their associated $\overline{\dot{\omega}}[i]$ and $\nabla \cdot \tau[i]$ values are drawn. 20 values of c^* uniformly distributed between 0 and 1 ($\Delta c^* = 0.05$) are defined. For each value of c^*, 50 images are selected randomly such that $\tilde{c}(\underline{x}, t) \in [c^* - \Delta c^*/2; c^* + \Delta c^*/2]$ ($\underline{x}$ is at the center of the test box, see Fig. 11.9).
- To avoid overfitting, uncorrelated random perturbations are added to the images, $\dot{\omega}(\tilde{c}(\underline{x}_j, t))$ and $\nabla \cdot (\overline{\rho} D_c(\tilde{c}) \nabla \tilde{c})\, (\underline{x}_j, t)$ for $j = 1, \ldots, N = 27$, as 10% of their

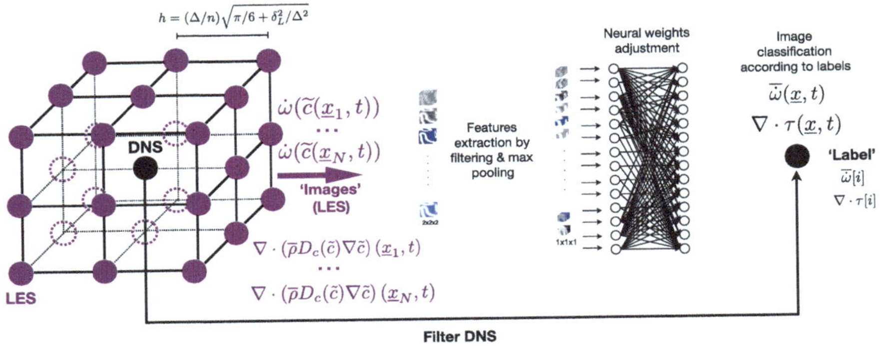

Fig. 11.9 CNN training from DNS, sketch of the database construction. Reprinted with permission [39]

maximum in the test box, thus a second image is available for each label. $N_L =$ 2000 images then exist for 1000 labels.

Using two filter sizes, the database to train the network contains 4000 images and 2000 labels and for every quantity learned, a set of $27 \times 4000 = 108000$ data $(\dot{\omega}(\tilde{c}(\underline{x}_j, t))$ and $\nabla \cdot (\overline{\rho} D_c(\tilde{c}) \nabla \tilde{c})\ (\underline{x}_j, t))$ is involved, data which are associated to the 2000 reference labels ($\overline{\dot{\omega}}[i]$ and $\nabla \cdot \tau[i]$ for $i = 1, \ldots, N_L$).

During the training phase, a series of convolution/sampling operations are performed iteratively, in which the neural weights are adjusted until a satisfying minimal error is found between the value of the labels and the neural network prediction. A series of convolution/sampling operations are applied to the database, to extract its features using a number of different kernels [41, 42]:

- Each image is first convoluted with 32 different filter kernels obtained with random values from a truncated normal distribution. Meaningful values of the returned features are extracted with a max pooling non-linear function to avoid excessive computational costs.
- The process is repeated with 64 filters, decomposing the image into several most meaningful features, needed for seeking out the inner properties of the fluxes and sources.
- Two fully connected layers are built to process the 64 obtained features, and to classify the image by giving the probability it belongs to each label.

The training of the network was conducted using the TensorFlow (www.tensorflow.org) library and breakdown of the network structure is given in Fig. 11.10. The training error function is based on cross entropy [43] and this training is controlled by the Adam optimizer [44] for stochastic gradient descent, with a learning rate of 10^{-4}.

In using the networks, the $N = 27$ values of the chemical sources and of the divergence of the fluxes computed from the resolved progress variable field in the test box surrounding the LES cell (Fig. 11.9) are the input. For each filter size, 1000 filtered DNS fields are used for a priori tests (the noised images of the training phase do not enter these tests).

Figure 11.11 shows the comparison between the conditional means of the predicted divergence of the fluxes and their DNS counterpart. For the trained filter size, the CNN returns the expected value, for the untrained one some departure is observed, but compared to what can be observed in some case using a physical model, results are more than encouraging. (Compare for instance these results against those of Fig. 11.5 with the Clark model.)

The filtered chemical source is also very well predicted; both the shape and the amplitudes are recovered by the CNN (Fig. 11.12). Interestingly, even the data corresponding to the filter sizes that were not used for training (0.4 mm and 0.45 mm) are quite well reproduced.

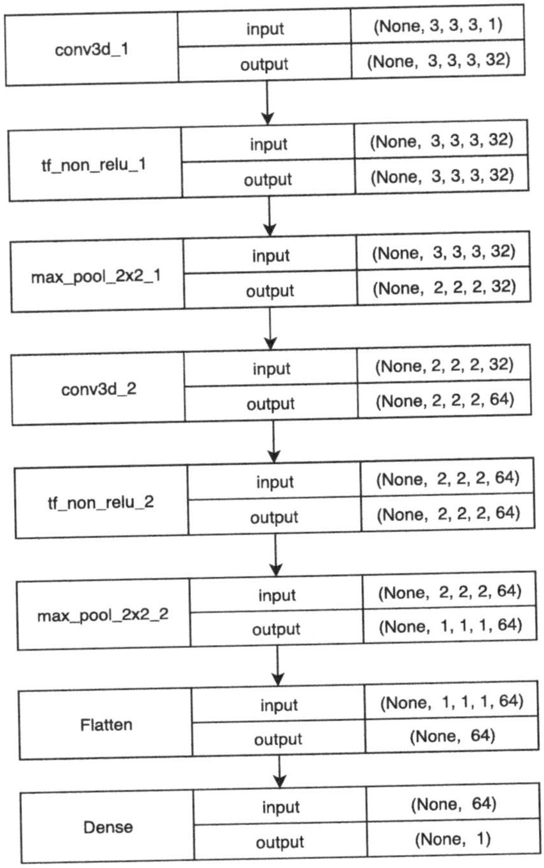

Fig. 11.10 Typical structure of a convolutional network for scalar deconvolution. More details on the TensorFlow functions used may be found in [39]. Reprinted with permission [39]

11.4 Conclusion

Large datasets from direct numerical simulation have been mostly used so far to explore reacting flow physics and to test well-established or novel modeling ideas. With rapidly growing computing facilities, followed by the development of well-resolved large eddy simulations for real combustion systems, the possibility of relying on modeling based on a direct analysis of the signals is now emerging.

Instead of expressing unclosed terms from algebraic relations involving quantities resolved by the mesh, deconvolution may be used to directly reconstruct the non-linear terms to close the scalar balance equations, then some form of sub-grid scale interpolation is required. Both discretized and iterative deconvolutions have been found valuable along these lines.

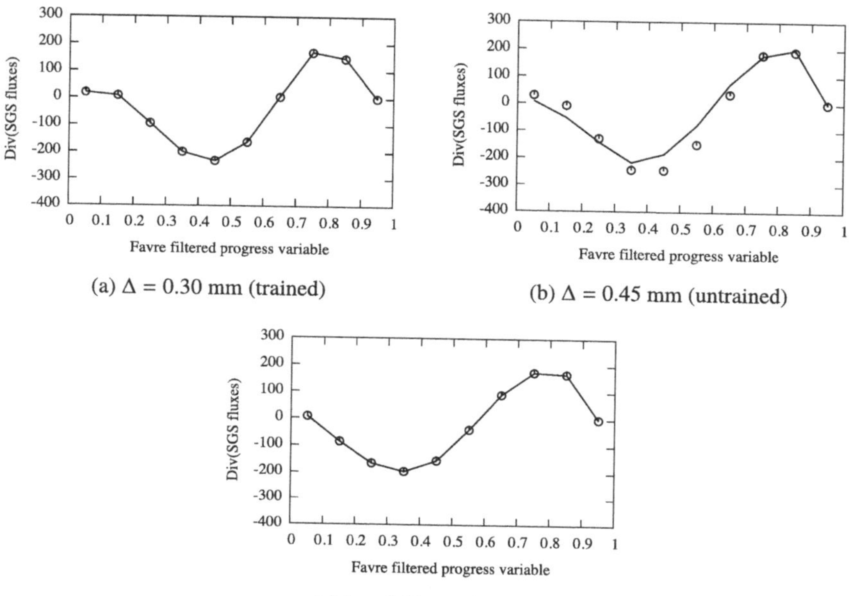

(a) Δ = 0.30 mm (trained)

(b) Δ = 0.45 mm (untrained)

(c) Δ = 0.90 mm (trained)

Fig. 11.11 $\langle \nabla \cdot \tau \mid \tilde{c} \rangle$ vs $\tilde{c}$. Symbols: DNS reference. Line: CNN prediction. Reprinted with permission [40]

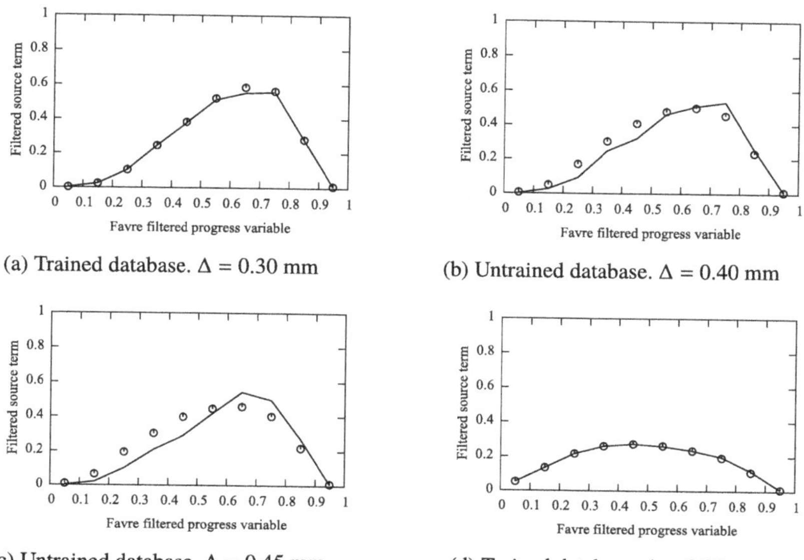

(a) Trained database. Δ = 0.30 mm

(b) Untrained database. Δ = 0.40 mm

(c) Untrained database. Δ = 0.45 mm

(d) Trained database. Δ = 0.90 mm

Fig. 11.12 $\langle \overline{\dot{\omega}^{+}} \mid \tilde{c} \rangle$ vs $\tilde{c}$. Symbols: DNS reference. Line: CNN prediction

In addition, the capabilities of machine learning tools to estimate unclosed terms from the resolved ones have been demonstrated, which opens many perspectives such as the coupling between advanced computational flame dynamics tools and artificial intelligence.

The deconvolution of scalar signals and the direct mapping form LES resolved scales to unknown filtered terms from CNN do not operate at the same level (signal reconstruction or direct prediction) and could thus be combined as complementary tools.

References

1. R.W. Bilger, S.B. Pope, K.N.C. Bray, J.F. Driscoll, Paradigms in turbulent combustion research. Proc. Combust. Inst **30**(1), 21–42 (2005)
2. L.Y.M. Gicquel, G. Staffelbach, T. Poinsot, Large Eddy simulations of gaseous flames in gas turbine combustion chambers. Prog. Energy Combust. Sci. **38**(6), 782–817 (2012)
3. T. Poinsot, Prediction and control of combustion instabilities in real engines. Proc. Combust. Inst. **36**(1), 1–28 (2017)
4. E. Mastorakos, Forced ignition of turbulent spray flames. Proc. Combust. Inst. **36**(2), 2367–2383 (2017)
5. C. Locci, L. Vervisch, B. Farcy, N. Perret, Selective non-catalytic reduction (SNCR) of nitrogen oxide emissions: a perspective from numerical modeling. Flow Turbul. Combust. **100**(2), 301–340 (2018)
6. K.N.C. Bray, The challenge of turbulent combustion. Symp. (Int.) Combust. **26**, 1–26 (1996)
7. N. Peters, *Turbulent Combustion* (Cambridge University Press, Cambridge, 2000)
8. D. Veynante, L. Vervisch, Turbulent combustion modeling. Prog. Energy Combust. Sci. **28**, 193–266 (2002)
9. H. Pitsch, Large Eddy simulation of turbulent combustion. Annu. Rev. Fluid Mech. **38**, 453–482 (2006)
10. M. Lesieur, O. Métais, P. Comte, *Large-Eddy Simulations of Turbulence* (Cambridge University Press, Cambridge UK, 2005)
11. Z. Nikolaou, L. Vervisch, A priori assessment of an iterative deconvolution method for les sub-grid scale variance modelling. Flow Turbul. Combust. **101**(1), 33–53 (2018)
12. F. Katopodes, R.L. Street, M. Xue, J.H. Ferziger, Explicit filtering and reconstruction turbulence modeling for large-eddy simulation of neutral boundary layer flow. J. Atmos. Sci. **62**(7), 2058–2077 (2004)
13. P. Domingo, L. Vervisch, Large Eddy simulation of premixed turbulent combustion using approximate deconvolution and explicit flame filtering. Proc. Combust. Inst. **35**(2), 1349–1357 (2015)
14. Q. Wang, M. Ihme, Regularized deconvolution method for turbulent combustion modeling. Combust. Flame **176**, 125–142 (2017)
15. C. Mehl, J. Idier, B. Fiorina, Evaluation of deconvolution modelling applied to numerical combustion. Combust. Theory Model. **22**(1), 38–70 (2018)
16. A.W. Vreman, R.J.M. Bastiaans, B.J. Geurts, A similarity sub-grid model for premixed turbulent combustion. Flow Turbul. Combust. **82**(2), 233–248 (2009)
17. Y.-C. Chen, N. Peters, G.A. Schneemann, N. Wruck, U. Renz, M.S. Mansour, The detailed flame structure of highly stretched turbulent premixed methane-air flames. Combust. Flame **107**(3), 223–244 (1996)
18. L. Bouheraoua, P. Domingo, G. Ribert, Large-eddy simulation of a supersonic lifted jet flame: Analysis of the turbulent flame base. Combust. Flame **179**, 199–218 (2017)

19. O. Gicquel, N. Darabiha, D. Thevenin, Laminar premixed hydrogen/air counterflow flame simulations using flame prolongation of ILDM with differential diffusion. Proc. Combust. Inst. **28**, 1901–1908 (2000)
20. J.A. van Oijen, F.A. Lammers, L.P.H. de Goey, Modeling of complex premixed burner systems by using flamelet-generated manifolds. Combust. Flame **127**(3), 2124–2134 (2001)
21. G.P. Smith, D.M. Golden, M. Frenklach, N.W. Moriarty, B. Eiteneer, M. Goldenberg, C.T. Bowman, R.K. Hanson, S. Song, W.C. Gardiner, V.V. Lissianski, Z. Qin, Technical report (1999). http://www.me.berkeley.edu/gri-mech/
22. G. Godel, P. Domingo, L. Vervisch, Tabulation of nox chemistry for large-eddy simulation of non-premixed turbulent flames. Proc. Combust. Inst. **32**, 1555–1561 (2009)
23. F. Ducros, F. Laporte, T. Soulères, V. Guinot, P. Moinat, B. Caruelle, High-order fluxes for conservative skew-symmetric-like schemes in structured meshes: application to compressible flows. J. Comput. Phys. **161**, 114–139 (2000)
24. G. Lodato, P. Domingo, L. Vervisch, Three-dimensional boundary conditions for direct and large-eddy simulation of compressible viscous flows. J. Comput. Phys **227**(10), 5105–5143 (2008)
25. M. Klein, A. Sadiki, J. Janicka, A digital filter based generation of inflow data for spatially developing direct numerical or large eddy simulations. J. Comput. Phys. **186**(2), 652–665 (2002)
26. P. Domingo, L. Vervisch, Dns and approximate deconvolution as a tool to analyse one-dimensional filtered flame sub-grid scale modeling. Combust. Flame **177**, 109–122 (2017)
27. P.H. Van Cittert, Zum einfluss der spaltbreite auf die intensitätverteilung in spektralinien. II, Z. Physik **69**, 298–308 (1931)
28. P.A. Jansson, in *Deconvolution with Applications in Spectroscopy* (Academic, New York, 1984), pp. 67–134
29. Z.M. Nikolaou, N. Swaminathan, Direct numerical simulation of complex fuel combustion with detailed chemistry: physical insight and mean reaction rate modeling. Combust. Sci. Tech. **187**, 1759–1789 (2015)
30. R.S. Cant, Senga2 user guide. cued/a?thermo/tr67. Technical report (2012)
31. Z. Nikolaou, R.S. Cant, L. Vervisch, Scalar flux modelling in turbulent flames using iterative deconvolution. Phys. Rev. Fluids. **3**(4), 043201 (2018)
32. R.A. Clark, Evaluation of sub-grid scalar models using an accurately simulated turbulent flow. J. Fluid Mech. **91**(1) (1979)
33. D. Veynante, A. Trouvé, K.N.C. Bray, T. Mantel, Gradient and counter-gradient scalar transport in turbulent premixed flames. J. Fluid Mech. **332**, 263–293 (1997)
34. Z.M. Nikolaou, Y. Minamoto, L. Vervisch, Unresolved stress tensor modeling in turbulent premixed v-flames using iterative deconvolution: An a priori assessment. Phys. Rev. Fluids **4**, 063202 (2019)
35. L. Cifuentes, C. Dopazo, J. Martin, P. Domingo, L. Vervisch, Local volumetric dilatation rate and scalar geometries in a premixed methane-air turbulent jet flame. Proc. Combust. Inst. **35**(2), 1295–1303 (2015)
36. L. Cifuentes, C. Dopazo, J. Martin, P. Domingo, L. Vervisch, Effects of the local flow topologies upon the structure of a premixed methane-air turbulent jet flame. Flow Turbul. Combust. **96**(2), 535–546 (2016)
37. P. Domingo, L. Vervisch, D. Veynante, Large-Eddy Simulation of a lifted methane-air jet flame in a vitiated coflow. Combust. Flame **152**(3), 415–432 (2008)
38. A.W. Vreman, An eddy-viscosity subgrid-scale model for turbulent shear flow: Algebraic theory and applications. Phys. Fluids. **16**(10), 3670–3681 (2004)
39. A. Seltz, P. Domingo, L. Vervisch, Z. Nikolaou, Direct mapping from LES resolved scales to filtered-flame generated manifolds using convolutional neural networks. Combust. Flame **210**, 71–82 (2019)
40. Z. Nikolaou, C. Chrysostomou, L. Vervisch, R.S. Cant, Progress variable variance and filtered rate modelling using convolutional neural networks and flamelet methods. Flow Turbul. Combust. (2019). https://doi.org/10.1007/s10494-019-00028-w

41. J. Schmidhuber, Deep learning in neural networks: an overview. Neural Netw. **61**, 85–117 (2015)
42. Y. Lecun, Y. Bengio, G. Hinton, Deep learning. Nature **521**, 436–444 (2015)
43. P.-T. de Boer, D.P. Kroese, S.S. Mannor, R.Y. Rubinstein, A tutorial on the cross-entropy method. Ann. Oper. Res. **134**(1), 19–67 (2005)
44. D.P. Kingma, J.L. Ba, ADAM: a method for stochastic optimization (2017). https://arxiv.org/pdf/1412.6980

Chapter 12
Analysis of Turbulent Reacting Jets via Principal Component Analysis

Giuseppe D'Alessio, Antonio Attili, Alberto Cuoci, Heinz Pitsch, and Alessandro Parente

Abstract The interpretation of high-dimensional data, like those obtained from Direct Numerical Simulations (DNS) of turbulent reacting flows, constitutes one of the biggest challenges in science and engineering. Although these simulations are a source of key information to advance the knowledge of turbulent combustion, as well as to develop and validate modeling approaches, the dimensionality of the data often limits the full opportunity to leverage the detailed and comprehensive information stored in datasets. The Principal Component Analysis (PCA) and its local formulation (LPCA) are widely used in many fields, including combustion. During the last 20 years, they have been used in combustion for the identification of low-dimensional manifolds, data analysis, and development of reduced-order models. Lower dimensional structures, either global or local, can provide better insights on the underlying physical phenomena, and lead to the formulation of high-fidelity models. This chapter aims to offer to the reader a comprehensive introduction of the PCA potential for data analysis, firstly introducing the main theoretical concepts, and then going through all the required computational steps by means of a MATLAB®

G. D'Alessio (✉) · A. Parente
Aero-Thermo-Mechanics Laboratory, Université Libre de Bruxelles, Bruxelles, Belgium
e-mail: giuseppe.dalessio@ulb.ac.be

A. Parente
e-mail: alessandro.parente@ulb.be

G. D'Alessio · A. Cuoci
CRECK Modeling Lab, Department of Chemistry, Materials and Chemical Engineering, Politecnico di Milano, Piazza Leonardo da Vinci 32, 20131 Milano, Italy
e-mail: alberto.cuoci@polimi.it

A. Attili · H. Pitsch
Institute for Combustion Technology, RWTH Aachen University, 52056 Aachen, Germany
e-mail: a.attili@itv.rwth-aachen.de

H. Pitsch
e-mail: h.pitsch@itv.rwth-aachen.de

A. Parente
Combustion and Robust Optimization Group (BURN), Université Libre de Bruxelles and Vrije Universiteit Brussel, Brussels, Belgium

H. Pitsch and A. Attili (eds.), *Data Analysis for Direct Numerical Simulations of Turbulent Combustion*, https://doi.org/10.1007/978-3-030-44718-2_12

code. Finally, the methodology is applied to data obtained from a DNS of a turbulent reacting non-premixed n-heptane jet in air. The latter can be regarded as an optimal case for data analysis because of the complex physics characterized by turbulence–chemistry interaction and soot formation.

12.1 Theory

12.1.1 Building the Dataset and Dataset Preprocessing

In order to apply any kind of statistical tool, data must be organized as matrices. The matrix $\mathbf{X}$, representing the original dataset, consists of n rows, which represent the statistically equivalent observations of a phenomenon, i.e., the different samples of an experiment, or the grid points of a numerical simulation, and p columns, which represent the variables of the problem, i.e., chemical species, velocity, temperature, and pressure. Since the variables are characterized by different units and ranges, preprocessing in the form of centering and scaling is a mandatory operation [1, 2]. Data centering consists of subtracting the mean value of each variable to all dataset observations: in this way, all the observations can be seen as fluctuations from a mean value. Scaling is achieved by dividing each variable by a given scaling factor, which can be different depending on the adopted scaling criterion. Therefore, the ith observation of the jth variable, $x_{i,j}$, from the original dataset matrix $\mathbf{X}$ can be centered and scaled by means of Eq. 12.1, where $\bar{x}_j$ and d_j are the centering and scaling factors for the considered jth variable, respectively.

$$\tilde{x}_{i,j} = \frac{x_{i,j} - \bar{x}_j}{d_j} \tag{12.1}$$

The way data are preprocessed can have a strong influence on the data analysis and the reduced-order modeling for combustion applications [3], as the scaling technique can be more or less sensitive to the presence of outliers or it can highlight a specific pattern in the data:

1. **Auto scaling**: the standard deviation of each variable, $\sigma(\mathbf{x})$, is used as a scaling factor. After Auto scaling, all the variables are characterized by a standard deviation equal to one. It is also one of the most used scaling criteria as it allows to consider evenly all the variables.
2. **Pareto scaling**: it uses the square root of the standard deviations for each variable, $\sqrt{\sigma(\mathbf{x})}$, as a scaling factor. More importance is given to variables with a very high standard deviation and with high numerical values. Importantly, the variables do not become dimensionless after scaling.
3. **Range scaling**: the difference between the minimum and the maximum value is adopted as scaling factor. It results to be more sensitive, if compared to the

Table 12.1 Scaling criteria and scaling factors for multivariate datasets

Scaling criterion	Scaling factor ($\mathbf{d}$)
Auto	$\sigma(\mathbf{x})$
Pareto	$\sqrt{\sigma(\mathbf{x})}$
Range	$\max(\mathbf{x}) - \min(\mathbf{x})$
Vast	$\sigma(\mathbf{x}) \frac{\sigma(\mathbf{x})}{\mathrm{mean}(\mathbf{x})}$

other scalings, to outliers, which can significantly change the numerical values of minimum and maximum.

4. **Vast scaling**: the scaling factor is the product between a variable's standard deviation and the coefficient of variation, i.e., the ratio: $\frac{\sigma(\mathbf{x})}{\mathrm{mean}(\mathbf{x})}$. It has been proven to focus on the variables which do not show strong variation.

Auto scaling is the optimal option for combustion applications if the main objective is the reconstruction of the overall state space, with no major differences between the major and minor state variables [3]. The other scalings, such as Range scaling and Vast scaling, on the other hand, are more focused on the stable and major species [3]. In Table 12.1, the aforementioned scaling criteria, as well as their associated scaling factors, are summarized.

12.1.2 Principal Component Analysis

The Principal Component Analysis (PCA) is a statistical technique used to find a reduced set of uncorrelated variables, starting from a larger set of interdependent variables, losing only a small amount of information [4, 5]. Starting from a centered and scaled data matrix $\tilde{\mathbf{X}}$, consisting of n observations and p variables, it is possible to compute the associated covariance matrix $\mathbf{S}$ and decompose it by means of an eigenvalue decomposition:

$$\mathbf{S} = \frac{1}{n-1}\tilde{\mathbf{X}}^T\tilde{\mathbf{X}} \tag{12.2}$$

$$\mathbf{S} = \mathbf{A}\mathbf{L}\mathbf{A}^T. \tag{12.3}$$

The columns of the matrix $\mathbf{A}$, whose size is $p \times p$, are an orthonormal basis of eigenvectors (Principal Components), while the diagonal elements of $\mathbf{L}$ correspond to their associated eigenvalues. Each eigenvalue represents a fixed percentage of information, in terms of variance of the original dataset, accounted by the associated Principal Component (PC). As the eigenvalues are ordered in descending order of magnitude $l_1 > l_2 > \cdots > l_p$, the PCs are also ordered in descending order of importance. The matrix $\tilde{\mathbf{X}}$ can be expressed as a function of the Principal Components by means of the scores matrix, $\mathbf{Z}$:

$$\mathbf{Z} = \tilde{\mathbf{X}}\mathbf{A}. \tag{12.4}$$

With the linear transformation described in Eq. 12.4, the original variables are recasted into a new set of uncorrelated variables. From a geometrical point of view, the axes of the new variables are represented by the columns of the matrix **A**. Moreover, given the orthonormality of the latter, it results that $\mathbf{A^T} = \mathbf{A^{-1}}$.

Thus, it is possible to uniquely recover the values of the original variables from **Z**:

$$\tilde{\mathbf{X}} = \mathbf{Z}\mathbf{A^T}. \tag{12.5}$$

The dimensionality reduction comes by considering only a q-dimensional subset, $\mathbf{A}_q$, from the original p-dimensional full set of PCs, **A**. If the cumulative variance, t_q, for the truncated q-dimensional basis of eigenvectors is within a desired accuracy, the basis can be considered as representative of the problem, and the original dataset $\tilde{\mathbf{X}}$ can be correctly compressed to the chosen reduced dimensionality finding the matrix of the scores, $\mathbf{Z_q}$. Equations 12.6 and 12.7 report the expressions for the cumulative variance, t_q, and the representation of the original data by means of the truncated basis of PCs, $\mathbf{A_q}$:

$$t_q = \frac{\sum_{j=1}^{q} \lambda_j}{\sum_{j=1}^{p} \lambda_j}, \tag{12.6}$$

$$\mathbf{Z_q} = \tilde{\mathbf{X}}\mathbf{A_q}. \tag{12.7}$$

The data matrix $\tilde{\mathbf{X}}$ can be reconstructed from the reduced dimensionality space by means of Eq. 12.8:

$$\tilde{\mathbf{X}} \approx \tilde{\mathbf{X}}_q = \mathbf{Z}_q\mathbf{A}_q^T. \tag{12.8}$$

The difference between the data matrix $\tilde{\mathbf{X}}$ and the reconstructed $\tilde{\mathbf{X}}_q$ is defined as the low-rank approximation error, and it can be used to evaluate the quality of the dimensionality reduction. The low-rank approximation error, ϵ, can be defined as follows:

$$\epsilon = \sum_{j=q+1}^{p} \lambda_j = \sum_{i=1}^{n}\sum_{j=1}^{p}(\tilde{\mathbf{x}}_{q,ij} - \tilde{\mathbf{x}}_{ij})^2, \tag{12.9}$$

where $\tilde{\mathbf{x}}_{q,ij}$ and $\tilde{\mathbf{x}}_{ij}$ correspond to the lower dimensional and the original observation, respectively.

The possibility to use PCA as a data analysis tool comes by considering that each PC is a linear combination of the original variables of the dataset. The *j*th variable will be characterized by a weight $w_{i,j}$, indicating how much it is represented by the *i*th PC [6, 7]. Thus, analyzing the distribution of the weights on the retained PCs of $\mathbf{A_q}$, it is possible to gain a better insight about the features of the system. As PCA is particularly sensitive to the presence of outliers, an outlier removal procedure is recommended in the preprocessing step, before applying the algorithm, in case of analysis on data obtained from experimental setups [3].

In the case of datasets with a large number of variables, it could be sometimes difficult to perform the analysis via visual inspection of the weights, because many variables could have comparable weights. Thus, the PCs physical interpretation can be aided by *rotation* methods, a class of statistical tools often coupled to PCA and other similar techniques such as Factor Analysis [7, 8]. The Varimax rotation, firstly developed by Kaiser [9], is an orthogonal rotation method which rigidly rotates the PCs over a fixed angle, while keeping the components orthogonal. When rotated, the subset of PCs spanning the lower dimensional space accounts for the same amount of cumulative variance as the unrotated, but it is redistributed within the components. Therefore, the information regarding the relative importance of the PCs, if the latter are rotated, is lost [4].

12.1.3 Local Principal Component Analysis

PCA is a linear technique, so the dimensionality reduction is limited when dealing with datasets obtained from non-linear systems such as those in combustion, since large reconstruction errors are obtained.

One option to overcome the intrinsic limitation of the PCA is to adopt a piecewise linear, local formulation for the dimensionality reduction (LPCA). Partitioning the data in k groups (clusters) and then performing the dimensionality reduction in each of them separately can lead to a drastic decrease of the reconstruction error.

Two methods [10, 11] are available to perform the dataset partitioning: an iterative unsupervised algorithm based on the minimization of the reconstruction error, the Vector Quantization Principal Component Analysis (VQPCA), or a supervised partitioning based on an a priori conditioning, by means of a selected variable which is known to be important for the process (FPCA). As the latter is not an iterative algorithm, it allows for a faster clustering in comparison with VQPCA, even if the choice of the optimal variable could constitute a difficult task for some applications, as it requires prior knowledge on the process, and the choice must be assessed case-by-case. For non-premixed, turbulent combustion applications, the mixture fraction Z is an optimal variable for the data conditioning, leading to excellent results both for data compression and interpretation tasks [11]. In the present approach, the FPCA algorithm groups the data in k bins: $k/2$ are allocated for all the observations under the condition of Z being lower than the stoichiometric mixture fraction Z_{st}, and the remaining $k/2$ for the observations at $Z > Z_{st}$.

The iterative VQPCA algorithm, instead, is based on the following steps:

1. *Initialization*. The cluster centroids $\tilde{\mathbf{r}}^{(k)}$ are initializated: a random allocation, a uniform distribution between all the observations of the dataset or a previous clustering solution can be chosen to compute the $\tilde{\mathbf{r}}^{(k)}$ initial values. The eigenvectors in each cluster, $\mathbf{A}^{(\mathbf{k})}$, are initialized as identity matrices.
2. *Partition*. Each observation is assigned to a cluster k such that the local reconstruction error is minimized:

$$\epsilon(\tilde{\mathbf{x}}_i, \tilde{\mathbf{r}}^{(k)}) = (\tilde{\mathbf{x}}_i - \tilde{\mathbf{r}}^{(k)})^T \mathbf{A}_q^{(k)T} \mathbf{A}_q^{(k)} (\tilde{\mathbf{x}}_i - \tilde{\mathbf{r}}^{(k)}), \tag{12.10}$$

3. *PCA*. The Principal Component Analysis is performed in each of the clusters found in the previous step. A new set of centroids is computed after the new partitioning step: their coordinates are calculated as the mean of all the observations in each cluster.
4. *Iteration*. All the previous steps are iterated until convergence is reached.

   The available convergence criteria are the following:

   i. The global mean reconstruction error, i.e., the averaged reconstruction error taking into account all the clusters, is below a fixed threshold.
   ii. All the clusters centroids positions are not changing between two consecutive iterations.
   iii. The variation of the global mean reconstruction error between two consecutive iterations is below a fixed threshold.

12.2 Methods and MATLAB® Code for Data Analysis with PCA

In this section, a simple procedure for data analysis via PCA with the corresponding MATLAB® code is presented, in order to show in details the application of the theory explained in Sect. 12.1. To perform data analysis with PCA, the following steps must be followed:

i. Standardize the initial dataset by means of centering and scaling.
ii. Perform PCA (global or local algorithm).
iii. Evaluate how many PCs are necessary to properly describe the system using Eq. 12.6.
iv. Evaluate the variables' weights on the retained PCs, and possibly apply a rotation to increase interpretability.

After the data have been organized as a matrix and loaded in MATLAB®, the first operation to accomplish is the standardization. The code for centering and scaling a generic $\mathbf{X}$ matrix to obtain $\tilde{\mathbf{X}}$ is reported below:

Data Pre-processing

```
[n_obs, n_var] = size(X);
tolerance = 1e-08;

% First of all, the mean value for each variable
% is calculated.
xbar = mean(X);
```

```
% Depending on the scaling criterion, the
% scaling factor is calculated:
switch upper(scaling)
  case {'NONE' ''}
    d = ones(1,nvar);
  case {'AUTO' 'STD'}
    d = zeros(1,n_var);
    for i=1:n_var
      d(i) = std(X(:,i),1);
    end
  case 'VAST'
    d = zeros(1,n_var);
    for i=1:n_var
      d(i) = std(X(:,i),1).^2 ./ (xbar(i));
    end
  case 'RANGE'
    d = max(X)-min(X);
  case 'PARETO'
    d = zeros(1,n_var);
    for i=1:n_var
      d(i) = sqrt(std(X(:,i),1));
    end
  otherwise
    error('Unsupported scaling option');
end

% All the observations can be now centered and scaled.
% In case of big matrices it is convenient to preallocate
% the memory.
X_tilde = zeros(size(X));
for i = 1: n_var
X_tilde(:,i) = (X(:,i) - xbar(i)) / (d(i) + tolerance);
end
```

After these operations, it is possible to perform PCA.

Principal Component Analysis

```
% The covariance matrix of the centered and scaled data
% must be computed.
cov_data = cov(X_tilde, 1);

% The eigenvectors and the eigenvalues are calculated
% from the covariance matrix.
% The eigenvalues, originally returned as a diagonal matrix,
% are stored into a vector (lambda).
[eigenvectors, eigenvalues] = eig(cov_data);
lambda = diag(eigenvalues);

% The eigenvalues must be now sorted in descending order.
```

```
% Their original indeces (sort_index) are also stored, as
% they are later used to arrange the eigenvectors in order
% of descending importance, thus building the matrix of
% the Principal Components (PCs).
[sort_eigval, sort_index] = sort(lambda, 'descend'); PCs =
zeros(n_var, n_var);
for i = 1 : n_var
    PCs(:,i) = eigenvectors(:, sort_index(i));
end

% Two different kinds of principal component scores can be
% now computed.
% 1) U-scores: obtained by projecting the matrix X_tilde of
% the centered and scaled data on the PCs.
% The resulting U-scores are uncorrelated and have variances
% equal to the corresponding eigenvalues.
% 2) W-scores: obtained by projecting the matrix X_tilde of
% the centered and scaled data on the PCs previously
% scaled by the inverse of the eigenvalues square root.
% The W-scores are still uncorrelated and have variances
% equal to 1.
U_vec = PCs;
W_vec = zeros(n_var, n_var);
for j = 1 : n_var
    W_vec(:, j) = (PCs(:, j)/sort_eigval(j)^0.5);
end
U_scores = X_tilde*U_vec;
W_scores = X_tilde*W_vec;
```

In alternative, a built-in function is also already available in MATLAB®:

Principal Component Analysis

```
[PCs, U_scores, Eigenvalues, ~, Explained] = ...
pca(X_tilde, 'Centered', false);
```

In this function, as well as providing the $n \times p$ standardized matrix in input, it is specified that the data have already been centered and scaled. In the output, the function returns:

i. **PCs**: the PCs matrix $\mathbf{A}$ containing all the principal components, whose size is $p \times p$.
ii. **U_scores**: the scores matrix, $\mathbf{Z}$, which consist of the projection of the input matrix on the full set of PCs: $\mathbf{Z} = \tilde{\mathbf{X}}\mathbf{A}$.
iii. **Eigenvalues**: the eigenvalues vector ($p \times 1$), containing the eigenvalues associated to each PC.

iv. **Explained**: the explained variance vector ($p \times 1$), which consists of the percentage of explained variance by each PC depending on the magnitude of the associated eigenvalue.

This last output is important for the next step, the choice of the number of PCs. As already explained in Sect. 12.1, a good approximation of the original problem requires selecting a basis that can explain a large amount of cumulative variance, i.e, from 95 to 100%. In this way, it is possible to have an initial guess for the dimensionality needed by the reduced basis to be representative. In the following code, it is required that more than 99% of the global variance is explained by the retained Principal Components:

Choice of the Reduced Dimensionality

```
cumulative_explained = cumsum(Explained)/sum(Explained);
variance_cut = find(cumulative_explained > 0.99);
required_number = variance_cut(1);
```

Once the required number of PCs is calculated, the reduced eigenvectors basis, $\mathbf{A_q}$, can be built as follows:

Selecting the PCs

```
A_q = PCs(:, 1:required_number);
```

Sometimes, the explained variance criterion might not be enough to assess the number of required PCs, especially if the data are not standardized with the Auto scaling criterion. Thus, another method must be taken into account to verify if the choice of the number of PCs is appropriate, checking the reconstruction error for the single variables. If the number of PCs is correctly determined, the variables' reconstruction from the reduced dimensionality is characterized by a low error, otherwise the number of the retained PCs must be increased. A large reconstruction error can have a negative impact on the analysis, as the feature extraction process from the data could also be compromised. In fact, the distribution of the weights on the modes could be too noisy or some important processes might not be extracted. The code to reconstruct the original matrix from the reduced dimensionality space is reported below, assessing the quality of the reconstruction using parity plots of the original and the reconstructed variables, after uncentering and unscaling the data.

Reconstruction Error for the Variables

```
% The original matrix must be reconstructed using the set of
% truncated modes calculated in the previous steps, first.
recovered_X_fromPCA = X_tilde*A_q*A_q';
```

```
% This matrix must be then unscaled and uncentered
% (in this order), with the same scaling and centering
% factors used in the previous  steps, to make a proper
% comparison with the original matrix, X.

% Unscaling
for i = 1:n_var
X_unscal(:, i) = recovered_X_fromPCA(:,i)*d(i);
end

% Uncentering
for i = 1:n_var
X_recovered(:,i) = X_unscal(:,i) + xbar(i);
end

% The parity plots between the original and the reconstructed
% variables can then be drawn to evaluate the reconstruction:
% the more the scatter points are aligned with the red solid
% line, the more precise is the PCA reconstruction. In order
% to have  a quantitative indication, it is also possible to
% use error metrics such as the Mean Square Error or the
% Root Mean Square Error.
for i = 1:n_var
figure, plot(X(:,i), X(:,i), 'r', 'LineWidth', 2);
hold on
scatter(X(:,i), X_recovered(:,i), 15, 'filled');
xlabel('Original variable');
ylabel('Reconstructed variable');
end
```

For a local, supervised, partitioning in k clusters, using mixture fraction as a conditioning variable, the following code can be implemented:

Local Partitioning via FPCA

```
% Initialization of bin data matrices
bin_data = cell(k, 1);
idx_clust = cell(k, 1);
idx = zeros(size(X_tilde,1), 1);

% Number of intervals
n = k + 1;
min_z = min(Z);
max_z = max(Z);

ints_1 = linspace(min_z, z_stoich, ceil(n/2));
ints_2 = linspace(z_stoich, max_z, ceil((n+1)/2));
ints = [ints_1(1:ceil(n/2-1)) ints_2];
```

```
% Partition
for bin = 1 : k
    idx_clust{bin} = find((Z>=ints(bin))&(Z<=ints(bin+1)));
    bin_data{bin} = X_tilde(idx_clust{bin}, :);
    idx(idx_clust{bin}) = bin;
end

% Perform PCA in each bin after the centroid has been removed
PCs_clusters = cell(k,1);
for i = 1:k
[rows, columns] = size(bin_data{i});
mean_var = mean(bin_data{i}, 1);
X_ave = repmat(mean_var, rows, 1);
X0 = bin_data{i} - X_ave;
[PCs, Scores, Eigenvalues, ~, Explained] = ...
pca(X0, 'Centered', false);
PCs_clusters{i} = PCs;
end

% Plot the weights on the PCs to analyze the data
for j = 1:k
A_q = PCs_clusters{j}(:, 1:required_number);
for i = 1: required_number
figure,bar(A_q(:,i),'FaceColor',[0, 0, 0],'LineWidth',1.5)
xlabel('Variables');
ylabel('Weights on the PC');
end
end
```

12.3 Application: PCA of a Non-premixed Sooting Flame DNS

12.3.1 Data Description

Data obtained from a 2D slice of a 3D temporally evolving DNS simulation of an n-heptane turbulent jet [12] are here considered for the analysis by means of PCA. The jet is non-premixed with a Reynolds number of 15,000. The fuel has an initial temperature of 400 K, while the oxidizer stream (air) is at 800 K. The kinetic mechanism used for the n-heptane flame consists of 47 species, including naphthalene and other soot precursors, with 290 reactions in total. Additional information on the mechanism and the gas phase hydrodynamics can be found in [12–15]. This can be considered as an optimal case for a data analysis task since it includes a large number of available observations and chemical species in the mechanism, and the physics is characterized by many complex phenomena such as turbulence–chemistry interaction and soot formation. The dataset considered here consists of the full thermo-chemical space and it is organized as a matrix of 1,048,576 observations (grid points of the 2D slice) and 48 variables (temperature and mass fractions of all the chemical species).

12.3.2 Analysis

12.3.2.1 Principal Component Analysis

The matrix **X** containing the input data was scaled with the four scalings methods discussed in Sect. 12.1 (Auto, Vast, Pareto, Range) to test their effect on the analysis. A different relative importance given to the PCs is observed in the four cases, due to the different eigenvalues magnitude distribution. As it is shown in Fig. 12.1, for Auto and Range scaling criteria, the curve representing the cumulative explained variance has a more moderate slope, starting from small values (i.e., about 0.4 and 0.65, respectively) and then asymptotes to one for a relatively large number of PCs, while for Pareto and Vast scaling criteria, the first principal component already explains an almost unitary cumulative variance. With Auto scaling, the number of eigenvectors to retain in order to explain at least 99% of the data cumulative variance is 19, while with Pareto scaling 1 PC is already enough to explain 99% of the total data variation. The examination of the cumulative explained variance is not always optimal for the assessment of the number of PCs to analyze, so the reconstruction error for the variables must be also investigated.

If the data are reconstructed from nineteen-dimensional compressed space and from the one-dimensional compressed space, respectively, the results in terms of accuracy of the reconstruction are totally different. Even if the amount of explained cumulative variance is the same, the coefficient of determination, R^2, for the parity

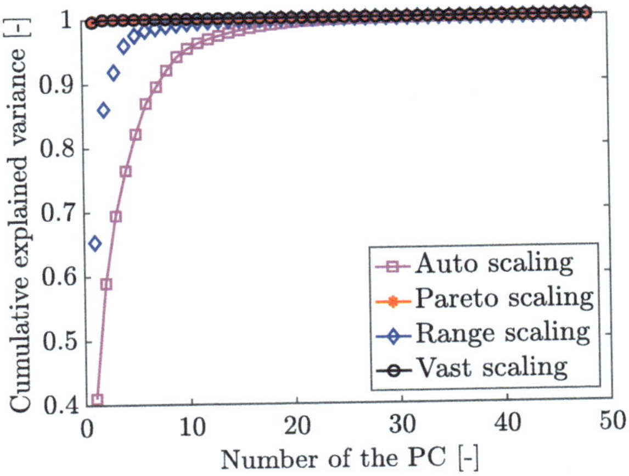

Fig. 12.1 Cumulative explained variance for the PCs for different scaling criteria

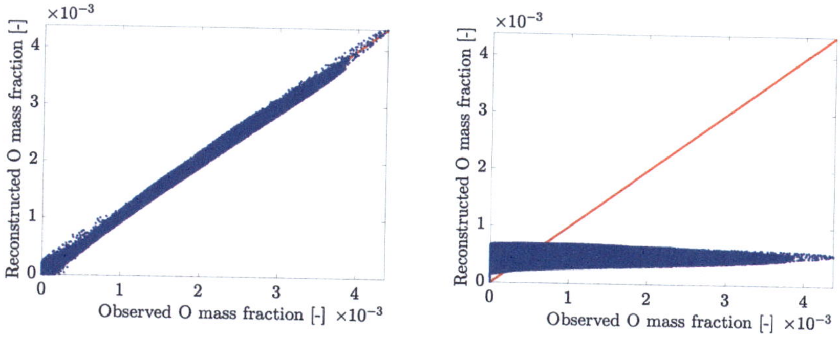

Fig. 12.2 Left: Parity plot for the reconstruction of the oxygen radical from the compressed space adopting Auto scaling with 19 retained PCs; Right: Parity plot for the reconstruction of the oxygen radical from the compressed space adopting Pareto scaling with 1 retained PC; on equal terms of cumulative explained variance ($t_q > 0.99$)

plots obtained from the variables' reconstruction are completely different: for the reconstruction of the oxygen radical, in case of Auto scaling with 19 PCs, the R^2 amounts to 0.996, while in case of Pareto scaling with 1 PC it amounts to 0.218. The parity plots for the reconstructed variable using the two different scalings are shown in Fig. 12.2. The only variable which has an acceptable reconstruction error with only 1 PC, in case of Pareto scaling, is the temperature with a R^2 equal to 1.

As highlighted in Sect. 12.1, the analysis is performed via examination of the weights' distribution on the PCs, and in many cases they tend to represent one or more physical quantities. For example, examining the sixth rotated PC obtained from the data scaled with Auto scaling reported in Fig. 12.3, it is possible to notice that nitrogen and oxygen have the largest negative weights, while n-heptane has the largest positive weight. This PC clearly represents the mixture fraction, as also confirmed by its correlation coefficient with the mixture fraction itself, which is equal to 0.72. This high correlation is particularly interesting because the mixture fraction was not included in the variables of the dataset, which consisted only of temperature and species mass fractions.

Other global important features extracted via PCA were, for example, the most important radicals involved in the branching reactions (O, OH, H on one rotated PC and HO_2 on another) and in the soot formation mechanism (C_6H_5, C_7H_8, C_3H_3 with highest weights on the third rotated PC).

12.3.2.2 Local Principal Component Analysis

The quality of the data analysis can be enhanced if a local formulation is considered. A piecewise linear local formulation for PCA has several advantages with respect to a global analysis: a lower reconstruction error, a lower intrinsic reduced dimensionality, and the possibility to highlight local processes. The algorithm has only one

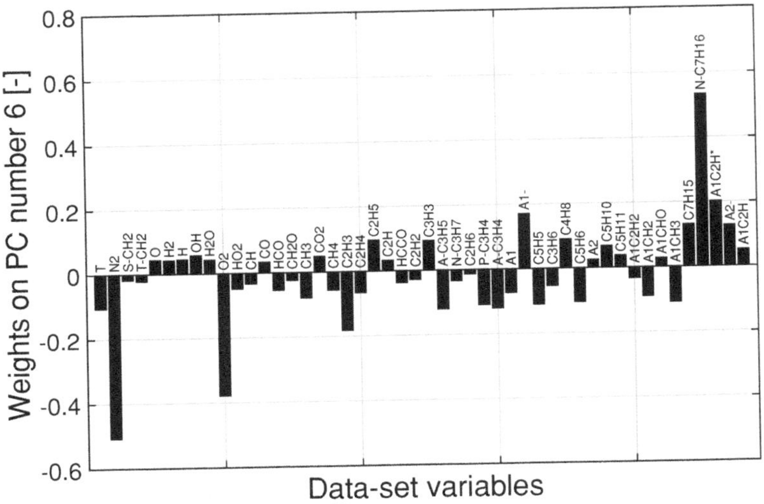

Fig. 12.3 Weights distribution on the sixth Principal Component, Auto scaling criterion

hyperparameter, namely, the choice of the number of bins of mixture fraction (the number of clusters) to use to partition the data, k. Despite the fact that a defined way to properly set the number of variables does not exist in literature, the total number of clusters can be retrieved from a trade-off between the accuracy of the reconstruction and the feasibility of the data analysis. In fact, as the number of clusters grows, the reconstruction error decreases, but the analysis with an excessively large k could be infeasible.

In Fig. 12.4, the reconstruction error obtained with FPCA for an increasing number of clusters, starting from the global PCA ($k = 1$), is reported. The error decreases in total of one order of magnitude, and it starts asymptoting from $k = 16$. The latter could also be ideally chosen as a good k to perform the analysis as it is a reasonable number of clusters to examine, not being too large for the manual weights inspection and interpretation, as $k = 64$ or $k = 128$ could be. LPCA is also robust to underfitting and overfitting, as the reconstruction error does not depend on the dataset dimensions, as also shown in the aforementioned figure where the errors using the full dataset and only the 50% of the observations are compared.

As discussed in Sect. 12.2, for the FPCA algorithm the mixture fraction space is divided in k/2 bins for all the points below the stoichiometric mixture fraction and in k/2 above Z_{st}, while VQPCA assigns the cluster index to one particular observation on the basis of the reconstruction error minimization criterion. In Figs. 12.5 and 12.6, the flame partitionings for $k = 16$ for the two different LPCA algorithms are reported. As expected, the results are different; even if the VQPCA partitioning is completely unsupervised, it gives better results in terms of data compression, as its average reconstruction error for $k = 16$ amounts to 0.0121, while with FPCA, the reconstruction error for the variables is 0.0256 on equal terms of number of clusters

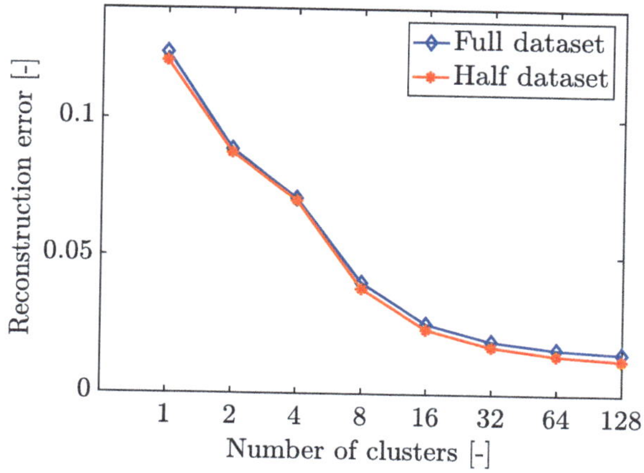

Fig. 12.4 Reconstruction error for an increasing number of bins of mixture fraction for the full dataset (all the observations) and a dataset consisting only of half of the total observations. The value for number of clusters equal to 1 is the error with global PCA

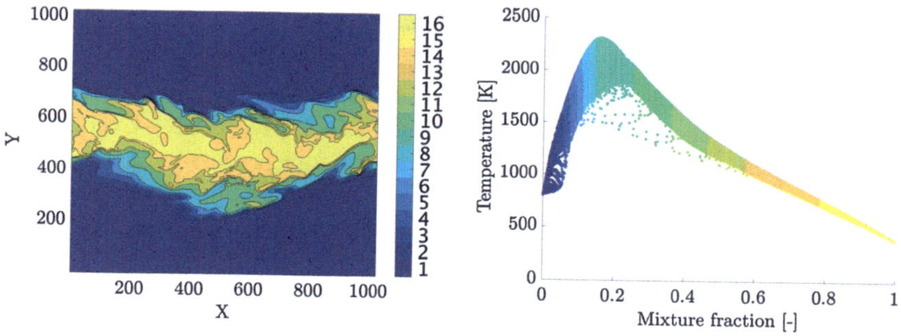

Fig. 12.5 Left: DNS flame partitioning via FPCA with $k = 16$; Right: mixture fraction partitioning via FPCA with $k = 16$. The colorbar indicates the index of the cluster assignment

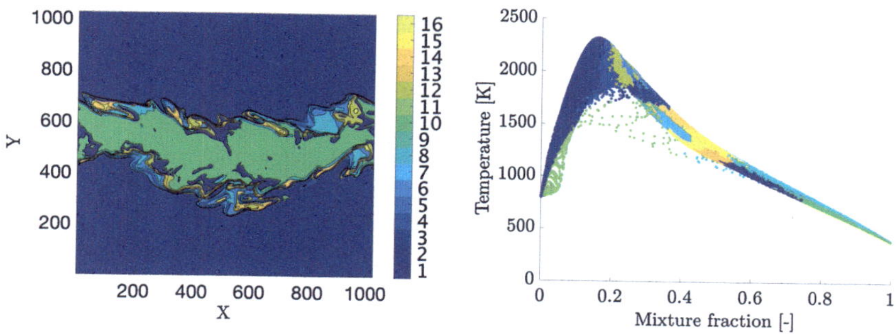

Fig. 12.6 Left: DNS flame partitioning via VQPCA with $k = 16$; Right: mixture fraction partitioning via VQPCA with $k = 16$. The colorbar indicates the index of the cluster assignment

and number of retained principal components. In each of the sixteen clusters, the main features of the flame can be visualized plotting the weights of the variables on the modes as done in the previous paragraph.

As shown in the mixture fraction—temperature plot, colored by means of the LPCA partitioning, the unsupervised algorithm allocates only two clusters for the lean conditions (from zero up to stoichiometric), while with the supervised approach, eight clusters were intentionally allocated for low values of Z. One of these two clusters found via VQPCA for the lean conditions contains all the points below the curve (in light green, cluster number 11), and it groups all the points of local extinction, characterized by quasi-stoichiometric and stoichiometric mixture fraction, low temperature, and OH concentration almost equal to zero. In the right branch of the mixture fraction—temperature plot, corresponding to rich conditions, the clusters found via the unsupervised algorithm have a larger extension in terms of mixture fraction range if compared to the supervised ones. In this case, separate clusters are assigned to points at the same mixture fraction range, but different temperature. Examining the plot on the right of Fig. 12.6, indeed, it is possible to see that cluster number 9 goes from $Z \sim 0.5$ to $Z \sim 0.95$, the same mixture fraction covered by clusters number 1 and number 10, but with the latter being at a lower temperature. This partitioning is totally in line with the physics of the flame, as these clusters are representative of different chemical features. In fact, while clusters number 1 and 10 are representative for the fuel jet and its decomposition, as the highest weights on the first modes are representative for species such as n-C_7H_{16}, C_7H_{15}, C_5H_{10}, C_4H_8, C_2H_5, CH_2O, cluster number 9 is representative of soot precursors, as in the first two modes the species characterized by highest weights are all aromatics involved in the soot formation such as naphthalene and its naphthyl radical $C_{10}H_8$, $C_{10}H_7^-$, benzyl radical and ethynyl benzene C_7H_7, C_8H_7. Thus, the FPCA algorithm, despite the conditioning variable proven to be optimal for turbulent non-premixed reacting jets, was not capable to be competitive with VQPCA for data analysis purposes as the physics of the jet were too complex, involving local extinction phenomena and the dynamics of soot precursors.

These data analysis methods can also be coupled to other techniques, such as Principal Variables (PV) [4, 16, 17]. Its purpose is to find a relationship between the Principal Components and a subset of the original variables by means of the maximization of the variance of the original data. Many strategies to find the PVs are available in the literature: in the present work, the B2 backward method [4] was tested. This technique has been successfully used for reduced-order modeling in combustion applications [17–20]. The variables extracted by the latter can be grouped in three main categories: radicals involved in branching reactions, such as H, HO_2, CH_3, C_7H_{15}, stable species such as O_2, CO_2, and species involved in the soot formation mechanism, like aromatic compounds $C_{10}H_8$, C_8H_7, C_7H_7, C_7H_6O and propargyl C_3H_3. Many of these species were also extracted by means of a direct analysis of the clusters' weights found via VQPCA, as a proof of the effectiveness of the local PCA for data analysis tasks. Moreover, the possibility to know the spatial position of the cluster where these features are important constitutes a relevant property of the partitioning algorithm. On the other hand, the PV algorithm offers the possibility

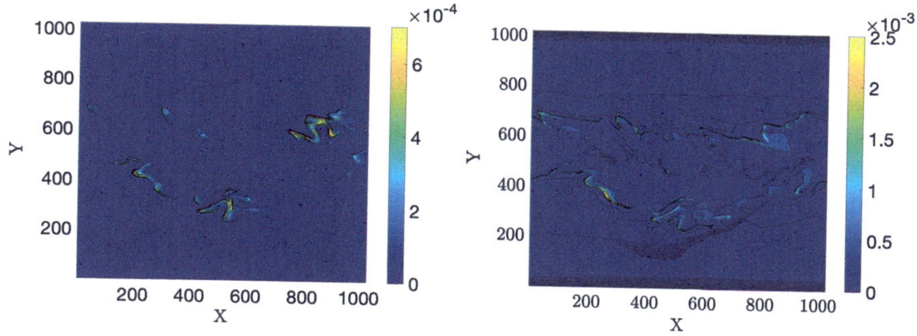

Fig. 12.7 Left: mass fraction contours of naphthalene; Right: mass fraction contours of propargyl

Fig. 12.8 Cluster number 9 obtained via VQPCA partitioning (in yellow)

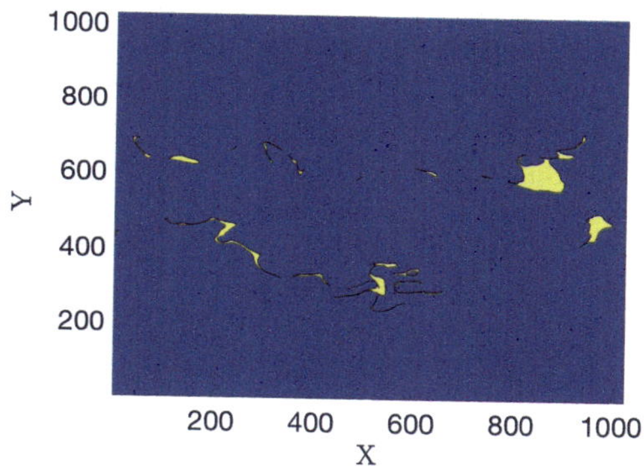

to identify features in an automated fashion, without having to visually inspect the weights of the LPCA modes, which could be unfeasible in case of a high number of clusters or retained modes for each cluster.

An optimal solution for data analysis could be represented by a hybrid method VQPCA-PV. The first algorithm can be used to partition the original dataset in k groups according to the reconstruction error minimization, and then the principal variables can be found in each cluster. In this case, applying the PV algorithm in each cluster found with the previous VQPCA partitioning led to similar results in terms of extracted features, but in a totally unsupervised fashion, without a visual inspection of the weights to be required. In Fig. 12.7, the contours of the mass fractions of naphthalene and propargyl, two species identified as principal variables in cluster number 9, are reported, and it can be observed that the highest concentrations for these species are observed in the region of the geometrical domain corresponding to that cluster, which is reported, colored in yellow, in Fig. 12.8. In fact, this cluster resulted again to be associated to soot formation, with the chemical species C_6H_5, C_6H_7, C_7H_8, $C_{10}H_7$, C_2H_3 identified as *local* principal variables.

12.4 Conclusions

The present work investigates the potential of the Principal Component Analysis to analyze data obtained from Direct Numerical Simulations of turbulent reacting flows.

The Principal Component Analysis is widely used in many fields for dimensionality reduction, but it can also be exploited for data analysis tasks. In fact, if the original variables' weights distribution on the principal components is examined, it is possible to obtain a physical interpretation for the latter, and an insight about the system features can be gained. Moreover, two local formulations of the PCA are available to overcome the limitations due to the linearity of the method: an iterative unsupervised algorithm, based on the minimization of the reconstruction error (VQPCA), and a supervised partitioning algorithm, based on an a priori conditioning by means of a selected variable which is known to be important for the process (FPCA). With the last two algorithms, the local phenomena, which could have been overlooked by a global analysis, can be highlighted.

The aforementioned techniques were tested on the analysis of data obtained in a 3D temporally evolving DNS of an n-heptane turbulent jet in air. The dataset consisted of the full thermo-chemical space, organized as a matrix of 1,048,576 observations (grid points of the simulation) and 48 variables (temperature and mass fractions of all the chemical species).

The global PCA was able to recognize the key role covered by the mixture fraction in the process even if it was not included in the variables of the dataset, as well as to highlight the most important radicals involved in the branching reactions and in the soot formation mechanism. The analysis done using the local algorithms was more effective both in terms of reconstruction error and feature extraction, as more physical processes such as decomposition of the fuel jet, local extinction phenomena, branching reactions, and soot formation were highlighted. In particular, VQPCA was more effective than FPCA because the conditioning variable, the mixture fraction, although proven to be optimal for turbulent non-premixed reacting jets, was not capable to deal with the complex physics of the system, characterized by local extinction phenomena and the dynamics of soot precursors. Finally, an hybrid algorithm coupling VQPCA with the Principal Variables method was proposed. This led to similar results as with VQPCA in terms of extracted features, but in a totally automated fashion, without requiring a visual inspection of the weights.

Acknowledgements The first author acknowledges the support of the Fonds National de la Recherche Scientifique (FRS-FNRS) through a FRIA fellowship.

A.A. and H.P. acknowledge funding from the European Research Council (ERC) under the European Unions Horizon 2020 research and innovation program under grant agreement No 695747.

A.P. acknowledges funding from the European Research Council (ERC) under the European Union's Horizon 2020 research and innovation program, grant agreement No 714605.

References

1. R. Bro, A.K. Smilde, Centering and scaling in component analysis. J. Chemom. **17**(1), 16–33 (2003)
2. R.A. van den Berg, H.C. Hoefsloot, J.A. Westerhuis, A.K. Smilde, M.J. van der Werf, Centering, scaling, and transformations: improving the biological information content of metabolomics data. BMC Genomics **7**(1), 142 (2006)
3. A. Parente, J.C. Sutherland, Principal component analysis of turbulent combustion data: data pre-processing and manifold sensitivity. Combust. Flame **160**(2), 340–350 (2013)
4. I. Jolliffe, *Principal Component Analysis* (Springer, Berlin, 2011)
5. C.M. Bishop, *Pattern Recognition and Machine Learning* (Springer, Berlin, 2006)
6. A. Parente, J.C. Sutherland, B.B. Dally, L. Tognotti, P.J. Smith, Investigation of the MILD combustion regime via principal component analysis. Proc. Combust. Inst. **33**(2), 3333–3341 (2011)
7. A. Bellemans, G. Aversano, A. Coussement, A. Parente, Feature extraction and reduced-order modelling of nitrogen plasma models using principal component analysis. Comput. Chem. Eng. **12**(115), 504–514 (2018)
8. M.B. Richman, Rotation of principal components. J. Climatol. **6**(3), 293–335 (1986)
9. H.F. Kaiser, The varimax criterion for analytic rotation in factor analysis. Psychometrika **23**(3), 187–200 (1958)
10. N. Kambhatla, T.K. Leen, Dimension reduction by local principal component analysis. Neural Comput. **9**(7), 1493–1516 (1997)
11. A. Parente, J.C. Sutherland, L. Tognotti, P.J. Smith, Identification of low-dimensional manifolds in turbulent flames. Proc. Combust. Inst. **32**(1), 1579–1586 (2009)
12. A. Attili, F. Bisetti, M.E. Mueller, H. Pitsch, Formation, growth, and transport of soot in a three-dimensional turbulent non-premixed jet flame. Combust. Flame **161**(7), 1849–1865 (2014)
13. A. Attili, F. Bisetti, M.E. Mueller, H. Pitsch, Effects of non-unity Lewis number of gas-phase species in turbulent nonpremixed sooting flames. Combust. Flame **1**(166), 192–202 (2016)
14. A. Attili, F. Bisetti, M.E. Mueller, H. Pitsch, Damkohler number effects on soot formation and growth in turbulent nonpremixed flames. Proc. Combust. Inst. **35**(2), 1215–1223 (2015)
15. A. Attili, F. Bisetti, Application of a robust and efficient Lagrangian particle scheme to soot transport in turbulent flames. Comput. Fluids **15**(84), 164–175 (2013)
16. G.P. McCabe, Principal variables. Technometrics **26**(2), 137–144 (1984)
17. B.J. Isaac, A. Coussement, O. Gicquel, P.J. Smith, A. Parente, Reduced-order PCA models for chemical reacting flows. Combust. Flame **161**(11), 2785–2800 (2014)
18. B.J. Isaac, A. Parente, C. Galletti, J.N. Thornock, P.J. Smith, L. Tognotti, A novel methodology for chemical time scale evaluation with detailed chemical reaction kinetics. Energy Fuels **27**(4), 2255–2265 (2013)
19. A. Coussement, B.J. Isaac, O. Gicquel, A. Parente, Assessment of different chemistry reduction methods based on principal component analysis: comparison of the MG-PCA and score-PCA approaches. Combust. Flame **1**(168), 83–97 (2016)
20. A. Coussement, O. Gicquel, A. Parente, MG-local-PCA method for reduced order combustion modeling. Proc. Combust. Inst. **34**(1), 1117–1123 (2013)

Chapter 13
Application of an Evolutionary Algorithm to LES Modelling of Turbulent Premixed Flames

M. Schöpplein, J. Weatheritt, M. Talei, M. Klein, and R. D. Sandberg

Abstract Gene Expression Programming (GEP) has been used successfully for modelling the unclosed terms in the context of Reynolds Averaged Navier–Stokes (RANS) and Large Eddy Simulation (LES)-based turbulence modelling. In contrast to deep-learning-based methodologies, this approach has the advantage that the model can be documented in the form of a mathematical expression; it can be interpreted and easily implemented in existing solvers. Recently, application of GEP to a priori LES modelling has demonstrated the efficiency of the approach to find high fidelity LES closures. The present contribution explains the methodology, reviews recent work in the field and focuses on the robustness of the method and the scope for future efficiency improvements, by applying it to the modelling of the unclosed stress tensor in turbulent premixed statistically planar flames.

13.1 Introduction

Evolutionary programming has been extensively used for numerous optimization problems with great success in the past years. It mimics nature's process of evolution to find solutions to mathematical problems. Populations of individual solutions (individuals) are genetically modified and evolve to new generations by the survival of the fittest principle. One class of evolutionary programming is known as *Gene expression programming* (GEP), which describes the mathematical function of a possible solution as a symbol string, the chromosome. Each symbol represents either a function, a variable or a constant from a predefined set of possible symbols. The GEP algorithm randomly initializes an initial population of individuals and attempts to enhance

M. Schöpplein · M. Klein (✉)
Department of Aerospace Engineering, Bundeswehr University Munich, Neubiberg, Germany
e-mail: markus.klein@unibw.de

J. Weatheritt · M. Talei · R. D. Sandberg
Department of Mechanical Engineering, University of Melbourne, Parkville, VIC 3010, Australia

H. Pitsch and A. Attili (eds.), *Data Analysis for Direct Numerical Simulations of Turbulent Combustion*, https://doi.org/10.1007/978-3-030-44718-2_13

the solution using genetic operations. A fitness function determines the probability of an individual being picked for reproduction which can happen through cloning, mutation and recombination of different chromosomes. The basic idea behind evolutionary algorithms is that better solutions pass on their genetic information (parts of mathematical expressions) with a higher probability than bad solutions.

Using GEP for turbulence modelling is an emerging field of research. This method is advantageous relative to other machine learning models such as artificial neural networks (ANNs) as it can produce a turbulence model as a function of key physical parameters. The obtained function can be readily implemented, and is also repeatable and provides insight into the phenomenon of interest.

Large Eddy Simulation (LES) is one of the numerical tools that can benefit from using GEP for subgrid-scale modelling. The objective of this chapter is to demonstrate the application of this technique to find models for the LES subgrid-scale (SGS) stress tensor explicitly for the combustion zone of statistically planar, turbulent premixed flames. In particular, we aim to highlight the robustness of the GEP algorithm in finding an appropriate SGS model. The relevant study to this chapter is presented in [1]. The algorithm uses a pre-existing GEP code that was used successfully before in RANS applications [2, 3]. A priori testing is chosen to rate a model's performance, which in this case means the model is tested on an explicitly filtered DNS flow field. To limit the number of independent tensor functions as well as invariants, a tensor basis is derived based on the previous work by Pope [4] but applied to LES SGS stress tensor modelling.

13.2 Evolutionary Programming

Two classes of evolutionary programming are discussed first, genetic algorithms (GAs) and genetic programming (GP). GEP which has inherited characteristics from these two techniques will then be described.

13.2.1 Genetic Algorithms

John Holland introduced GAs in the 1960s to study nature's process of adaption through evolution and learn how to apply similar mechanisms to computer systems. He encoded the chromosomes as bit strings and applied operators based on genetic processes (such as reproduction, mutation, crossover and inversion) to these chromosomes [5].

The idea behind GAs is that individuals which are closer to the optimal solution can only survive when generations are reproduced. The following terminology is commonly used in GAs (see Fig. 13.1):

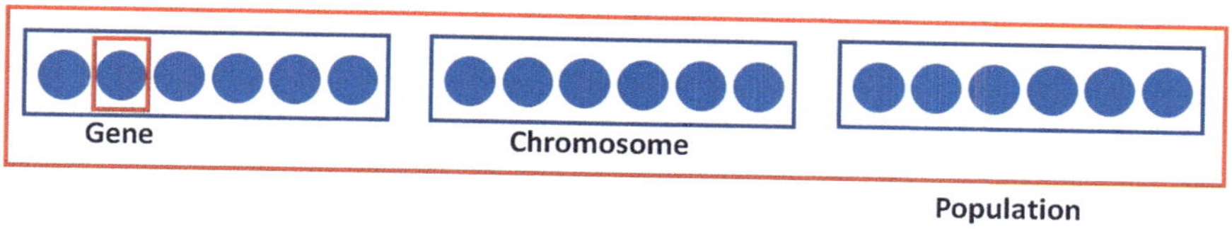

Fig. 13.1 Schematic of different elements in genetic algorithm, reproduced from [6]

- A **Gene** is the encoding of one basic characteristic of a solution.
- A **Chromosome (Individual)** is one of the candidates for the optimal solution and is described as a string of characters and numbers. This is referred to as chromosome.
- The **Population** is the set of a number of individuals.
- A **Generation** is the renewed population, obtained by replacing some of old individuals by those which are closer to the optimal solution.
- The **Fitness** of an individual is a representation of its quality that allows comparison with other individuals within the population. In other words, it shows how close each individual is to the optimal solution.

A simple genetic algorithm looks for a solution by generating new generations of genomes according to the flowchart presented in Fig. 13.2. At the beginning of every run, n_{pop} individuals are randomly created according to the problem-specific rules. The fitness value is then calculated with the specified fitness function *fit*, which

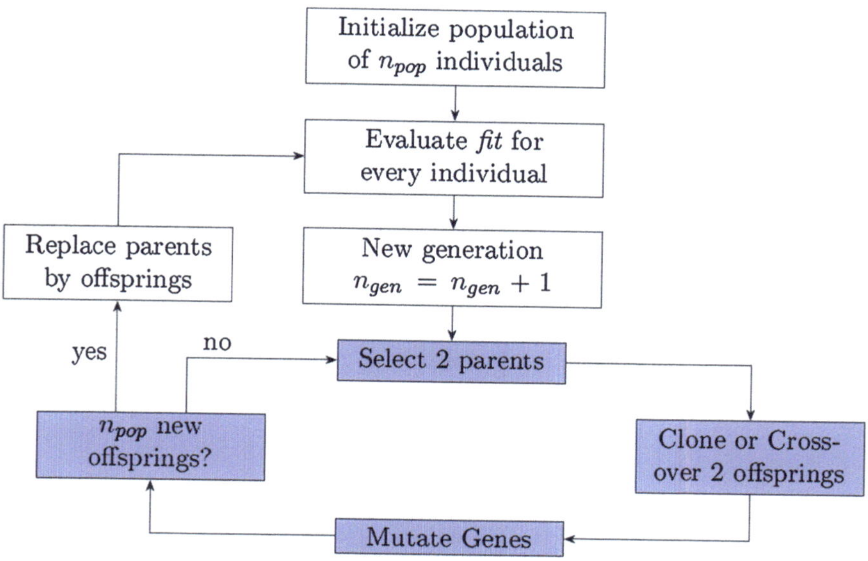

Fig. 13.2 General form of a genetic algorithm

determines the probability for an individual to be picked for reproduction. Two parents are selected and they reproduce two offsprings using different genetic operators, for example:

- **Cloning** the parents by copying their chromosomes into the offspring. This method is often used for the best solution in each generation (simple elitism) to ensure that it is passed on to the next generation.
- **Single point crossover** reproduces offspring by exchanging gene sequences between both parent chromosomes. The crossover operator has to ensure that the resulting offspring is always valid.
- **Mutation** is the random change of genes. A mutation operator could be a simple swap operator.

The reproduction loop (marked blue in Fig. 13.2) repeats until n_{pop} new offspring are created and the old population is subsequently replaced by the new generation. The iteration can be stopped either after a specified number of generations or when the best individual meets a certain fitness.

13.2.2 Genetic Programming

In GP, programs are expressed in parse trees by writing the program in a top to bottom structure [7, 8]. The operators precede their arguments, i.e. $a + b$ encodes $+ab$. Figure 13.3 shows a parse tree for the Pythagorean theorem (squared hypotenuse). The corresponding expression is $Q + \times \times aabb$, where Q represents the square-root. The objective of a GP algorithm is to find a functional expression of unknown complexity and form, that successfully solves a given problem, for example, attaining a relation between the two legs (a, b) and the hypotenuse (c) in a right-angled triangle. In GP, every candidate solution is expressed as a parse tree that can incorporate a predefined set of symbols (functions and terminals). During the evolutionary process, the parse trees of individuals are changed with respect to their form and the symbols on their nodes. The search process is very similar to the GA algorithm, shown in Fig. 13.2, as it uses a fitness value to evaluate the quality of every individual and applies genetic operations on them according to their fitness.

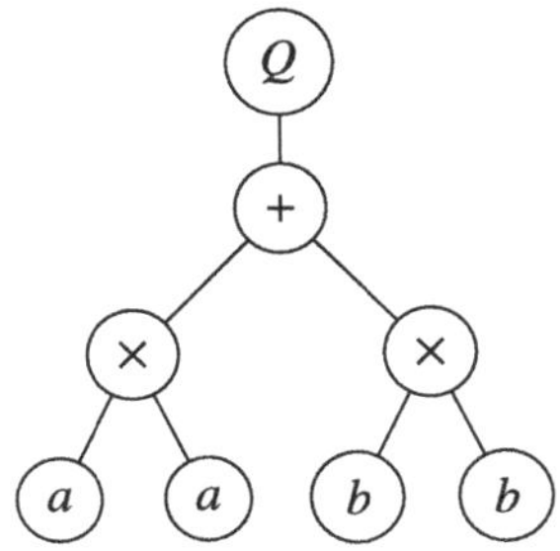

Fig. 13.3 Pythagorean theorem as parse tree

13.2.3 Gene Expression Programming

GEP was developed by Ferreira [9] using the simplicity of chromosomes used in GA and the functional complexity inherent to GP. Chromosomes are linear, symbolic strings of fixed length that can code expression trees (ETs) of various length. The structure of these strings ensures syntactically correct ETs under any genetic modification. The string sequence that is translated into an ET is called open reading frame (ORF), because like in molecular biology, an ORF is the sequence of the DNA that potentially encodes genetic information. A GEP ORF always starts at the first position of the gene but terminates at various positions.

In GEP, the chromosome's string is divided into two parts: a head with symbol length h and a tail with length t, subject to the relation:

$$t = h(n_a - 1) + 1. \tag{13.1}$$

Therein, n_a is the maximal number of arguments of all the functions in the function set defined for a given problem. Consider the function set $\mathscr{F} = \{+, -, \times, Q\}$ and the terminal set $\mathscr{T} = \{a, b, c, 1\}$. For a gene with $h = 6, t = 7$, the function with the highest arity has an arity of two, hence $n_a = 2$. The head may contain any symbol of both the function set $\mathscr{F}$ and the terminal set $\mathscr{T}$, while the tail must only consist of symbols from the terminal set. The full GEP gene of the expression described in Fig. 13.3 can then be described as

$$\begin{array}{cccccc|ccccccc} \scriptstyle 0 & \scriptstyle 1 & \scriptstyle 2 & \scriptstyle 3 & \scriptstyle 4 & \scriptstyle 5 & \scriptstyle 6 & \scriptstyle 7 & \scriptstyle 8 & \scriptstyle 9 & \scriptstyle 10 & \scriptstyle 11 & \scriptstyle 12 \\ Q & + & \times & \times & a & a & b & b & a & a & b & c & 1 \end{array} \tag{13.2}$$

where the head precedes the tail separated by a vertical line. In this case, the coding region starts at position 0 and ends at position 7, but the gene ends at position 12. We now assume, that a genetic mutation occurs in the head at position 4 and changes the symbol to a plus sign:

$$\begin{array}{cccccc|ccccccc} \scriptstyle 0 & \scriptstyle 1 & \scriptstyle 2 & \scriptstyle 3 & \scriptstyle 4 & \scriptstyle 5 & \scriptstyle 6 & \scriptstyle 7 & \scriptstyle 8 & \scriptstyle 9 & \scriptstyle 10 & \scriptstyle 11 & \scriptstyle 12 \\ Q & + & \times & \times & + & a & b & b & a & a & b & c & 1 \end{array}. \tag{13.3}$$

The coding area now ends at position 9 and the ET has changed to the sparse tree presented in Fig. 13.4. Despite the fixed length of the chromosome string, each gene can encode expression trees of different complexity. Furthermore, the simple form of the genotype allows any genetic modification without restrictions while the chromosome will always translate to a syntactically correct expression.

There is however one issue with the terminal set: while the set of variables can be estimated for the problem a priori, the constants are in general unknown. This is especially the case for turbulence modelling, where the model constants can depend heavily on the flow situation (e.g. Smagorinsky constant, refer to [10]). GEP solves this issue by defining an array of random numerical constants (RNCs) at the start of each run with various length in the range $[r_{min}, r_{max}]$. The chromosome string is

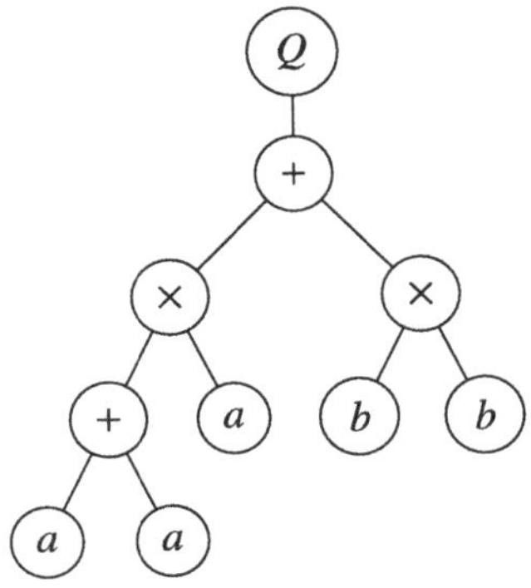

Fig. 13.4 ET for Eq. (13.3)

extended by a third part with length equal to the tail's length t, that consists of array indices to the RNC array. A symbol "?" is added to the terminal set, that accesses the next unused index pointing to an RNC. For example, consider the terminal set of $\mathscr{T} = \{a, b, c, 1, -1, ?\}$ and the RNC array of $\mathscr{R} = \{0.7, 0.1, 0.6, 0.9\}$ and the individual of

$$\begin{array}{cccccc|ccccccc|ccccccc} {\scriptstyle 0} & {\scriptstyle 1} & {\scriptstyle 2} & {\scriptstyle 3} & {\scriptstyle 4} & {\scriptstyle 5} & {\scriptstyle 6} & {\scriptstyle 7} & {\scriptstyle 8} & {\scriptstyle 9} & {\scriptstyle 10} & {\scriptstyle 11} & {\scriptstyle 12} & {\scriptstyle 13} & {\scriptstyle 14} & {\scriptstyle 15} & {\scriptstyle 16} & {\scriptstyle 17} & {\scriptstyle 18} & {\scriptstyle 19} \\ Q & + & \times & ? & - & + & ? & a & ? & b & b & c & 1 & 3 & 4 & 1 & 3 & 1 & 2 & 3 \end{array}. \tag{13.4}$$

Because ? is used for three positions, the first three indices of the third part of the chromosome will be used to replace the symbol ? in the head and tail as follows:

$$\begin{array}{cccccc|ccccccc} {\scriptstyle 0} & {\scriptstyle 1} & {\scriptstyle 2} & {\scriptstyle 3} & {\scriptstyle 4} & {\scriptstyle 5} & {\scriptstyle 6} & {\scriptstyle 7} & {\scriptstyle 8} & {\scriptstyle 9} & {\scriptstyle 10} & {\scriptstyle 11} & {\scriptstyle 12} \\ Q & + & \times & R_3 & - & + & R_4 & a & R_1 & b & b & c & 1 \end{array}. \tag{13.5}$$

These indices are then replaced by their associated numbers in the RNC array, leading to

$$\begin{array}{cccccc|ccccccc} {\scriptstyle 0} & {\scriptstyle 1} & {\scriptstyle 2} & {\scriptstyle 3} & {\scriptstyle 4} & {\scriptstyle 5} & {\scriptstyle 6} & {\scriptstyle 7} & {\scriptstyle 8} & {\scriptstyle 9} & {\scriptstyle 10} & {\scriptstyle 11} & {\scriptstyle 12} \\ Q & + & \times & 0.6 & - & + & 0.9 & a & 0.7 & b & b & c & 1 \end{array}. \tag{13.6}$$

13.3 An Example of GEP for Symbolic Regression

In this section, the application of GEP for symbolic regression is illustrated through an example. In this example, we aim to find an analytical expression for an (unknown) function:

$$z = x^2 + y^2 + xy \tag{13.7}$$

based on the training data which is sampled as $-1.0 < x < 1.0$, $-1.0 < y < 1.0$, $\Delta x = \Delta y = 0.02$ ($x_i = -1.0, -0.98, \ldots, 0.98, 1.0$; $y_i = -1.0, -0.98, \ldots, 0.98, 1.0$).

The objective is to find a functional form $f(x, y)$ which can describe the sample data within 0.01 absolute error. In an ideal case, GEP will find the exact form, i.e. Eq. 13.7. We consider the function set of $\mathscr{T} = \{+, -, *, /\}$: and a terminal set of $\mathscr{F} = \{x, y\}$. To evaluate the performance of GEP, the following cost function is used:

$$fitness = \sum_{i=1}^{n}(f_i - z_i)^2 \tag{13.8}$$

where n is the number of the training data points, z_i is the exact value at data point i which is given by Eq. 13.7, and f_i represents the predicted value by the trained function $f(x, y)$.

The following parameters need to be chosen to start the algorithm:

- Length of the head, h: it is recommended to start with a single gene chromosome and gradually increase the length.
- Population size: it is usually set as hundreds or up to 1000 individuals in one population, depending on the complexity of the problem.
- Number of generations: it is set as 1000 to ensure convergence of the training process.

To demonstrate the performance of GEP in finding a solution, one can examine the sensitivity of the results to the chosen parameters such as the length of the head, the population size and the number of generations. For brevity, we only examine the success rate of GEP in finding the functional form as the chromosome length increases (see Fig. 13.5).

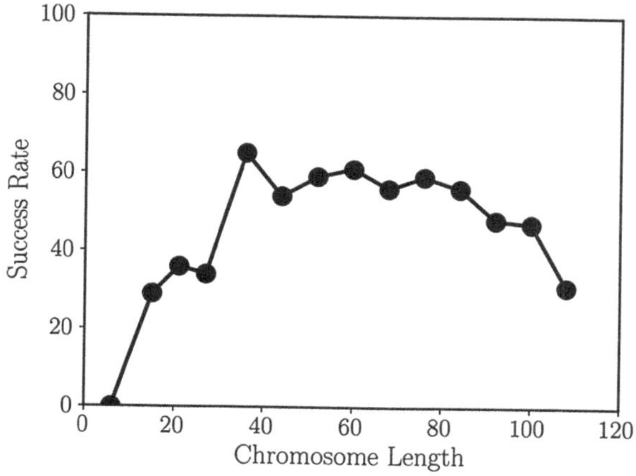

Fig. 13.5 Success rate as a function of the chromosome length

For a run to be successful, the fitness needs to be less than 0.01. We noted that for most successful runs, the fitness is as low as 10^{-30}, meaning that the exact function was found by GEP. There were solutions that have a higher fitness values like 10^{-4}. For these cases, the obtained function from GEP is quite similar to the target function except for a small deviation in the constant.

As can be seen in Fig. 13.5, once the length of chromosome is about 40, a success rate of more than 60% can be achieved. It is interesting to note that the performance does not improve as the chromosome length increases above 40.

13.4 Application of GEP to RANS Turbulence Modelling

GEP was first used for turbulence modelling in [2] and subsequent studies can be found in [3, 11]. The initial work focused on improving a shear stress transport (SST) Reynolds stress closure for RANS calculations of separated flows, in particular by testing a model developed on a backward facing step case on the different configurations of periodic hills. The Boussinesq approximation was extended by

$$\tau_{ij}^{Re} = \frac{2}{3}\delta_{ij}k - 2\nu_t S_{ij} + a_{ij}^{X} \tag{13.9}$$

and the extra anisotropy tensor a_{ij}^{X} was trained on a DNS flow field. Despite the non-determinism of the used M-GEP algorithm (an extension of the GEP algorithm to matrix functions), models with similar forms were found, highlighting the potential of evolutionary training. Furthermore, the model's fitness was evaluated using an a priori approach, but still showed good performance in a posteriori cases. The authors also observed that the trained models can perform inadequately when applied on flow regimes, which were not present in the training data (e.g., positive pressure gradients and walls). This issue can be alleviated using multiple datasets with different flow characteristics for the fitness evaluation.

In a subsequent study [3], a framework for explicit algebraic Reynolds stress model (EARSM) development applicable for industrial applications was introduced. The fitness function was evaluated using a priori testing on a hybrid RANS/LES solution of a rectangular duct. Again, the GEP algorithm found terms of similar form although every run started with different initial populations, proving the capability of GEP to find functional expressions (building blocks) for turbulence modelling. The resulting models were evaluated a priori on hybrid RANS/LES solutions of rectangular ducts different from the training case and demonstrated good predictions for the Reynolds stress using high-fidelity data. The work indicates that GEP is indeed capable of finding models that feature better predictions beyond the cases used in the evolutionary process. The authors argued that high-fidelity data which can capture the important features of the flow field could be sufficient to regress successful models and perhaps no DNS or even LES data is needed. This expands the scope of this method to more complex geometries in industrial applications for which performing DNS or LES is not feasible.

13.5 Application of GEP to a Priori LES Modelling

The current analysis considers a statistically planar turbulent premixed flame using a generic single-step Arrhenius type irreversible chemical mechanism for the purpose of computational economy, as the present analysis focuses mostly on the fluid-dynamical aspects of flame turbulence interaction. The relevant study to this analysis is presented in [1].

13.5.1 Computational Configuration

The well known compressible combustion DNS code SENGA [12] has been used, where the conservation equations of mass, momentum, energy and reaction progress variables are solved in non-dimensional form. The spatial discretisation scheme corresponds to a 10th-order central differencing scheme but the order of accuracy drops to a second-order one-sided scheme at non-periodic boundaries. The time advancement has been carried out using a low-storage third-order Runge–Kutta scheme. The flame is initialized with an unstrained laminar flame solution with superimposed synthetic turbulence using a pseudo spectral method [13] together with the model spectrum of [14]. The values of turbulent Reynolds number, normalized turbulent root-mean-square (rms) velocity fluctuation, longitudinal integral length scale to thermal flame thickness ratio, Damkoehler number, and Karlovitz number at the start of the simulation are given by $Re_t = u'l/\nu = 87.5$, $u'/S_L = 7.5$, $l/\delta_{th} = 4.58$, $Da = lS_L/\delta_{th}u' = 0.61$ and $Ka = (u'/S_L)^{3/2}(l/\delta_{th})^{-1/2} = 9.6$, respectively. By the time statistics are taken, i.e. after two eddy turnover times, turbulence has decayed by about 50% but this does not affect the current analysis and a detailed discussion in this regard can be found in [15]. The heat release parameter $\tau = (T_{ad} - T_0)/T_0$ and the Zel'dovich number β are taken to be 4.5 and 6.0 where T_{ad} and T_0 are the adiabatic flame temperature and the fresh gas temperature, respectively. Standard values of Prandtl number ($Pr = 0.7$) and ratio of specific heats ($\gamma_g = 1.4$) have been used. The simulation domain is a cube with edge length $45.75\delta_{th}$ which is discretized using a uniform Cartesian grid of 512^3 points, to ensure resolution of both the Kolmogorov length scale and the flame thickness. Navier–Stokes characteristic boundary conditions are used in the direction of mean flame propagation and periodic boundaries are used in both spanwise directions. Figure 13.6 shows a snapshot of the flame structure at the time when statistics are taken. The flame is part of a considerably larger database similar to those reported in the past [10, 16, 17], but featuring larger scale separation and higher turbulent Reynolds number. For more details in regards of the computational setup the reader is referred to these references. For the purpose of a priori analysis, the DNS data has been explicitly filtered with three convoluted 1D Gaussian filter kernels with filter size $\Delta/\Delta_{DNS} = 4, 16, 28$ corresponding to $\Delta/\delta_{th} \approx 0.36, 1.45, 2.54$.

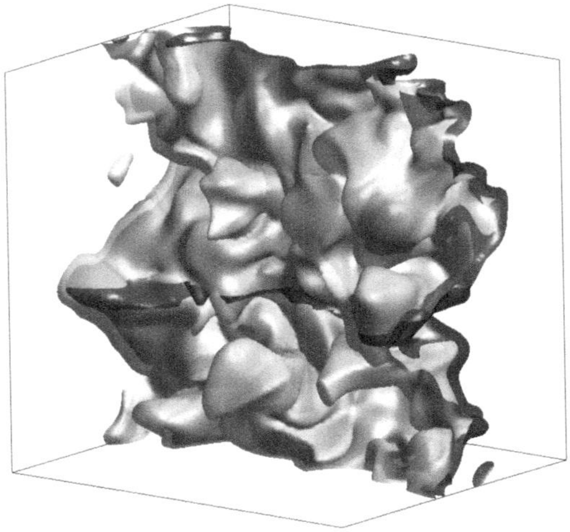

Fig. 13.6 Isosurfaces of reaction progress variable c. The flame burns from right to left. The light (dark) grey isosurface is located towards the fresh (burned) gas side of the reaction zone

13.5.2 Methodology

The density weighted momentum conservation equation is given by

$$\frac{\partial \bar{\rho}\tilde{u}_i}{\partial t} + \frac{\partial}{\partial x_j}\left(\bar{\rho}\tilde{u}_i\tilde{u}_j\right) = -\frac{\partial}{\partial x_j}\left(\overline{\rho u_i u_j} - \bar{\rho}\tilde{u}_i\tilde{u}_j\right) \tag{13.10}$$

$$+\frac{\partial}{\partial x_j}\left(\bar{\rho}\tilde{\nu}\left(\frac{\partial \tilde{u}_i}{\partial x_j} + \frac{\partial \tilde{u}_j}{\partial x_i}\right) - \frac{2}{3}\bar{\rho}\tilde{\nu}\frac{\partial \tilde{u}_k}{\partial x_k}\delta_{ij}\right) - \frac{\partial \bar{p}}{\partial x_i}, \tag{13.11}$$

where u_i is the ith component of the velocity vector, ν is the kinematic viscosity and p denotes pressure. The filtering operation gives rise to the unknown term:

$$\tau_{ij}^{SGS} = \overline{\rho u_i u_j} - \bar{\rho}\tilde{u}_i\tilde{u}_j = \bar{\rho}(\widetilde{u_i u_j} - \tilde{u}_i\tilde{u}_j). \tag{13.12}$$

In the following the assumption is made that τ_{ij}^{sgs} can be modelled as

$$\tau_{ij}^{sgs} = \tau_{ij}^{sgs}(\tilde{S}_{ij}, \tilde{\Omega}_{ij}, \Delta, \bar{\rho}) \tag{13.13}$$

where Δ is the LES filter width and

$$\tilde{S}_{ij} = \frac{1}{2}\left(\frac{\partial \tilde{u}_i}{\partial x_j} + \frac{\partial \tilde{u}_j}{\partial x_i}\right) \quad \tilde{\Omega}_{ij} = \frac{1}{2}\left(\frac{\partial \tilde{u}_i}{\partial x_j} - \frac{\partial \tilde{u}_j}{\partial x_i}\right). \tag{13.14}$$

For the GEP optimization dimenionless quantities are considered and the dimensionless strain rate and rotation rate tensor are denoted henceforth as s_{ij} and ω_{ij}. Details of the non-dimensionalisation and its inversion at the end of the optimisation process are omitted here for the sake of brevity.

Using the symmetry properties $\tau_{ij}^{sgs} = \tau_{ji}^{sgs}, s_{ij} = s_{ji}$ and $\omega_{ij} = -\omega_{ji}$, one obtains 11 basis functions and 7 invariants [4, 18] given by[1]

$$II \quad s \quad s^2 \quad \omega^2 \quad s\omega - \omega s \quad s^2\omega - \omega s^2 \quad \omega^2 s + s\omega^2 \quad s^2\omega^2 + \omega^2 s^2 \quad s\omega s^2 - s^2\omega s$$
$$\omega s\omega^2 - \omega^2 s\omega \quad \omega s^2\omega^2 - \omega^2 s^2\omega \tag{13.15}$$

where II is the identity matrix. The invariants are

$$tr(s) \quad tr(s^2) \quad tr(\omega^2) \quad tr(s^3) \quad tr(s\omega^2) \quad tr(s^2\omega^2) \quad tr(s\omega s^2\omega^2). \tag{13.16}$$

In the context of GEP the unknown τ_{ij}^{SGS} is approximated by a set of basis functions as

$$\tau_{ij}^{SGS} = \sum_{\alpha=1}^{n} G_\alpha \cdot T_{ij}^{\alpha}; \quad G_\alpha = G_\alpha(I_1, I_2, \ldots I_m). \tag{13.17}$$

The role of GEP is now to find the coefficients G, so that ultimately we are combining machine learning (GEP) with classical turbulence modelling (basis functions). For the current analysis only the first four basis functions and invariants are considered $(m = 4, n = 4)$ for the sake of computational economy:

$$T_{ij}^1 = s_{ij}, \quad T_{ij}^2 = s_{ik}\omega_{kj} - \omega_{ik}s_{kj}, \quad T_{ij}^3 = s_{ik}s_{kj}, \quad T_{ij}^4 = \omega_{ik}\omega_{kj}, \tag{13.18}$$

$$I_1 = s_{mn}s_{nm}, \quad I_2 = \omega_{mn}\omega_{nm}, \quad I_3 = s_{km}s_{mn}s_{nk}, \quad I_4 = \omega_{km}\omega_{mn}s_{nk}. \tag{13.19}$$

The performance of the model is evaluated using the following cost function

$$fit(\tau_{ij}^{mod}) = \frac{1}{n_p}\sum_{\alpha=1}^{n_p} \frac{\left|\tau_{ij}^{sgs}(\alpha) - \tau_{ij}^{mod}(\alpha)\right|_F}{\left|\tau_{ij}^{sgs}(\alpha)\right|_F} \qquad \left|\tau_{ij}\right|_F = \sqrt{\sum_{i=1}^{3}\sum_{j=1}^{3}\tau_{ij}^2} \tag{13.20}$$

where α denotes an arbitrary sample point used for averaging.

13.5.3 Results

Figure 13.7 indicates the strain tensor T_{ij}^1 has very low and frequently negative correlations with the unknown stress. This behaviour is consistent with the results reported in [10, 19]. Nevertheless, because T_{ij}^1 is a fundamental tensor in LES modelling it has been considered during the analysis. The correlations for T_{ij}^2, T_{ij}^3, T_{ij}^4 are reasonably high but in general they decrease with increasing filter size. It is further worth noting that T_{ij}^3 (T_{ij}^2 and T_{ij}^4) depict clearly positive (negative) correlations with the

[1] The indices are omitted for better readability.

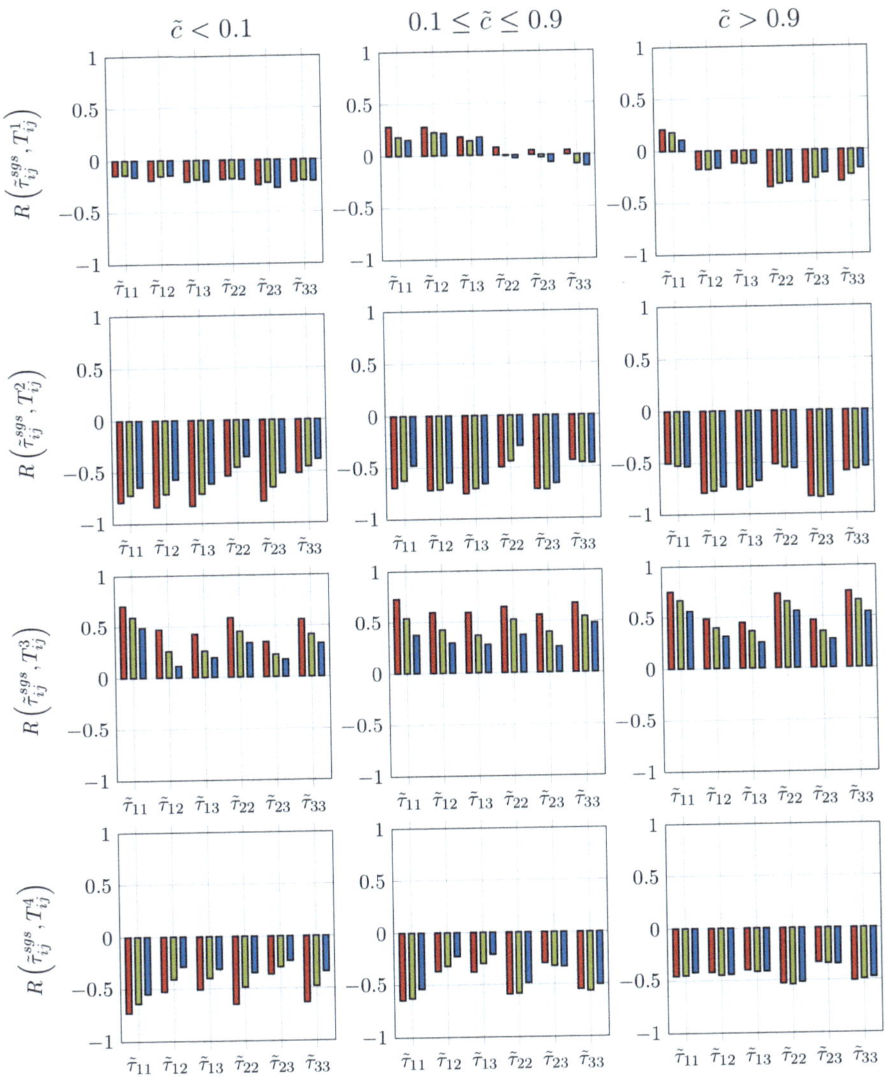

Fig. 13.7 Pearson correlation coefficient between tensor components of the first four basis functions and SGS stresses calculated from DNS conditionally averaged on three regimes (unburned side, reaction zone and burned gas side) of reaction progress variable $\tilde{c}$. The colours represent three different filter width with $\Delta/\Delta_{DNS} = 4, 16, 28$. 04, 16, 28

unknown SGS stress for all tensor components. There is no pronounced difference in correlation strength depending on the region of the flame brush. This indicates that all basis functions except T_{ij}^1 are able to account for the effects of gradient and counter-gradient transport which can be typically found in premixed flames when the effects of heat release are strong relative to the turbulent velocity perturbations [20]. This behaviour is discussed in detail in [10, 19] for a similar database. Figure 13.8 shows the correlation coefficients for the first four invariants with individual tensor components. The correlations are in general reasonably high for the diagonal components and rather low for the off-diagonal components of the stress tensor. In particular I_3 and I_4 show very low correlations within the reaction zone and towards the burned gas side.

The framework outlined before has been applied to modelling the unknown tensor $\tilde{\tau}_{ij}^{SGS}$ using the first four basis functions and the first four invariants and running the GEP algorithm for 500 generations. The fitness value of all 100 models is shown in Fig. 13.9. Setting the model expression to zero would give rise to a fitness value of unity. Hence it is reasonable to discard all models that have a fitness value larger than 0.5 and a complete list of the remaining models is given in Table 13.1.

The best model found in generation 77 is given by

$$\tilde{\tau}_{ij}^{SGS} \approx \bar{\rho}\Delta^2 C \cdot \left(T_{ij}^3 - T_{ij}^2 - T_{ij}^4\right) \quad C = 8.21 \times 10^{-2} \,. \tag{13.21}$$

Using the definitions for strain and vorticity rate together with the basis functions, Eq. 13.21 can be also written in the following manner

$$\tilde{\tau}_{ij}^{SGS} \approx \bar{\rho}\Delta^2 C \cdot \frac{\partial \tilde{u}_i}{\partial x_k}\frac{\partial \tilde{u}_j}{\partial x_k} \tag{13.22}$$

and the constant can be estimated by

$$C = 8.21 \times 10^{-2} \approx \frac{1}{12.18} \approx \frac{1}{12} = 8.333 \times 10^{-2} \,. \tag{13.23}$$

In other words the model found by the GEP algorithm corresponds exactly to the model by Clark [21], including its model parameter. The performance of Clark's model has been discussed in detail in [10, 19] and this will not be repeated here. It is worth mentioning that it is very surprising (but also encouraging) that GEP did find this well known model from the literature including the model coefficient based on a suitable non-trivial combination of the basis functions.

It is interesting to observe from Table 13.1 that very often the constant coefficient functions G_α are selected by GEP, which implies that they result in small fitness values. Apart from the four basic arithmetic operations the $tanh$ and $power$ functions can be found in the best models, whereas other functions, that are as well used with a certain probability, obviously do no result in useful model combinations and hence do not occur in the best model at the end of evolution.

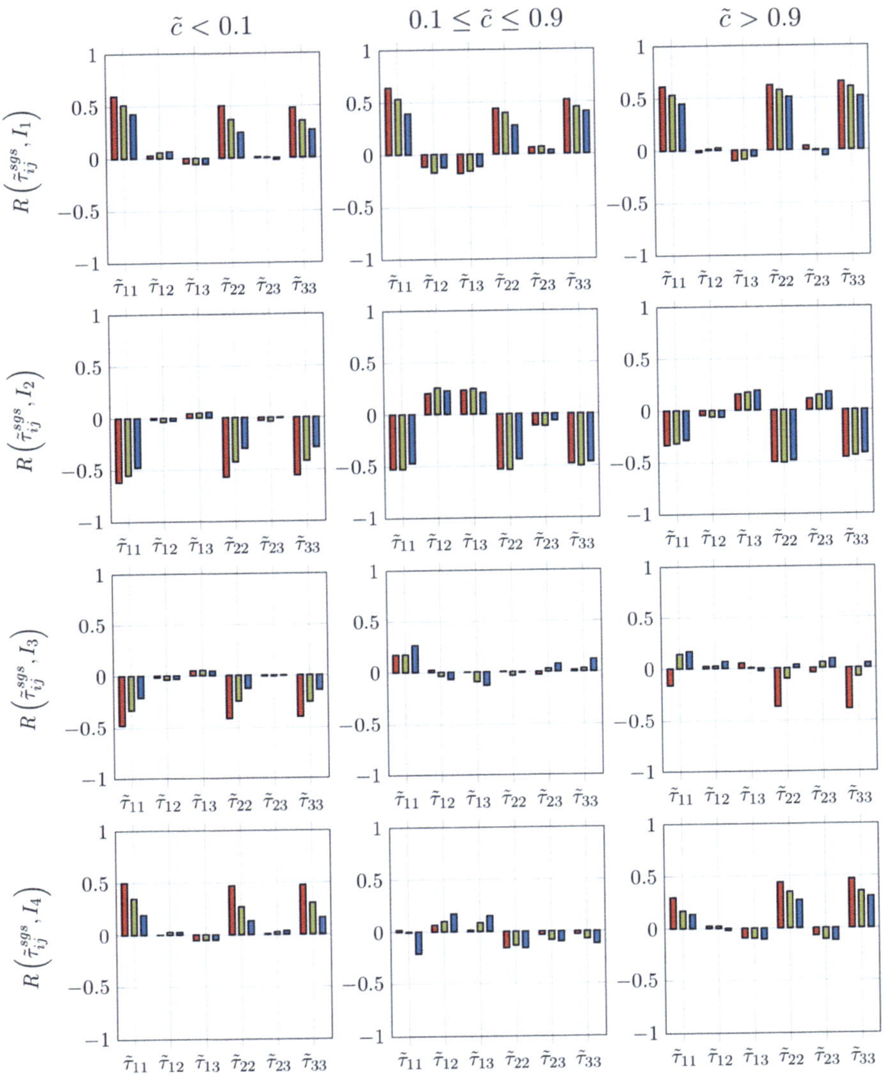

Fig. 13.8 Pearson correlation coefficient between the first four invariants and SGS stresses calculated from DNS conditionally averaged on three regimes (unburned side, reaction zone and burned gas side) of reaction progress variable $\tilde{c}$. The colours represent three different filter width with $\Delta/\Delta_{DNS} = 4, 16, 28$. 04, 16, 28

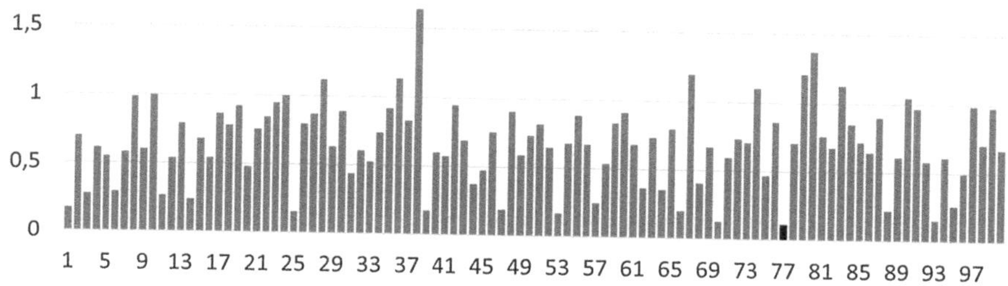

Fig. 13.9 Fitness distribution for the 100 models using the first four basis functions

In the previous example the GEP algorithm was run 100 times for 500 generations with a population size of 75, resulting in 100 model proposals. In each generation about 20 new models are tested which implies that the cost function has to be evaluated $20 \times 500 \times 100 = 10^6$ times which can be quite expensive. In order to assess the robustness of the GEP algorithm and in order to understand the required amount of raw data, the GEP algorithm has been run in the next step for only 10 model generations using, for the purpose of this simple test, only the first three basis functions and the first four invariants. The best model out of 10 is given by

$$\tilde{\tau}_{ij}^{SGS} \approx \bar{\rho}\Delta^2 \tilde{S}_{mn}\tilde{S}_{nm} \cdot \left[0.11 \cdot T_{ij}^3 - 0.07 \cdot T_{ij}^2\right] . \quad (13.24)$$

Subsequently, the amount of raw data has been reduced by considering only every 2nd and 4th data point. Table 13.2 indicates that the first 10 model expressions found by GEP are identical but only the fitness values change slightly. When considering only every 32th or 128th data point (out of a 512 cube), the models change slightly but the best model (i.e. the one found in the third generation) remains the same (see Tables 13.3 and 13.4). This demonstrates that there is scope for improving the computational efficiency by subsampling the data. For instance, this would allow the use of more complex functions and even consideration of the complete set of basis functions. The results of this exercise suggest that a careful selection of data for training is necessary for both enhancing the computational efficiency and developing a more accurate model.

13.6 Conclusions

This chapter presented an application of genetic expression programming (GEP) for subgrid-scale (SGS) modelling in statistically planar, turbulent premixed flames. The chapter introduced the basic concepts of genetic algorithms (GAs), genetic programming (GP) and GEP, which inherit characteristics from both GAs and GP. Application of GEP for RANS turbulence modelling was then discussed. Next, the

Table 13.1 List of models with fitness value smaller than 0.5 found during the training run

	Fitness	Model expression
1	0.165	$0.0776 \cdot T_{ij}^3 - 0.071\,578 \cdot T_{ij}^2 - 0.0776 \cdot T_{ij}^4$
3	0.2728	$0.087\,974 \cdot T_{ij}^3 - 0.059 \cdot T_{ij}^2 - 0.118 \cdot T_{ij}^4 + 0.013\,924 \cdot I_3 \cdot I_4 \cdot T_{ij}^1$
6	0.2847	$0.0596 \cdot T_{ij}^3 - 0.080\,605 \cdot T_{ij}^2 - 0.084 \cdot T_{ij}^4$
11	0.2616	$0.116 \cdot T_{ij}^3 - T_{ij}^2 \cdot (0.035\,276 - 0.048\,275) - 0.0291 \cdot T_{ij}^4 - T_{ij}^1 \cdot (0.0291 \cdot I_3 + 0.0291 \cdot I_4)$
14	0.2341	$0.0784 \cdot T_{ij}^3 - 0.057\,268 \cdot T_{ij}^2 - 0.0784 \cdot T_{ij}^4 - 0.0262 \cdot I_4 \cdot T_{ij}^1$
20	0.4787	$0.11 \cdot T_{ij}^3 - 0.022\,644 \cdot T_{ij}^2 - 0.037 \cdot T_{ij}^4 + T_{ij}^1 \cdot (0.011\,692 \cdot I_2 - 0.011\,692 \cdot I_3)$
25	0.1491	$0.095\,625 \cdot T_{ij}^3 - T_{ij}^2 \cdot 0.097\,96i - 0.085\,68 \cdot T_{ij}^4$
31	0.4342	$T_{ij}^1 \cdot tanh(I_3 \cdot I_4) - 0.077 \cdot T_{ij}^4 - 0.0871 \cdot T_{ij}^2 - T_{ij}^3 \cdot (I_3 \cdot I_4 - 0.0871)$
39	0.1697	$0.011\,292 \cdot T_{ij}^3 \cdot (200.0 \cdot I_4 - 53.0)^{1/2} - 0.0722 \cdot T_{ij}^4 - 0.0722 \cdot T_{ij}^2 + T_{ij}^1 \cdot tanh(0.051 \cdot I_4)$
44	0.3696	$0.0198 \cdot T_{ij}^1 - 0.076\,44 \cdot T_{ij}^2 + 0.059\,099 \cdot T_{ij}^3 + 0.0198 \cdot I_2 \cdot T_{ij}^4$
45	0.4712	$0.049\,123 \cdot T_{ij}^1 - 0.0544 \cdot T_{ij}^2 + 0.031\,93 \cdot T_{ij}^3 - 0.0544 \cdot T_{ij}^4$
47	0.187	$0.099\,579 \cdot T_{ij}^3 - 0.102\,43 \cdot T_{ij}^2 - 0.102\,43 \cdot T_{ij}^4$
53	0.1642	$0.0715 \cdot T_{ij}^3 - T_{ij}^2 \cdot 0.074\,825i - 0.0715 \cdot T_{ij}^4 + 0.014\,861 \cdot T_{ij}^1 \cdot (-tanh(I_3))^{1/2}$
57	0.2435	$0.0901 \cdot T_{ij}^3 - 0.115\,52 \cdot T_{ij}^2 - 0.085\,67 \cdot T_{ij}^4$
62	0.3618	$0.172\,97 \cdot T_{ij}^3 \cdot tanh(I_3) - 0.078\,273 \cdot T_{ij}^4 - 0.0856 \cdot T_{ij}^2$
64	0.3535	$0.085\,28 \cdot T_{ij}^3 - 0.0615 \cdot T_{ij}^2 - 0.746 \cdot I_3 \cdot T_{ij}^1$
66	0.1963	$0.094 \cdot T_{ij}^3 - 0.0644 \cdot T_{ij}^2 - 0.064\,311 \cdot T_{ij}^4$
67	1.197	$0.5 \cdot I_4 \cdot T_{ij}^3 - 0.227 \cdot T_{ij}^4 - 0.906 \cdot I_3 \cdot I_4 \cdot T_{ij}^1$
70	0.1264	$0.0949 \cdot T_{ij}^3 - 0.0949 \cdot T_{ij}^2 - 0.003\,796 \cdot T_{ij}^1 - 0.0949 \cdot T_{ij}^4$
75	0.4674	$0.088\,369 \cdot T_{ij}^3 - 0.119 \cdot T_{ij}^2$
77	0.115	$0.0821 \cdot T_{ij}^3 - 0.0821 \cdot T_{ij}^2 - 0.0821 \cdot T_{ij}^4$
88	0.2175	$T_{ij}^3 \cdot (0.1726 \cdot I_4 + 0.112\,19) - 0.0863 \cdot T_{ij}^4 - 0.0863 \cdot T_{ij}^2 - 0.0863 \cdot I_3 \cdot I_4^{3/2} \cdot T_{ij}^1$
95	0.2562	$0.0358 \cdot I_3 \cdot T_{ij}^1 \cdot tanh(0.761\,59 \cdot I_3) - 0.0358 \cdot T_{ij}^4 - T_{ij}^3 \cdot (0.0358 \cdot I_4 - 0.0716) - 0.079\,911 \cdot T_{ij}^2$
96	0.4923	$0.078 \cdot T_{ij}^3 - 0.087\,93 \cdot T_{ij}^4 - T_{ij}^2 \cdot (I_3 - tanh(I_3))$

Table 13.2 Models for data sample size variation. Every (second column), every 2nd (third column), every 4th (fourth column) data point is used for the analysis

	Fitness	Fitness	Fitness	Model expression
1	0.6133	0.6132	0.613	$0.0776 \cdot T_{ij}^3$
2	1.799	1.799	1.798	$I_3^2 \cdot T_{ij}^1 \cdot abs(I_2) + I_2 \cdot I_3 \cdot T_{ij}^3$
3	0.358	0.358	0.3578	$0.105\,85 \cdot T_{ij}^3 - 0.069\,705 \cdot T_{ij}^2$
4	1.26	1.26	1.26	$I_3 \cdot T_{ij}^3 \cdot (I_3 - 0.568) - 0.047 \cdot I_3 \cdot T_{ij}^1$
5	0.9436	0.9436	0.9433	$0.0368 \cdot T_{ij}^2 + 0.0368 \cdot T_{ij}^3$
6	0.6771	0.677	0.6766	$0.0596 \cdot T_{ij}^1 - 0.039\,69 \cdot T_{ij}^2 + 0.0361 \cdot T_{ij}^3$
7	0.6122	0.6121	0.6119	$0.000\,235 \cdot T_{ij}^1 - 0.000\,235 \cdot T_{ij}^2 + 0.077 \cdot T_{ij}^3$
8	0.509	0.5089	0.5086	$0.027\,105 \cdot T_{ij}^1 - 0.0432 \cdot T_{ij}^2 + 0.0453 \cdot T_{ij}^3$
9	2.256	2.256	2.255	$I_3^2 \cdot T_{ij}^3 - T_{ij}^1 \cdot (0.563 \cdot I_3 - 0.091\,769)$
10	1.282	1.282	1.281	$0.01 \cdot abs(I_2 \cdot I_3)^{1/2} \cdot (72.18 \cdot T_{ij}^3 - 55.0 \cdot I_3 \cdot T_{ij}^1)$

Table 13.3 Models for data sample size variation. Only every 32th data point is used for the analysis

	Fitness	Model expression
1	0.6129	$0.0776 \cdot T_{ij}^3$
2	1.8	$I_3^2 \cdot T_{ij}^1 \cdot abs(I_2) + I_2 \cdot I_3 \cdot T_{ij}^3$
3	0.3574	$0.105\,85 \cdot T_{ij}^3 - 0.069\,705 \cdot T_{ij}^2$
4	1.259	$I_3 \cdot T_{ij}^3 \cdot (I_3 - 0.568) - 0.047 \cdot I_3 \cdot T_{ij}^1$
5	0.6566	$0.0368 \cdot T_{ij}^1 - 0.051\,156 \cdot T_{ij}^2 + 0.014\,941 \cdot T_{ij}^3$
6	0.6773	$0.0596 \cdot T_{ij}^1 - 0.039\,69 \cdot T_{ij}^2 + 0.0361 \cdot T_{ij}^3$
7	0.6118	$0.000\,235 \cdot T_{ij}^1 - 0.000\,235 \cdot T_{ij}^2 + 0.077 \cdot T_{ij}^3$
8	0.5758	$0.0453 \cdot T_{ij}^1 - 0.0432 \cdot T_{ij}^2 + 0.0453 \cdot T_{ij}^3$
9	2.255	$I_3^2 \cdot T_{ij}^3 - T_{ij}^1 \cdot (0.563 \cdot I_3 - 0.091\,769)$
10	3.812	$0.7218 \cdot T_{ij}^3 - 0.396 \cdot T_{ij}^1 + T_{ij}^2 \cdot (I_3 - abs(I_3)^{1/2})$

DNS dataset used for LES a priori testing was introduced. This was followed by description of the methodology and presentation of the results.

The GEP algorithm was run 100 times for 500 generations with a population size of 75 resulting in 100 model proposals. A key observation was that the found model with the best fitness corresponds exactly to the model proposed by Clark [21], even including its model parameter. This is a very promising result demonstrating

Table 13.4 Models for data sample size variation. Only every 128th data point is used for the analysis

	Fitness	Model expression
1	0.611	$0.0776 \cdot T_{ij}^3$
2	1.062	$-0.401 \cdot I_3 \cdot T_{ij}^2 \cdot (I_3 - 0.591)$
3	0.3555	$0.105\,85 \cdot T_{ij}^3 - 0.069\,705 \cdot T_{ij}^2$
4	1.209	$I_3 \cdot T_{ij}^3 \cdot (I_3 - 0.568) - 0.07 \cdot I_3^2 \cdot T_{ij}^1 - 0.13 \cdot T_{ij}^2$
5	0.6555	$0.0368 \cdot T_{ij}^1 - 0.051\,156 \cdot T_{ij}^2 + 0.014\,941 \cdot T_{ij}^3$
6	0.6735	$0.0596 \cdot T_{ij}^1 - 0.039\,69 \cdot T_{ij}^2 + 0.0361 \cdot T_{ij}^3$
7	0.6099	$0.000\,235 \cdot T_{ij}^1 - 0.000\,235 \cdot T_{ij}^2 + 0.077 \cdot T_{ij}^3$
8	0.7329	$0.0453 \cdot T_{ij}^1 - 0.0453 \cdot T_{ij}^2$
9	2.241	$I_3^2 \cdot T_{ij}^3 - T_{ij}^1 \cdot (0.563 \cdot I_3 - 0.091\,769)$
10	3.798	$0.7218 \cdot T_{ij}^3 - 0.396 \cdot T_{ij}^1 + T_{ij}^2 \cdot (I_3 - abs(I_3)^{1/2})$

the capabilities of GEP for SGS modelling. In order to assess the robustness of the GEP algorithm, only 10 model generations were used in the next stage. Running the GEP algorithm on several sets of subsampled raw data resulted in identical models with small changes in the fitness values. This demonstrates that there is scope for efficiency improvements in future work which might allow using the complete set of basis functions.

Acknowledgements The stay abroad of M.S. was supported by the Federal Ministry of Defence.

References

1. M. Schoepplein, J. Weatheritt, R. Sandberg, M. Talei, M. Klein, Application of an evolutionary algorithm to LES modelling of turbulent transport in premixed flames. J. Comput. Phys. **374**, 1166–1179 (2018)
2. J. Weatheritt, R. Sandberg, A novel evolutionary algorithm applied to algebraic modifications of the RANS stress strain relationship. J. Comput. Phys. **325**, 22–37 (2016)
3. J. Weatheritt, R.D. Sandberg, The development of algebraic stress models using a novel evolutionary algorithm. Int. J. Heat Fluid Flow **68**, 298–318 (2017)
4. S.B. Pope, A more general effective-viscosity hypothesis. J. Fluid Mech. **75**, 331–340 (1975)
5. J.H. Holland, *Adaptation in Natural and Artificial Systems* (1975)
6. Genetic Algorithms, https://www.geeksforgeeks.org/genetic-algorithms/. Accessed 16 Jun 2019
7. J.R. Koza, *Genetic Programming II: Automatic Discovery of Reusable Programs* (1994)
8. J.R. Koza, *Genetic Programming: On the Programming of Computers by Means of Natural Selection* (1992)

9. C. Ferreira, Gene expression programming: a new adaptive algorithm for solving problems. Complex Syst. **13**, 87–129 (2001)
10. M. Klein, C. Kasten, N. Chakraborty, Y. Gao, A-priori direct numerical simulation assessment of sub-grid scale stress tensor closures for turbulent premixed combustion. Comput. Fluids **122**, 1–11 (2015)
11. J. Weatheritt, R.D. Sandberg, Hybrid Reynolds-averaged/large-eddy simulation methodology from symbolic regression: formulation and application. AIAA J. (2017)
12. K. Jenkins, R. Cant, DNS of turbulent flame kernels, in *Proceedings of 2nd AFOSR Conference on DNS and LES*, ed. by C. Liu, L. Sakell, T. Beautner (Kluwer Academic Publishers, 1999), pp. 192–202
13. R.S. Rogallo, NASA Technical Memorandum 81315 (NASA Ames Research Center, 1981)
14. B.P. Pope, *Turbulent Flows* (Cambridge University Press, Cambridge, 2000)
15. M. Klein, N. Chakraborty, S. Ketterl, A comparison of strategies for direct numerical simulation of turbulence chemistry interaction in generic planar turbulent premixed flames. Flow Turbul. Combust. 1–17 (2017)
16. Y. Gao, N. Chakraborty, M. Klein, Assessment of sub-grid scalar flux modelling in premixed flames for large eddy simulations. Eur. J. Mech. - B/Fluids **52**, 97–108 (2015)
17. M. Klein, N. Chakraborty, Y. Gao, Scale similarity based models and their application to subgrid scale scalar flux modelling in the context of turbulent premixed flames. Int. J. Heat Fluid Flow **57**, 91–108 (2016)
18. A.J.M. Spencer, R.S. Rivlin, The theory of matrix polynomials and its application to the mechanics of isotropic continua, in *Collected Papers of R.S. Rivlin* (Springer, Berlin, 1997)
19. M. Klein, C. Kasten, N. Chakraborty, A-priori direct numerical simulation assessment of models for generalized sub-grid scale turbulent kinetic energy in turbulent premixed flames. Comput. Fluids **154**, 123–131 (2017)
20. D. Veynante, A. Trouve, K. Bray, T. Mantel, Gradient and counter-gradient scalar transport in turbulent premixed flames. J. Fluid Mech. **332**, 263–293 (1997)
21. R.A. Clark, J.H. Ferziger, W. Reynolds, Evaluation of subgrid-scale models using an accurately simulated turbulent flow. J. Fluid Mech. **91**, 1–16 (2017)

Chapter 14
Machine Learning of Combustion LES Models from Reacting Direct Numerical Simulation

Shashank Yellapantula, Marc T. Henry de Frahan, Ryan King, Marc Day, and Ray Grout

Abstract In this chapter we demonstrate how supervised deep learning techniques can be used to construct models for the filtered progress variable source term necessary for large eddy simulation (LES). The source data for the model is a direct numerical simulation (DNS) of a reacting flow in a low swirl burner configuration. Filtered quantities taken from the DNS data are used to train a deep neural network (DNN)-based model. An efficient data sampling strategy was devised to ensure that a uniform representation of all the states observed in the filtered DNS data are equally present in the training dataset. A-priori testing of the DNN-based model highlights the representative power of DNN to accurately reproduce the filtered reaction progress variable source term over a range of scales and various flame regimes as seen in an industrial burner.

S. Yellapantula (✉) · M. T. H. de Frahan · R. King · R. Grout
High Performance Algorithms and Complex Fluids, Computational Science Center, National Renewable Energy Laboratory, 15013 Denver W Pkwy, ESIF301, Golden, CO 80401, USA
e-mail: shashank.yellapantula@nrel.gov

M. T. H. de Frahan
e-mail: Marc.HenrydeFrahan@nrel.gov

R. King
e-mail: Ryan.King@nrel.gov

R. Grout
e-mail: Ray.Grout@nrel.gov

M. Day
Center for Computational Sciences and Engineering, Lawrence Berkeley National Laboratory, Berkeley, CA 94720, USA
e-mail: msday@lbl.gov

H. Pitsch and A. Attili (eds.), *Data Analysis for Direct Numerical Simulations of Turbulent Combustion*, https://doi.org/10.1007/978-3-030-44718-2_14

14.1 Introduction

The exponential increase in supercomputing capabilities has led to a phenomenal rise in direct numerical simulation (DNS) of reacting flows. These DNS studies, initially focused on 2D or simplified 3D cases such as temporal mixing layers, have significantly increased in complexity in terms of configurations, fuels, pressure, and temperature conditions [1–5]. Some of the recent DNS use upwards of 2 billion grid points with complex fuels such as *n*-dodecane and use millions of CPU hours on the leadership class supercomputing facilities [6–8]. All of these reacting flow DNSs have no doubt highlighted a number of key physical phenomena and have furthered knowledge in the community. These DNS studies have also led to a massive development in high-fidelity models for reacting flows.

In the past, typical model development exercises using DNS data have tested physical hypothesis against the data or used data to uncover model coefficients or constants. Strategies that rely on data to construct a model form or discover the most important parameters or underlying phenomenon have been scarce [9, 10] and have not been fully explored within the combustion modeling community. In the very recent past, there has been a massive surge in data-driven modeling strategies for modeling turbulent flows [11]. This surge in interest in data-driven models is being aided by the improvement in machine learning techniques especially in the area of deep learning. In this study, a deep learning-based strategy will be used to demonstrate the capability of these techniques in constructing a large eddy simulation (LES)-based model for reacting flow in an industrially relevant burner.

14.1.1 Deep Learning

In recent years, deep learning, relying on the deployment of deep neural networks, has enjoyed particular success in modeling tasks ranging from image processing [12–18] to text generation [19–21] and games [22]. Several reviews of the field give a summary of recent breakthroughs and developments [23–27]. The advent of deep neural networks has been driven in large part by the availability of large public training datasets and powerful computing platforms such as graphics processing units (GPUs) [26]. Additional advances in deep learning, specifically backpropagation, minibatch stochastic gradient, and novel architectures (e.g., convolutional and recurrent neural networks), have further accelerated the adoption of deep learning techniques.

DNNs are typically based on a sequence of learnable linear operators and non-linear activation functions that are tuned using a gradient-descent optimizer. It has been shown that DNNs are universal function approximators [28, 29] and, therefore, given enough layers and nodes; they can be used to model arbitrarily complex and highly non-linear functional forms. This raises the difference in how their performance is impacted by increasing quantities of input data compared to more traditional modes. While models constructed based on a small number of parameters, either physics-based or other analytic form, are inherently regularized in the face of insufficient data, they are also inherently limited by the degree to which the underlying model can capture variation in the data. Where the data is not self-consistent through the lens of the prescribed model, there is little recourse to discover what might be missing. Data-driven models such as DNN on the other hand can see their performance improve as they generalize to a model that captures all of the variation in the data. The representative power of deep neural networks offers an opportunity for regression-based tasks for modeling complex physics such as those found in large eddy simulations of reacting flows. Here we use the DNN not only to build a model, but to explore multiple sets of inputs to determine which are most important to model the desired output.

14.1.2 Filtered Progress Variable Source Term

A key unclosed term in large eddy simulation modeling of reacting flow is the chemical source term for reacting species. In some reacting LES formulations, a surrogate variable typically known as the progress variable is used to describe the progress of the reaction. In the current study, a progress variable formulation is chosen with progress variable, C, defined as the summation of mass fraction of major product species ($C = Y_{CO} + Y_{CO_2} + Y_{H_2O} + Y_{H_2}$). Equation 14.1 describes the transport equation of filtered progress variable, $\widetilde{C}$, where $\widetilde{\omega}_C$ is the filtered reaction progress variable source term:

$$\frac{\partial \bar{\rho}\widetilde{C}}{\partial t} + \frac{\partial \bar{\rho}\widetilde{u}_j\widetilde{C}}{\partial x_j} = \frac{\partial}{\partial x_j}\bar{\rho}\widetilde{D}\frac{\partial \widetilde{C}}{\partial x_j} + \frac{\partial \tau_{sgs,j}}{\partial x_j} + \bar{\rho}\widetilde{\omega}_C. \tag{14.1}$$

The primary goal of this study is to use DNN-based techniques to construct a model for filtered reaction progress variable source term, that is accurate over a range of scales and various flame regimes seen in an industrial burner. A secondary objective is to determine a suitable set of inputs to the model. In contrast to the existing modeling studies, a manifold form or a sub-grid PDF is not assumed a-priori and the DNN-based regression analysis using the filtered DNS data is used to derive the LES sub-grid model. A data-driven LES model of the form $\widetilde{\omega}_C = F\left(\widetilde{Z}, \widetilde{Z''^2}, \widetilde{C}, \widetilde{C''^2}, \widetilde{\chi}_Z, \widetilde{\chi}_C\right)$ will be constructed using the filtered DNS data. This requires that the DNN capture the influence of the filter, the chemistry, and the sub-filter fluctuations of the input variables. In the model form shown above, $\widetilde{Z}$ and $\widetilde{C}$ are the filtered mixture fraction and progress variable, respectively. Scalar dissipation rates of mixture fraction and progress variable are denoted by $\widetilde{\chi}_Z$ and $\widetilde{\chi}_C$, respectively.

The remainder of this chapter is organized as follows: Sect. 14.2 describes the DNS case used in this study. Sections 14.2.1 and 14.2.2 describe the procedure for preparing the training data for constructing the model and Sect. 14.3 describes the deep learning algorithm used in this study. Results from the a-priori tests to establish the accuracy of this data-driven model will be presented in Sect. 14.4. Finally a conclusion is provided in Sect. 14.5.

14.2 DNS of Low Swirl Burner

The direct numerical simulation (DNS) of the methane-based premixed low swirl burner [30, 31] provides the data for model development in this study. In this lean premixed low swirl burner configuration, a nozzle with circular vane imposes a swirl (geometric swirl number of 0.55) to a premixed CH_4 and air mixture with a fuel–air equivalence ratio of 0.7. A co-flow of air envelopes the nozzle region with an upward velocity of 0.25 $\frac{\mathrm{m}}{\mathrm{s}}$. The inflow velocity of the fuel–air mixture at the nozzle is 15 $\frac{\mathrm{m}}{\mathrm{s}}$. The simulation was performed using LMC, a low Mach number Navier–Stokes solver for turbulent reacting flows that leverages adaptive mesh refinement to resolve finer length scales [32]. The computational domain was 0.25 m in each dimension. Three levels of refinement were used, leading to an effective resolution of 100 μm in the flame region. For reference, at these conditions the laminar flame thickness computed from a laminar freely propagated premixed flame configuration is 600 μm. The DRM 19 chemical mechanism was used to model the finite rate kinetics [33]. The description of the flame and a detailed analysis of the flame are presented in [30] and will not be described here for brevity.

The choice of using this test case to demonstrate the capability of deep learning-based techniques is motivated by the presence of multiple regimes of combustion. In this case the primary reaction zone is set up by a premixed lifted flame followed by a secondary reaction zone where the products from the primary zone mix and react with the co-flow air. Identifying the most important input parameters and constructing a model that can accurately reproduce these reaction zones is a challenging exercise and thus this case acts as an ideal test case to highlight the power of deep learning-based techniques.

14.2.1 Model Inputs

The data-driven model proposed in this study is constructed using the input variables most commonly used in the combustion modeling community. This choice of input variables is made to create a direct analogy between data-driven models and the prevalent physics-based models. Additionally, all these input variables have well defined transport equations with validated and verified closure models. The following are the fields post processed from the DNS data to generate model inputs.

i. Mixture fraction, Z: In the current analysis, the mixture fraction, Z, is computed through a linear combination of the nitrogen mass fraction in the burner exit stream and the co-flow and is normalized such that it varies between 0 in the co-flow stream and 1 in the premixed burner exit stream. Figure 14.1a shows contours of mixture fraction, Z, on the mid-plane in the span-wise direction.

ii. Progress variable, C: The progress variable is defined in this work as $C = Y_{CO_2} + Y_{CO} + Y_{H_2} + Y_{H_2O}$ and varies between 0 and 0.21, where Y_i is the mass fraction of species i, shown in Fig. 14.1b.

iii. Mixture fraction dissipation rate, χ_Z: Using the mixture fraction field defined previously, the mixture fraction dissipation rate, χ_Z is computed as $D_Z|\nabla Z|^2$, where D_Z is the diffusivity of mixture fraction (in this case species N_2). Figure 14.1c shows the contours of χ_Z. In reacting LES equations, filtered mixture fraction dissipation rate, $\widetilde{\chi_Z}$, acts as the sink term in the conserved transport equation for the filtered mixture fraction variance, $\widetilde{Z''^2}$.

iv. Progress variable dissipation rate, χ_C: Progress variable dissipation rate, χ_C, is computed as $D_Z|\nabla C|^2$, where D_C is the diffusivity of progress variable, computed through summation of diffusivity of species CO_2, CO, H_2, and H_2O, shown in Fig. 14.1d. Similar to mixture fraction dissipation rate, filtered progress variable dissipation rate, $\widetilde{\chi_C}$, acts as a sink term along with the source term from progress variable reaction rate in the transport equation for filtered progress variable variance, $\widetilde{C''^2}$.

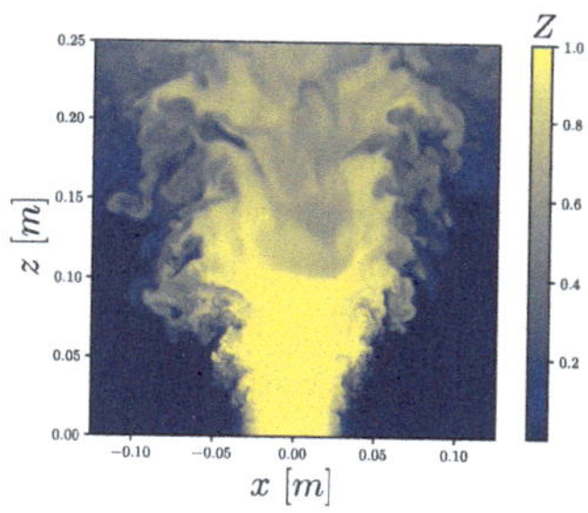

(a) Mixture fraction, Z

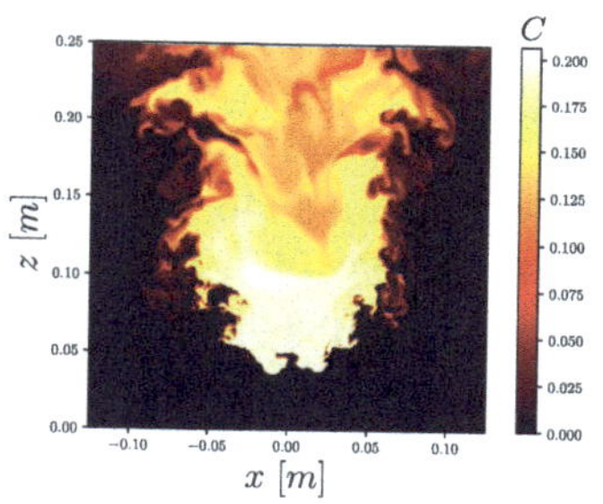

(b) Progress variable, C

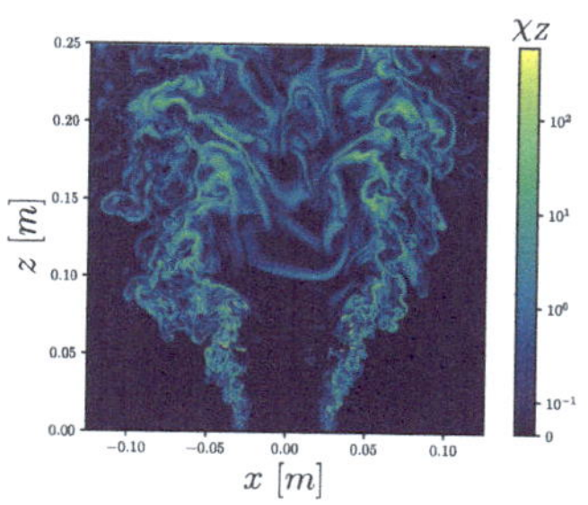

(c) Mixture fraction dissipation rate, χ_Z

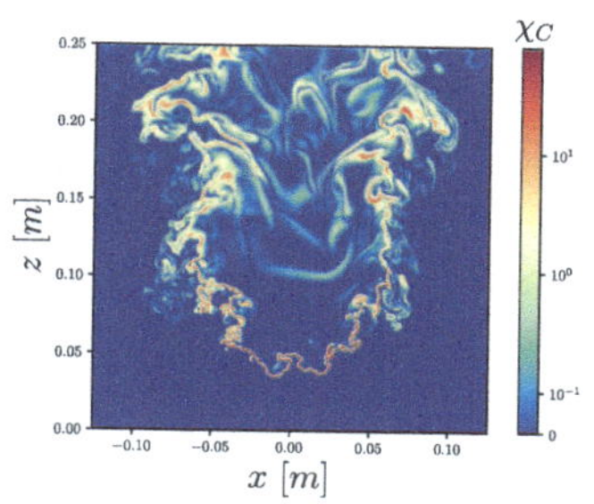

(d) Progress variable dissipation rate, χ_C

Fig. 14.1 DNS quantities used to construct inputs to the data-driven model. Instantaneous (at t = 0.0626 s) contours from DNS plotted on the span-wise mid-plane

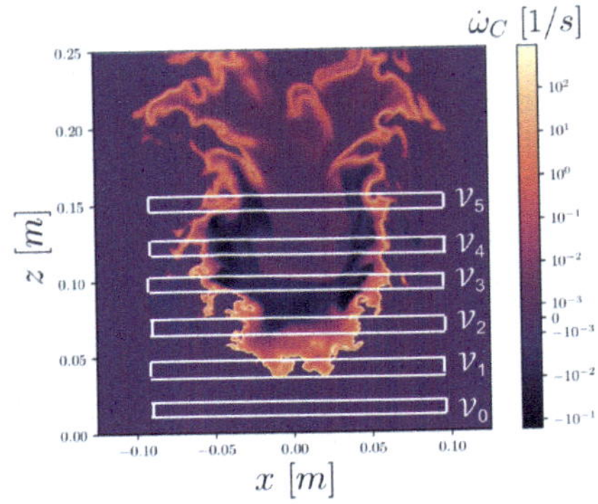

Fig. 14.2 Sub-volumes of DNS data used for training

14.2.2 Training Data

The 3D data from the DNS at a single time instance, $t = 0.0626$ s, is used to generate the training data for the data-driven model. The 3D data is first divided into a number of sub-volumes that span the initial region where mixing between the premixed jet from the nozzle and co-flowing air takes place, the base of the lifted flame, and downstream burning regions where mixing occurs between the products of the premixed flame and the co-flowing air. These volumes—denoted by $\mathcal{V}_i$, where $i = 0, \ldots, n_v - 1$ and $n_v = 6$ is the number of sub-volumes—are centered at $z_i = 0.0525\,\text{m} + (i - 1)0.0125\,\text{m}$, with height $0.00625\,\text{m}$ and width $0.14\,\text{m}$, composed of $1146 \times 1146 \times 51$ cells. The locations of several of these subregions are presented in Fig. 14.2. The premixed lifted flame, with high values of $\dot{\omega}_C$ and steep gradients corresponding to a thin flame, can be observed around $z = 0.05$ m. Farther downstream of the nozzle, the premixed flame products mix with the air coming from the co-flow and react to produce lower values of $\dot{\omega}_C$.

14.2.3 Filtering

For training and testing the proposed data-driven LES combustion model, the DNS data for source term of progress variable and the input variables are Favre filtered using a Gaussian filter kernel from the SciPy library [34]. The form for the Gaussian filter kernel used in this study is

$$G(x - y) = \sqrt{\frac{1}{\pi \Delta^2/6}} \exp\left(\frac{-|x - y|^2}{\Delta^2/6}\right). \tag{14.2}$$

The filter width, Δ, is varied as a multiple of the finest mesh spacing in the DNS. Throughout this paper, the ratio of filter width to the finest DNS mesh spacing is

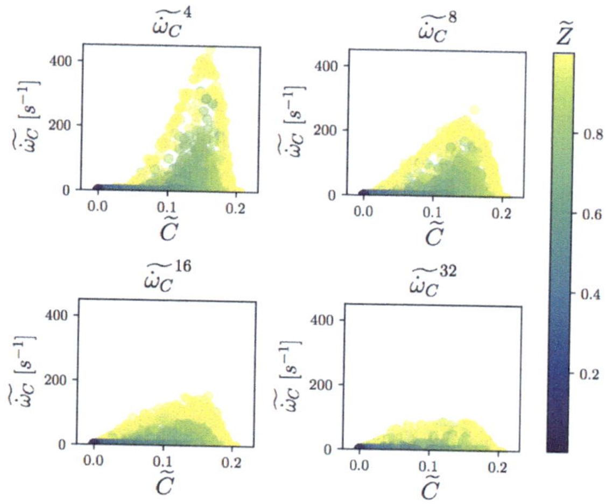

Fig. 14.3 Filtered source term of progress variable at varying filter widths, plotted on the span-wise mid-plane

referred to as the filter to grid ratio (FGR). All Favre filtered quantities will be denoted using a superscript, for example, a quantity filtered at filter length scale of FGR = 8 will be represented as $\widetilde{()}^8$. FGR values of 4, 8, 16, and 32 are used in this study. Figure 14.3 shows the impact of filtering on the peak values of progress variable source terms. As the filter width is increased, the sharp peaks in the progress variable source term associated with the flame are damped out and are observed to spread across a number of grid cells.

As can be observed from Fig. 14.2, a significant portion of the domain is non-reacting mixing with minuscule values of progress variable source term. A random sampling of point-wise data from sub-volumes will result in a dataset with only a small portion of the samples taken from the flame region, with high values of progress variable source term, and the rest of the data related to non-reacting mixing. Training a model with such a dataset would result in significant inefficient sampling. A sampling strategy based on k-means clustering [35] is used to generate a training and a validation sample. The k-means clustering algorithm is NP-hard and is prohibitively expensive needing $O(n^{k+1})$ time to cluster, where n is the number of observations and k is the number of clusters [36]. In this study, an alternative to k-means algorithm well suited to clustering massive datasets known as the minibatch k-means [37] is used. Minibatch k-means sample small random batches and compute clusters on those samples, iterating until the clusters are converged with new random samples being used to update the clusters on each iteration. In this study we use the minibatch k-means algorithm from scikit-learn [34] library.

The sampling strategy to prepare the training and validation data is broken down in the following steps:

i. The input and the output variable, $\widetilde{\dot{\omega}_C}$, from each of the sub-volume and all four filter sizes is collected in a large dataset.
ii. Using minibatch k-means, 40 clusters are created from the dataset.

iii. A random sampling of 50,000 samples are picked from each of the clusters and are randomly permuted, leading to a total of 2 million data points.
iv. This dataset with 2 million samples, consisting of input variables and output variables, is divided into 95% training and the other 5% is used for validation. The resulting training dataset constituted only 0.7% of the total data available for training.

This data sampling strategy ensures that a uniform representation of all the states observed in the filtered DNS data are equally present in the training and validation dataset. Prior to training the network, each input and output variables are independently scaled using the robust scaling technique from scikit-learn [34] library.

14.3 Machine Learning Algorithm

In the last decade, improvements in deep learning algorithms and computing technology have enabled phenomenal success in the areas of computer vision, image processing, speech recognition, and natural language processing [38–40]. Some of the network architectures highlighted in these review articles are very complex and are designed to accomplish specific goals. In this study a simpler feedforward, fully connected DNN is used to construct a model for progress variable source term. The network consists of 10 hidden layers and an output layer. Each hidden layer has 20 fully connected nodes, a leaky rectified linear unit activation function and batch normalization layer [41]. A number of training runs were performed to optimize these hyper-parameters to achieve both accuracy and efficiency. The total number of trainable parameters for this network architecture is 4800. The entire training data was cycled over 1500 epochs. In each epoch cycle training data was shuffled and divided into batches of 4096 samples. This choice of batch size was motivated by computational efficiency of the entire network training process. The Adam optimizer, an extension of the stochastic gradient descent algorithm, is used to update weights in the network. An initial learning rate of 10^{-4} was used for training. The weighted mean absolute error (WMAE), defined by Eq. 14.3, is used as the loss function.

$$l(t, p) = \frac{\sum_i^n |t_i - p_i|}{\sum_i^n t_i} \tag{14.3}$$

where, t and p are the vectors of true and predicted observations.The PyTorch [42] library was used to build the network and was trained on a single NVIDIA Tesla K80 GPU.

Two sets of deep learning models are constructed, DNN-6 and DNN-4. All the hyper-parameters listed above were optimized for the DNN-6 model. These models are evaluated and results are presented in Sect. 14.4.

i. DNN-4: $\widetilde{\omega}_C \approx G\left(\widetilde{Z}, \widetilde{Z''^2}, \widetilde{C}, \widetilde{C''^2}\right)$

ii. DNN-6: $\widetilde{\omega}_C \approx F\left(\widetilde{Z}, \widetilde{Z''^2}, \widetilde{C}, \widetilde{C''^2}, \widetilde{\chi}_Z, \widetilde{\chi}_C\right)$

Primary motivation behind the choice of input variables for DNN-4 is that some of the most commonly used physics-based manifold models for LES such as the Flamelet Generated Manifold (FGM) [43] typically rely on the same set of input variables to provide closure to the reacting flow LES equations.

14.4 Results

To establish the performance of the DNN-based model, results from a-priori testing, where the input parameters are generated from the filtered DNS data and output are compared against filtered progress variable source term from DNS, are presented in the following section. DNN-based models are stored in the form of weights associated with each of the nodes in each of the hidden layer. Model evaluation is performed as a series of matrix-vector multiplications, making them highly amenable to efficient computing architectures optimized for linear algebra operations such as the GPUs. In terms of memory footprint, the DNN-6 model occupied only 52 KB of memory, which highlights the efficiency of DNNs in storing information.

14.4.1 DNN-4 Predictions

Figure 14.4 shows the DNN-4 predictions of $\widetilde{\omega}_C$, computed using filtered mean and variance of mixture fraction and progress variable from sub-volume $\mathcal{V}_1$. The solid red line in Fig. 14.4 shows a perfect model and any deviation from the solid red line is the error in the model predictions. At $\Delta/\Delta_{DNS} = 4$, DNN-4 model under-predicts and at higher filter to DNS grid ratio, $\Delta/\Delta_{DNS} = 32$, over-predicts the $\widetilde{\omega}_C$ compared to the filtered DNS values. In order to quantify the modeling errors, an error metric is introduced

$$\epsilon = \frac{\sum\left(\widetilde{Y}_{model} - \widetilde{Y}_{DNS}\right)^2}{\sum\left(\widetilde{Y}_{DNS}\right)^2} \tag{14.4}$$

where $\widetilde{Y}_{model}$ and $\widetilde{Y}_{DNS}$ is the model prediction and filtered DNS on all sub-volumes. Table 14.1 lists the error metric, ϵ, calculated by comparing DNN-4 predictions and filtered DNS. The data from all the sub-volumes is used to calculate the error. From Fig. 14.4 and Table 14.1, it can be clearly observed that DNN-4 predictions correlate poorly with the filtered DNS predictions. As will be shown in the next section, Sect. 14.4.2, the primary reason behind this poor model prediction is the weak correlation of $\widetilde{\omega}_C$ on the input variables, $\widetilde{Z}$, $\widetilde{Z''^2}$, $\widetilde{C}$ and $\widetilde{C''^2}$. As mentioned previously, the main reason behind this choice of inputs variables is their common use in the

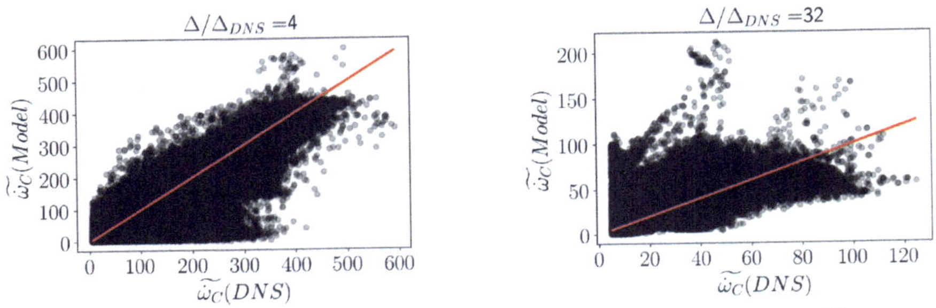

(a) $\Delta/\Delta_{DNS} = 4$ (b) $\Delta/\Delta_{DNS} = 32$

Fig. 14.4 Filtered progress variable source term, $\widetilde{\omega}_C$, from the sub-volume $\mathcal{V}_1$, filtered DNS compared against DNN-4 predictions

Table 14.1 Prediction error computed using Eq. 14.4. DNN-4 model predictions are compared against filtered DNS

Δ/Δ_{DNS}	ϵ(DNN-4)
4	0.397
8	0.627
16	0.788
32	1.328

manifold-based models such as the FGM [43] and well established transport equation and closure models. The correlation between output and input variables can be improved through a proper choice of input variables, which can be motivated through physical hypothesis or through algorithms such as principal component analysis (PCA) [9]. In the following section, Sect. 14.4.2, predictions from a deep neural network model (DNN-6), with filtered dissipation rates of mixture fraction and progress variable added to the input set, are presented. Inclusion of these additional variables is motivated by physical hypothesis supported by a number of studies that have shown the importance of dissipation rates to turbulence combustion modeling [44].

14.4.2 DNN-6 Predictions

Predictions from the DNN-6 model, with input variables $\widetilde{Z}$, $\widetilde{Z''^2}$, $\widetilde{C}$, $\widetilde{C''^2}$, $\widetilde{\chi}_Z$ and $\widetilde{\chi}_C$ are shown in Fig. 14.5. Model predictions and filtered DNS data in Fig. 14.5 are taken from sub-volume $\mathcal{V}_1$. The error metric defined by Eq. 14.4 computed by comparing model predictions from DNN-6 and filtered DNS is listed in Table 14.2. For sake of comparison error metric computed using DNN-4 model prediction is also shown. The data from all the sub-volumes is used to calculate this error metric. From Fig. 14.5 and Table 14.2 it can be clearly observed that inclusion of dissipation rates improves

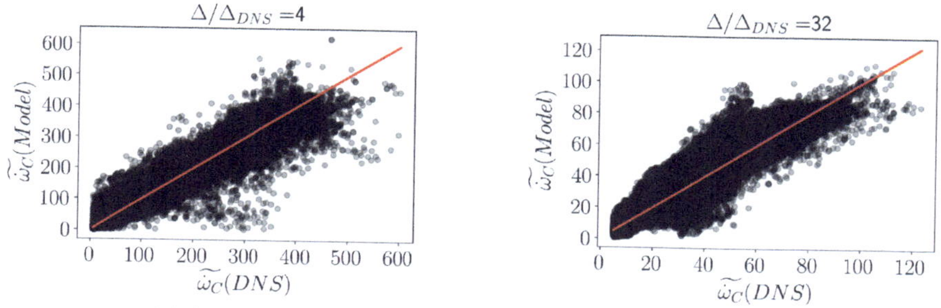

(a) $\Delta/\Delta_{DNS} = 4$ (b) $\Delta/\Delta_{DNS} = 32$

Fig. 14.5 Filtered progress variable source term, $\widetilde{\dot\omega}_C$, from the sub-volume $\mathcal{V}_1$, filtered DNS compared against DNN-6 predictions

Table 14.2 Prediction error computed using Eq. 14.4. Both DNN-6 and DNN-4 model predictions are compared against filtered DNS.

Δ/Δ_{DNS}	ϵ (DNN-6)	ϵ(DNN-4)
4	0.178	0.397
8	0.242	0.627
16	0.32	0.788
32	0.38	1.328

the model prediction significantly, highlighting the strong correlation between $\widetilde{\dot\omega}_C$ and $\widetilde{\chi}_C$. In a recent work [45], the authors analyzed a number of DNS studies, with methane, toluene, *n*-heptane, and *iso*-octane fuels, and arrived at the same conclusion that the burning rate is a strong function of the progress variable dissipation rate.

Figure 14.6 shows the deviation of DNN-6 model predictions from filtered DNS source term for progress variable. The data shown in this plot is from the mid-plane in the span-wise direction. The red line plotted in Fig. 14.6 shows the correlation derived from a perfect model. One thing to note here is that only a tiny subset of the data from this mid-plane was used to training the model. Overall the model predictions lie on the 45° line, denoting the perfect model correlation, with the significant portion of deviations in areas where the magnitude of the progress variable source term was two orders of magnitude lower than the peak progress variable around the lifted flame.

Figure 14.7 shows the contours, plotted on the mid-plane in the span-wise direction, of filtered $\widetilde{\dot\omega}_C$ from DNS and the predictions from the DNN-6 model. Figure 14.7 demonstrate that the data-driven model, derived using deep neural network (DNN-6), accurately reproduces the filtered DNS across multiple filter widths. Figure 14.7 only shows the plot at $\Delta/\Delta_{DNS} = 32$ but lower or similar errors are observed in model predictions at $\Delta/\Delta_{DNS} = 4, 8, 16$. The peak values of the filtered source term, observed near the premixed lifted flame are predicted accurately. In the regions downstream of the lifted premixed flame, where products of the primary reacting zone mix and burn with the co-flowing air, DNN-6 model over-predicts the filtered

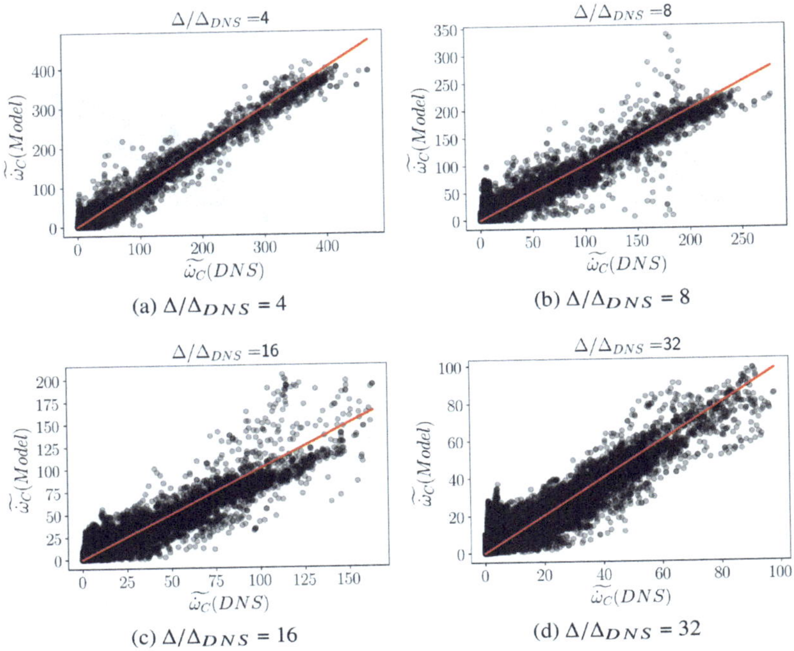

(a) $\Delta/\Delta_{DNS} = 4$

(b) $\Delta/\Delta_{DNS} = 8$

(c) $\Delta/\Delta_{DNS} = 16$

(d) $\Delta/\Delta_{DNS} = 32$

Fig. 14.6 Filtered $\widetilde{\dot{\omega}}_C$ from the midplane in the span-wise direction, filtered DNS compared against DNN-6 predictions

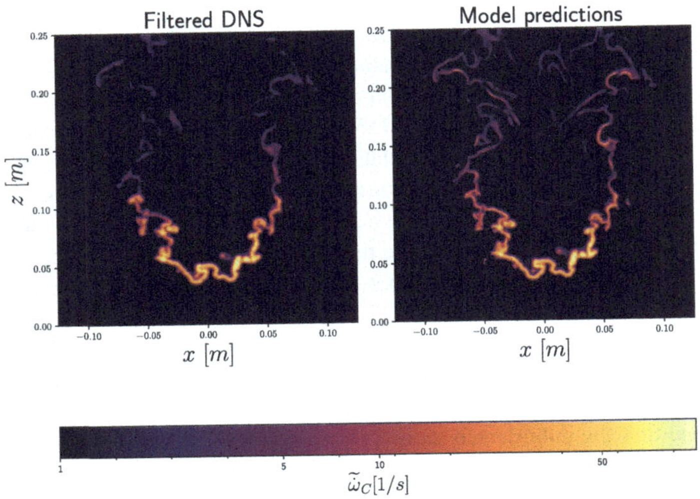

Fig. 14.7 Filtered $\widetilde{\dot{\omega}}_C$ at filter to DNS grid ratio, $\Delta/\Delta_{DNS} = 32$. Left image: DNS, Right image: DNN-6 predictions

progress variable. Peak values of filtered progress variable source term in these secondary reaction zones are much lower than the peak values observed around the premixed lifted flame.

14.4.2.1 Conditional Averages

Figures 14.8 and 14.9 show the conditional mean of filtered progress variable source term, conditioned on filtered mixture fraction, computed using data from $\mathcal{V}_1$, $\mathcal{V}_2$ sub-volumes, respectively. Figure 14.8 shows that in sub-volume $\mathcal{V}_1$ and at lower filter to grid ratios, peak values of progress variable source term are observed near Z = 1, corresponding to the premixed lifted flame established downstream of the nozzle. Conditional average from the sub-volume $\mathcal{V}_2$, shown in Fig. 14.9, shows that, due to mixing between the stream from the nozzle and the co-flow air, the peak values of $\widetilde{\dot{\omega}}_C$ transition toward leaner fuel–air equivalence ratio. From Figs. 14.8 and 14.9 it can be clearly observed that DNN-6 model predictions accurately reproduce the filtered DNS data. As observed in Fig. 14.4, DNN-4 model under predicts for lower Δ/Δ_{DNS} and over-predicts for higher values of Δ/Δ_{DNS}.

Figures 14.10, 14.11, and 14.12 show the conditional average of filtered progress variable source term conditioned on filtered progress variable, computed on the same sub-volumes, $\mathcal{V}_1$, $\mathcal{V}_2$ and $\mathcal{V}_4$, respectively. In the sub-volume $\mathcal{V}_1$, the filtered progress variable source term is well correlated with the filtered progress variable, $\widetilde{C}$, with the peak value observed at $\widetilde{C} \approx 0.15$. As the filter to grid ratio is increased, a flame thickening effect is observed which modifies the profile of the conditional mean filtered progress variable source term to a more symmetric profile. From Figs. 14.10 and 14.11 it can be clearly observed that DNN-6 model reproduces the conditional means from DNS very accurately. One key observation that can be made from these conditional averages plots is that $\widetilde{\dot{\omega}}_C$ is much more strongly correlated with filtered

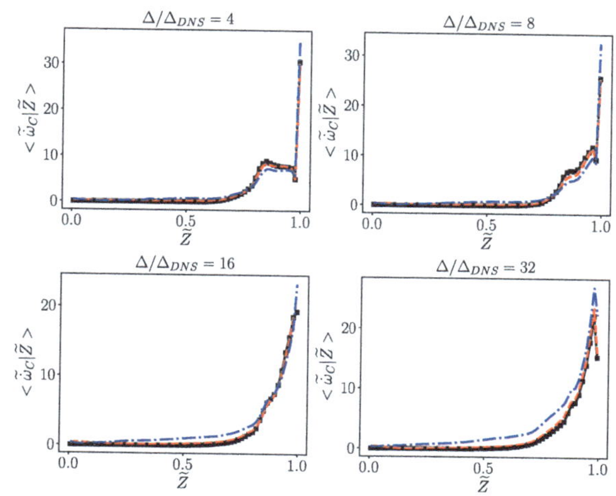

Fig. 14.8 Filtered $\widetilde{\dot{\omega}}_C$ conditioned on filtered mixture fraction, $\widetilde{Z}$. Data for conditional averages is taken from sub-volume, $\mathcal{V}_1$. Black solid line with symbols is DNS data and red dashed line is from DNN-6 model predictions and blue dashed dotted line is from DNN-4 predictions

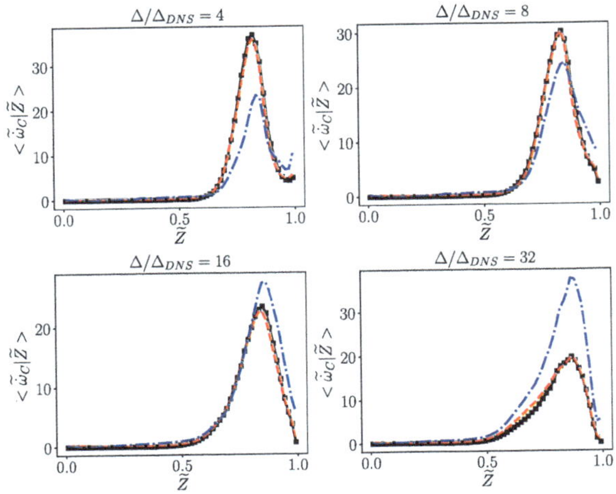

Fig. 14.9 Filtered $\widetilde{\dot{\omega}}_C$ conditioned on mixture fraction, $\widetilde{Z}$. Data for conditional averages is taken from sub-volume, $\mathcal{V}_2$. Black solid line with symbols is DNS data and red dashed line is from DNN-6 model predictions and blue dashed dotted line is from DNN-4 predictions

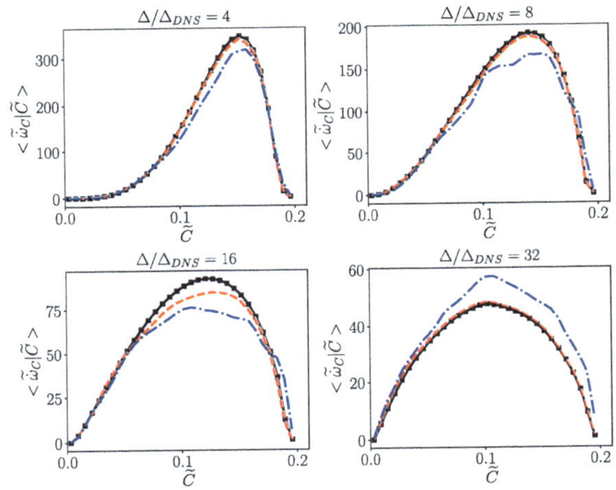

Fig. 14.10 Filtered $\widetilde{\dot{\omega}}_C$ conditioned on filtered progress variable, $\widetilde{C}$. Data for conditional averages is taken from sub-volume, $\mathcal{V}_1$. Black solid line is DNS data and red dashed line is from DNN-6 model predictions and blue dashed dotted line is from DNN-4 predictions

progress variable than filtered mixture fraction. This can be seen by comparing the peak magnitudes of $< \widetilde{\dot{\omega}}_C|\widetilde{C} >$ and $< \widetilde{\dot{\omega}}_C|\widetilde{Z} >$ in Figs. 14.8 and 14.10, and contours of input variables from the DNS data in Fig. 14.1. Near the lifted flame no variation in mixture fraction field is observed whereas the progress variable field has a very sharp gradient. Therefore in sub-volume $\mathcal{V}_1$ a major fraction of cells will show non-reacting, $\widetilde{\dot{\omega}}_C = 0$, in regions with $\widetilde{Z} = 1$, along with a tiny region with peak values of $\widetilde{\dot{\omega}}_C$ at $\widetilde{Z} = 1$, thus driving the magnitude of $< \widetilde{\dot{\omega}}_C|\widetilde{Z} >$ lower than $< \widetilde{\dot{\omega}}_C|\widetilde{C} >$. This dependence on filtered progress variable and relative independence of $\widetilde{\dot{\omega}}_C$ on filtered mixture fraction was embedded in the data and was correctly identified by the DNN-based model. In the DNN-6 model, one of the input variables is scalar dissipation rate of progress variable, $\widetilde{\chi}_C$. Figures 14.13 and 14.14 show the conditional mean filtered progress variable source term conditioned on filtered dissipation rate, $\widetilde{\chi}_C$.

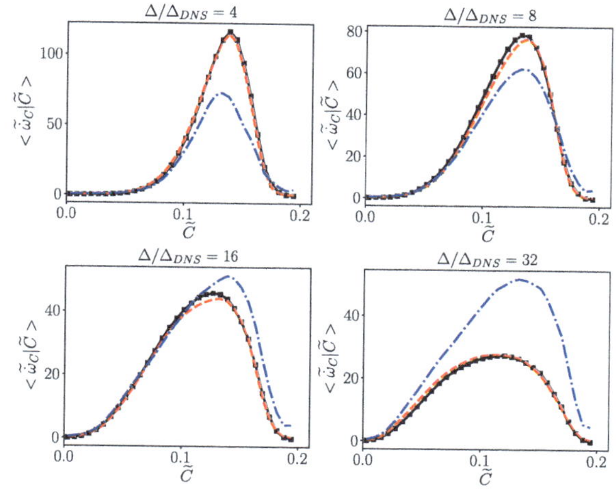

Fig. 14.11 Filtered $\widetilde{\dot{\omega}}_C$ conditioned on filtered progress variable, $\widetilde{C}$. Data for conditional averages is taken from sub-volume, $\mathcal{V}_2$. Black solid line is DNS data and red dashed line is from DNN-6 model predictions and blue dashed dotted line is from DNN-4 predictions

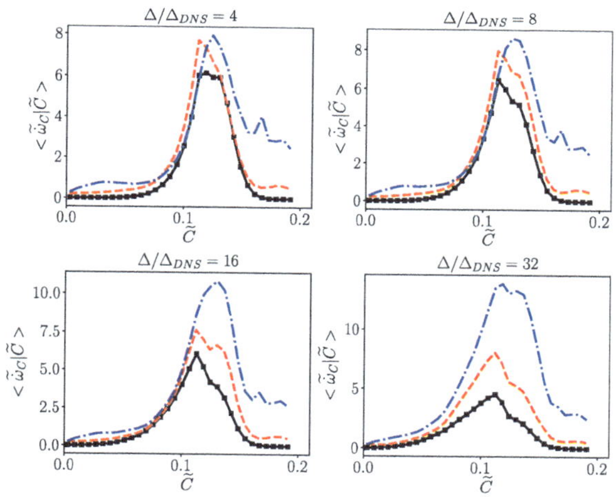

Fig. 14.12 Filtered $\widetilde{\dot{\omega}}_C$ conditioned on filtered progress variable, $\widetilde{C}$. Data for conditional averages is taken from sub-volume, $\mathcal{V}_4$. Black solid line is DNS data and red dashed line is from DNN-6 model predictions and blue dashed dotted line is from DNN-4 predictions

Figures 14.13 and 14.14 show an almost linear dependence of $\widetilde{\dot{\omega}}_C$ on $\widetilde{\chi}_C$. In these figures, large fluctuations are observed at higher values of filtered progress variable dissipation rate, $\widetilde{\chi}_C$, due to the reduced number of samples at those large values of $\widetilde{\chi}_C$. Similar to the previously observed trends, DNN-6 models accurately capture the dependence of $\widetilde{\dot{\omega}}_C$ on filtered progress variable dissipation rate, $\widetilde{\chi}_C$.

In terms of model input parameters, in the sub-volumes $\mathcal{V}_1$ and $\mathcal{V}_2$, a strong correlation was observed between $\widetilde{C''^2}$ and $\widetilde{\chi}_C$. This strong correlation was observed across different filter widths and not including both of these as input variables is redundant. However, in the downstream sub-volumes a relative weakening of this strong correlation between $\widetilde{C''^2}$ and $\widetilde{\chi}_C$ was observed. For this reason, both $\widetilde{C''^2}$ and $\widetilde{\chi}_C$ were considered in constructing the DNN-6 model to include all the available information. Table 14.3 lists out the Pearson correlation coefficient [46], r, between

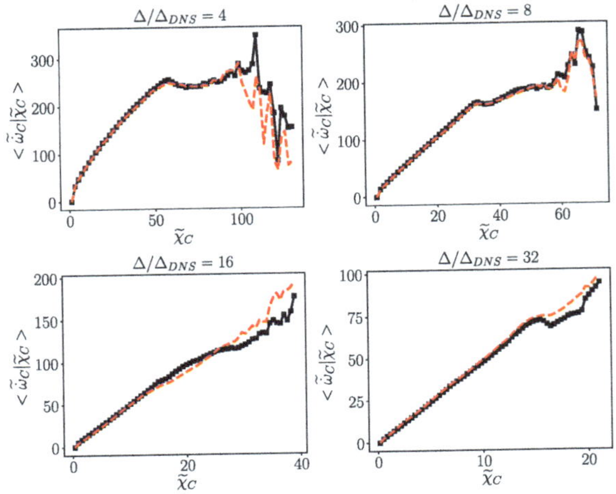

Fig. 14.13 Filtered $\widetilde{\dot{\omega}}_C$ conditioned on filtered progress variable, $\widetilde{\chi}_C$. Data for conditional averages is taken from sub-volume, $\mathcal{V}_1$. Black solid line is DNS data and red dashed line is from DNN-6 model predictions

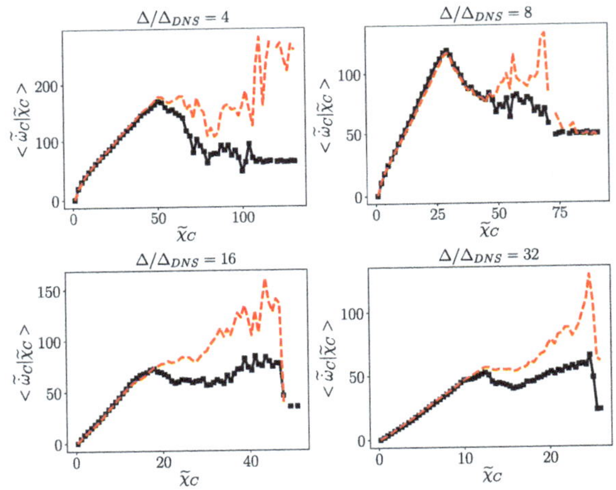

Fig. 14.14 Filtered $\widetilde{\dot{\omega}}_C$ conditioned on filtered progress variable, $\widetilde{\chi}_C$. Data for conditional averages is taken from sub-volume, $\mathcal{V}_2$. Black solid line is DNS data and red dashed line is from DNN-6 model predictions

Table 14.3 Pearson correlation coefficient between $\widetilde{C''^2}$ and $\widetilde{\chi}_C$

	Pearson coefficient, r			
Δ/Δ_{DNS}	$\mathcal{V}_1$	$\mathcal{V}_2$	$\mathcal{V}_3$	$\mathcal{V}_4$
4	0.919	0.722	0.858	0.915
8	0.914	0.877	0.890	0.905
16	0.912	0.892	0.778	0.847
32	0.928	0.922	0.754	0.770

$\widetilde{C''^2}$ and $\widetilde{\chi}_C$ in various sub-volumes and at various filter to DNS grid ratio. In the case of mixture fraction, a much weaker correlation coefficient ($\approx$0.5) was observed between $\widetilde{Z''^2}$ and $\widetilde{\chi}_Z$ and therefore both of these inputs variables were considered in the DNN-6 model formulation.

For a model of the form, $\widetilde{\dot{\omega}}_C = F\left(\widetilde{Z}, \widetilde{Z''^2}, \widetilde{C}, \widetilde{C''^2}, \widetilde{\chi}_Z, \widetilde{\chi}_C\right)$, conditional average conditioned on all the input variables is referred to as the optimal estimator [47]. The optimal estimator is the absolute best that any model with those corresponding inputs can perform [47] and the error made by the conditional averages is the irreducible error made by estimating $\widetilde{\dot{\omega}}_C$ using $\left(\widetilde{Z}, \widetilde{Z''^2}, \widetilde{C}, \widetilde{C''^2}, \widetilde{\chi}_Z, \widetilde{\chi}_C\right)$ as input parameters. From the conditional averages predicted by the DNN model, it can be clearly observed that by carefully considering the input data, optimizing the DNN architecture and the training procedure, models generated using neural networks can provide the best estimate for the optimal estimators.

14.4.3 Model Prediction at Time $t = 0.059$ s

As mentioned in Sect. 14.2.2, the training data for the DNN-6 and DNN-4 models was from an instantaneous snapshot of the DNS at time $t = 0.0626$ s. In order to test the model generality, the model was trained on data from one instant in time ($t = 0.0626$ s) and tested on data from the DNS solution at another instant in time ($t = 0.059$ s). The DNS data from $t = 0.059$ s is used to generate input variables for the model and 'truth' data for comparison to the model output (the filtered progress variable source term), but the no training was performed using this data. Figure 14.15 shows the DNN-6 model prediction on the span-wise mid-plane at time, $t = 0.059$ s, at filter width of $\Delta/\Delta_{DNS} = 32$. Similar or lower error was noticed for model predictions at filter width of $\Delta/\Delta_{DNS} = 4, 8, 16$ and therefore are not shown here. These results indicate that the model trained on data from time, $t = 0.0626$ s, accurately learned the functional relationship between input and output variables at dynamically steady state of the flame and is able to reproduce accurately the filtered DNS data at time $t = 0.059$ s.

14.5 Conclusion

A data-driven LES model for simulation of low swirl burner, with methane as fuel, was developed using deep learning techniques. Filtered DNS data was used to train the deep neural network (DNN) model for filtered progress variable reaction rate. An efficient strategy was devised to populate the training sample with uniform representation from various regions with varying magnitudes of filtered progress variable source term. A-priori testing of the proposed DNN models demonstrated the efficiency and accuracy of deep learning techniques in constructing a reacting flow

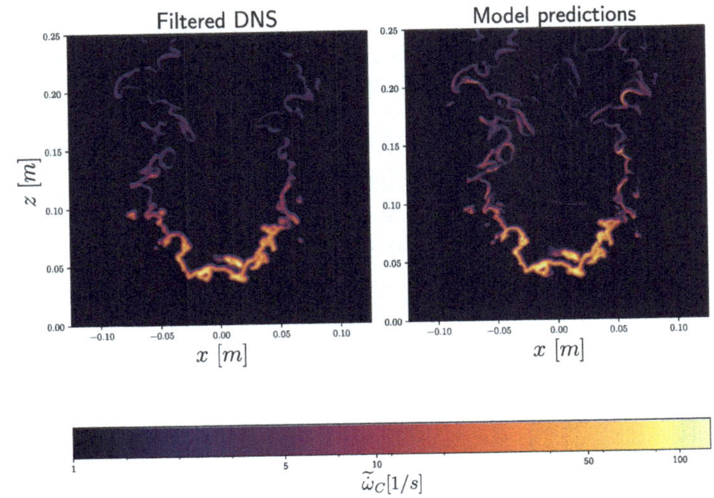

Fig. 14.15 Filtered $\widetilde{\dot{\omega}}_C$ at filter width $\Delta/\Delta_{DNS} = 32$. Left image: DNS, Right image: DNN-6

model valid across regimes and over a range of length scales. This indicates that (1) the DNN architecture used here was sufficiently flexible to reasonably capture the dependence of the filtered progress variable source term on the (filtered) input quantities and (ii) the expanded input set used in DNN-6 was necessary for good results. The latter observation, although not particularly surprising, illustrates how machine learning can be used to identify important physics—here the dissipation rates—and also provide a framework for potentially discarding extraneous model parameters. Future efforts in this area will be focused on development of general purpose models, using machine learning techniques, applicable across various configurations and regimes. Additionally, DNN-based model performance will be established using a posteriori analysis.

Acknowledgements This work was authored in part by the National Renewable Energy Laboratory, operated by Alliance for Sustainable Energy, LLC, for the U.S. Department of Energy (DOE) under Contract No. DE-AC36-08GO28308. Funding was provided by U.S. Department of Energy Office of Science and National Nuclear Security Administration. The views expressed in the article do not necessarily represent the views of the DOE or the U.S. Government. The U.S. Government retains and the publisher, by accepting the article for publication, acknowledges that the U.S. Government retains a nonexclusive, paid-up, irrevocable, worldwide license to publish or reproduce the published form of this work, or allow others to do so, for U.S. Government purposes.

References

1. P.A. McMurtry, W.H. Jou, J. Riley, R. Metcalfe, AIAA J. **24**(6), 962 (1986)
2. T. Poinsot, S. Candel, A. Trouvé, Prog. Energy Combust. Sci. **21**(6), 531 (1995)

3. L. Vervisch, T. Poinsot, Annu. Rev. Fluid Mech. **30**(1), 655 (1998)
4. E.R. Hawkes, R. Sankaran, J.C. Sutherland, J.H. Chen, J. Phys.: Conf. Ser. **16**, 65 (IOP Publishing, 2005)
5. J.H. Chen, A. Choudhary, B. De Supinski, M. DeVries, E.R. Hawkes, S. Klasky, W.K. Liao, K.L. Ma, J. Mellor-Crummey, N. Podhorszki et al., Comput. Sci. Discov. **2**(1), 015001 (2009)
6. K. Aditya, A. Gruber, C. Xu, T. Lu, A. Krisman, M.R. Bothien, J.H. Chen, Proc. Combust. Inst. **37**(2), 2635 (2019)
7. D. Dalakoti, E.R. Hawkes, M.S. Day, J.B. Bell, Direct numerical simulation of two-stage combustion and flame stabilisation in diesel engine-relevant conditions. in *International Colloquium on the Dynamics of Explosions and Reactive Systems (ICDERS)* (Boston USA. 2017), pp. 5835-5843
8. M. Rieth, M. Day, J. Chen, Bull. Am. Phys. Soc. **63** (2018)
9. J.C. Sutherland, A. Parente, Proc. Combust. Inst. **32**(1), 1563 (2009)
10. C.J. Lapeyre, A. Misdariis, N. Cazard, D. Veynante, T. Poinsot, Combust. Flame **203**, 255 (2019)
11. K. Duraisamy, G. Iaccarino, H. Xiao, Annu. Rev. Fluid Mech. **51**, 357 (2019)
12. I. Goodfellow, J. Pouget-Abadie, M. Mirza, B. Xu, D. Warde-Farley, S. Ozair, A. Courville, Y. Bengio, Adv. Neural Inf. Process. Syst. **27**, 2672–2680 (2014). https://doi.org/10.1017/CBO9781139058452, http://papers.nips.cc/paper/5423-generative-adversarial-nets.pdf
13. H.C. Burger, C.J. Schuler, S. Harmeling, in *Proceedings of the IEEE Computer Society Conference on Computer Vision and Pattern Recognition* (2012), pp. 2392–2399. https://doi.org/10.1109/CVPR.2012.6247952
14. A. Dosovitskiy, J.T. Springenberg, T. Brox, in *Proceedings of the IEEE Computer Society Conference on Computer Vision and Pattern Recognition, 07–12 June 2015*, pp. 1538–1546. https://doi.org/10.1109/CVPR.2015.7298761
15. S. Lefkimmiatis, in *Proceedings of the IEEE Conference on Computer Vision and Pattern Recognition* (2016), pp. 5882–5891. https://doi.org/10.1109/CVPR.2017.623, arXiv:1611.06757
16. C. Ledig, L. Theis, F. Huszar, J. Caballero, A. Cunningham, A. Acosta, A. Aitken, A. Tejani, J. Totz, Z. Wang, W. Shi, Conf. Comput. Vis. Pattern Recognit. (2016), pp. 1–14. https://doi.org/10.1109/CVPR.2017.19, arXiv:1609.04802
17. Y. Tai, J. Yang, X. Liu, in *Proceedings of the 30th IEEE Conference on Computer Vision and Pattern Recognition, CVPR 2017*, vol. 2017 (2017), pp. 2790–2798. https://doi.org/10.1109/CVPR.2017.298
18. W.S. Lai, J.B. Huang, N. Ahuja, M.H. Yang, in *Proceedings of the 30th IEEE Conference on Computer Vision and Pattern Recognition, CVPR 2017*, vol. 2017 (2017), pp. 5835–5843. https://doi.org/10.1109/CVPR.2017.618
19. A. Graves (2013). https://doi.org/10.1145/2661829.2661935, arXiv:1308.0850
20. Y. Wu, M. Schuster, Z. Chen, Q.V. Le, M. Norouzi, W. Macherey, M. Krikun, Y. Cao, Q. Gao, K. Macherey, J. Klingner, A. Shah, M. Johnson, X. Liu, Ł. Kaiser, S. Gouws, Y. Kato, T. Kudo, H. Kazawa, K. Stevens, G. Kurian, N. Patil, W. Wang, C. Young, J. Smith, J. Riesa, A. Rudnick, O. Vinyals, G. Corrado, M. Hughes, J. Dean (2016). https://doi.org/10.1038/nrn2258
21. W. Kwon (2017), arXiv:1706.03762 [cs]
22. D. Silver, J. Schrittwieser, K. Simonyan, I. Antonoglou, A. Huang, A. Guez, T. Hubert, L. Baker, M. Lai, A. Bolton, Y. Chen, T. Lillicrap, F. Hui, L. Sifre, G. van den Driessche, T. Graepel, D. Hassabis, Nature **550**(7676), 354 (2017). https://doi.org/10.1038/nature24270
23. Y. Lecun, Y. Bengio, G. Hinton, Nature **521**(7553), 436 (2015). https://doi.org/10.1038/nature14539
24. J. Schmidhuber, Neural Netw. **61**, 85 (2015). https://doi.org/10.1016/j.neunet.2014.09.003
25. A. Prieto, B. Prieto, E.M. Ortigosa, E. Ros, F. Pelayo, J. Ortega, I. Rojas, Neurocomputing (2016). https://doi.org/10.1016/j.neucom.2016.06.014
26. I. Goodfellow, Y. Bengio, A. Courville, *Deep Learning* (MIT Press, Cambridge, 2016), http://www.deeplearningbook.org

27. W. Liu, Z. Wang, X. Liu, N. Zeng, Y. Liu, F.E. Alsaadi, Neurocomputing (2017). https://doi.org/10.1016/j.neucom.2016.12.038
28. G. Cybenko, Math. Control Signals Syst. **2**(4), 303 (1989). https://doi.org/10.1007/BF02551274
29. K. Hornik, Neural Netw. **4**(2), 251 (1991). https://doi.org/10.1016/0893-6080(91)90009-T
30. M. Day, S. Tachibana, J. Bell, M. Lijewski, V. Beckner, R.K. Cheng, Combust. Flame **159**(1), 275 (2012). https://doi.org/10.1016/j.combustflame.2011.06.016, http://linkinghub.elsevier.com/retrieve/pii/S0010218011001969
31. R. Cheng, D. Yegian, M. Miyasato, G. Samuelsen, C. Benson, R. Pellizzari, P. Loftus, Proc. Combust. Inst. (2000). https://doi.org/10.1016/S0082-0784(00)80344-6
32. M.S. Day, J.B. Bell, Combust. Theory Model. **4**(4), 535 (2000). https://doi.org/10.1088/1364-7830/4/4/309
33. A. Kazakov, M. Frenklach, Reduced reaction sets based on GRI-Mech 1.2 (1994), http://www.me.berkeley.edu/drm/
34. F. Pedregosa, G. Varoquaux, A. Gramfort, V. Michel, B. Thirion, O. Grisel, M. Blondel, P. Prettenhofer, R. Weiss, V. Dubourg, J. Vanderplas, A. Passos, D. Cournapeau, M. Brucher, M. Perrot, E. Duchesnay, J. Mach. Learn. Res. **12**, 2825 (2011)
35. J. MacQueen et al., in *Proceedings of the Fifth Berkeley Symposium on Mathematical Statistics and Probability, Oakland, CA, USA*, vol. 1 (1967), pp. 281–297
36. M. Inaba, N. Katoh, H. Imai, in *Proceedings of the 10th Annual Symposium on Computational Geometry* (ACM, 1994), pp. 332–339
37. D. Sculley, in *Proceedings of the 19th International Conference on World Wide Web* (ACM, 2010), pp. 1177–1178
38. Y. LeCun, Y. Bengio, G. Hinton, Nature **521**(7553), 436 (2015)
39. J. Schmidhuber, Neural Netw. **61**, 85 (2015)
40. Y. Bengio et al., Found. Trends Mach. Learn. **2**(1), 1 (2009)
41. S. Ioffe, C. Szegedy (2015), arXiv:1502.03167
42. A. Paszke, S. Gross, S. Chintala, G. Chanan, E. Yang, Z. DeVito, Z. Lin, A. Desmaison, L. Antiga, A. Lerer., *Automatic differentiation in pytorch* (2017)
43. J. Van Oijen, A. Donini, R. Bastiaans, J. ten Thije Boonkkamp, L. De Goey, Prog. Energy Combust. Sci. **57**, 30 (2016)
44. D. Veynante, L. Vervisch, Prog. Energy Combust. Sci. **28**(3), 193 (2002)
45. B. Savard, G. Blanquart, Combust. Flame **180**, 77 (2017)
46. K. Pearson, Proc. R. Soc. Lond. **58**(347–352), 240 (1895)
47. A. Moreau, O. Teytaud, J.P. Bertoglio, Phys. Fluids **18**(10), 105101 (2006)

The manufacturer's authorised representative in the EU is Springer Nature Customer Service Centre GmbH, Europaplatz 3, 69115 Heidelberg, Germany. If you have any concerns regarding our products, please contact ProductSafety@springernature.com

Printed and bound by CPI Group (UK) Ltd, Croydon, CR0 4YY
22/09/2026
02221057-0013